ZAPOTEC SCIENCE

ZAPOTEC SCIENCE

Farming and Food
in the Northern Sierra
of Oaxaca

Roberto J. González

University of Texas Press
Austin

Copyright © 2001 by the University of Texas Press
All rights reserved
Printed in the United States of America
First edition, 2001

Requests for permission to reproduce material from this work should be sent to Permissions, University of Texas Press, P.O. Box 7819, Austin, TX 78713-7819.

♾ The paper used in this book meets the minimum requirements of ANSI/NISO Z39.48-1992 (R1997) (Permanence of Paper).

Library of Congress Cataloging-in-Publication Data

González, Roberto J. (Roberto Jesús), 1969–
Zapotec science : farming and food in the Northern Sierra of Oaxaca / Roberto J. González.— 1st ed.
    p.   cm.
Includes bibliographical references and index.
ISBN 978-0-292-72832-5

1. Zapotec Indians—Agriculture. 2. Zapotec Indians—Food. 3. Traditional farming—Mexico—San Miguel Talea de Castro. 4. Subsistence economy—Mexico—San Miguel Talea de Castro. 5. Sustainable development—Mexico—San Miguel Talea de Castro. 6. San Miguel Talea de Castro (Mexico)—Social life and customs. I. Title.
F1221.Z3 G66   2001
630'.972'74—dc21

00-012010

A los que han compartido sus conocimientos—

Roberto J. González, Jr., Imelda F. González,

Laura Nader,

Rodolfo Bautista Solís,

y *los bhni huen dxin ibá de Ralh'a* (los campesinos de Talea),

Juventino Chávez Méndez y su familia en especial.

CONTENTS

Acknowledgments	*xi*
1. The Conceptual Bases of Zapotec Farming and Foodways	*1*
2. Locating Talea: *Geography, History, and Cultural Contexts*	*32*
3. The Craft of the Campesino: *Measures, Implements, and Artifacts*	*70*
4. "Maize Has a Soul": *Rincón Zapotec Notions of Living Matter*	*102*
5. From Milpa to Tortilla: *Growing, Eating, and Exchanging Maize*	*130*
6. Sweetness and Reciprocity: *Sugarcane Work*	*175*
7. The Invention of "Traditional" Agriculture: *The History and Meanings of Coffee*	*195*
8. Agriculture Unbound: *Cultivating the Ground between Science Traditions*	*234*

APPENDIX A. Pronunciation of Rincón Zapotec Terms   263

APPENDIX B. Talean Food Plants   265

APPENDIX C. Talean Livestock and Game Animals   272

APPENDIX D. Selected Average Crop Yields   274

APPENDIX E. Recipes   276

Notes   279

References   295

Index   319

FIGURES

2.1. Map of the Sierra Juárez and the Rincón, *33*
2.2. Panoramic view of Talea in 1999, *35*
2.3. The municipal palace, *36*
2.4. Scene from a Talean fiesta, *39*
2.5. Aerial photo of Talea and its lands, *41*
2.6. Talean *danzantes*, *53*
3.1. Measurements used by Talean campesinos, *74*
3.2. Campesino using a short-handled hoe, *82*
3.3. Construction of the *timón* (plow beam), *84*
3.4. Construction of the *telera* (connecting beam), *85*
3.5. *Arado* (plow) measurements, *87*
3.6. Mother and daughters in the forest, *93*
4.1. San Isidro, the patron saint of campesinos, *108*
4.2. Campesino preparing to dress a chicken, *115*
4.3. Village band playing at a wedding party, *116*
5.1. Talean campesino plowing his field, *137*
5.2. Campesino planting maize, *142*
5.3. Woman and daughter spreading maize for drying, *154*

5.4. Electric *molino* for grinding maize into meal, *156*
5.5. Making tortillas in a Talean kitchen, *157*
5.6. Campesina cutting banana leaves, *160*
5.7. Talea's weekly market, *163*
6.1. Talean campesinos grinding sugarcane, *185*
6.2. Altar for Todos Santos, *191*
6.3. Village orchestra playing funeral march, *193*
7.1. Major coffee-producing regions of Mexico, *198*
7.2. Boy harvesting coffee, *211*
7.3. Rincón family sun-drying coffee beans, *217*
7.4. Talean *cafetal* with *guajinicuil* shade trees, *219*

TABLES

3.1. Agricultural Implements Used by Talean Campesinos, *81*
4.1. Religious Fiestas Celebrated in Talea, 1996, *107*
4.2. Average Yields and Calories per Hectare of Maize, Rice, and Wheat, *122*
5.1. Soil Criteria and Types Identified by Campesinos, *132*
5.2. Workday Schedule for a Campesino with Animals, *134*
5.3. Varieties of Maize Grown by Talean Campesinos, *141*
5.4. Foods and Beverages Prepared from Maize in Talea, *159*
6.1. Some Varieties of Sugarcane Cultivated in Talea, *180*
7.1. Coffee Production in Oaxaca, 1945–1963, *203*

ACKNOWLEDGMENTS

Many people and organizations helped to support this book. Research was made possible through a National Science Foundation Predoctoral Fellowship (September 1994–August 1997). The University of California, Berkeley (UC Berkeley) Institute for the Study of Social Change and the UC Berkeley Graduate Opportunity Program supported the writing phase of the project (September 1997–May 1998).

My greatest debt is to Laura Nader, who provided background information on the Sierra and made valuable comments on earlier drafts of this book. Nelson Graburn also provided helpful comments which improved the manuscript. I would like to thank June Nash, John Chance, Gerald Berreman, Stanley Brandes, Jonathan Fox, Rosemary Joyce, Beatriz Manz, Jay Ou, and Jessica Jerome. Suzanne Calpestri, of the George and Mary Foster Anthropology Library, collected some of the material on coffee.

In Oaxaca Alejandro de Avila, director of the Oaxaca Botanical Gardens, offered his assistance and friendship during my time in the field. Others, including Miguel Bartolomé and Alicia Barabas (at the Instituto Nacional de Antropología e Historia–Oaxaca), Arthur Murphy and Martha Rees (at the Cecil Welte Institute), and María de los Angeles Romero, provided hospitality and insight. I am especially grateful to Leonardo Tyrtania of the Universidad Autónoma Metropolitana-Iztapalapa for his collegiality and hospitality.

Fernando Guadarrama offered suggestions and encouragement during my time in the field. The members of several Sierra-based organizations were also supportive: Jaime Martínez Luna, Magdalena Andrade, Alejandro Ruiz, Salvador Aquino, Arturo Guerrero, and Tonatiuh Díaz (of Comunalidad A.C.); Gustavo Ramírez and Fernando Ramos (of Proyectos de Desarrollo Sierra Norte A.C.); and Aldo González, Prócuro Pascual, and Fernando Santillán (of Unión de Organizaciones de la Sierra Juárez de Oaxaca S.C.). Encouraging words and captivating photographs depicting life in the Sierra Norte and beyond sent to me by Gabriela Zamorano have been a great inspiration.

Rachel Chance and Theresa May, of the University of Texas Press, provided early comments on the manuscript and directed it to two reviewers, who provided extraordinarily thorough and insightful commentaries. I am especially grateful to Eugene Anderson for his constructive critique and his enthusiastic encouragement.

Roberto, Imelda, Jorge, Ernesto, and Veronica González deserve credit for taking an interest in my projects and, frequently, subjecting them to criticism. I thank them for their thoughtful letters and support.

My deepest appreciation goes to many Taleans: Rodolfo Bautista and Zoila Alonso helped make my adjustment to rural fieldwork smooth by opening their home to me and answering many questions. Over time, many others offered their hospitality and friendship: Reyna and Sara Olivera, Wilfrido Martínez, and Isabel Bautista; Rosendo Bautista, Candida Hernández, and their sons Jaime, Salvador, and Jacobo; Mario Bautista, Eva López, and Eli Bautista; Pedro Canseco; Pedro Hernández; Godofredo Rivera and Elisabeth Hernández; Reymundo González and Teresa Rivera; Epitasio Francisco Cruz; Pedro Chávez and Luminosa Miguel; Octavio Olivera, Estanislao Núñez, Andrés Pérez, and Gil Canseco; Cado Pérez, who patiently taught me the rudiments of Rincón Zapotec; and many, many others. Most of all, I am grateful to a man who taught me more in fifteen months than I would ever have thought possible: Juventino Chávez. He, his wife, Concepción Labastida, and the rest of the Chávez Labastida family (Rita, Elvira, Alfonso, Joaquina, and Abigail) fed me, cared for me, and taught me lessons in life that I will never forget.

ZAPOTEC SCIENCE

CHAPTER 1

THE CONCEPTUAL BASES OF ZAPOTEC FARMING AND FOODWAYS

The approach to the problems of farming must be made from the field, not from the laboratory... In this the observant farmer and labourer, who have spent their lives in close contact with nature, can be of greatest help to the investigator.
  Sir Albert Howard, *An Agricultural Testament* (1940)

The ability of these natives in cultivating the soil and making plantations would very shortly produce such abundance that great profit would accrue to the Imperial Crown...
  Hernán Cortés, October 15, 1524

The Zapotec people of Oaxaca have continuously cultivated maize, beans, squash, and other crops for more than 5,000 years. Evidence strongly indicates that their ancient Mesoamerican ancestors domesticated maize from a wild grass and that its propagation and improvement required a level of human intervention so sophisticated that it has been described as "the most remarkable plant breeding accomplishment of all time" (Galinat 1992:47). Over this period, techniques for producing, processing, preserving, and preparing maize and other crops evolved to the point that agricultural surpluses supported a series of civilizations, including those of the Olmec, Maya, Toltec, Aztec, Zapotec, and Mixtec. Maize, beans, and squash (the so-called American Trinity), when combined with small amounts of other foods and animal-based proteins, provided the nutritional base for the de-

1

velopment of complex societies like those of Monte Albán, Mitla, Tenochtitlán, Chichén Itzá, and others (Wolf 1959:63; Warman 1988:20).

In Talea de Castro, a Zapotec village in a region called the Rincón (literally "corner") of the Northern Sierra of Oaxaca, most households continue to depend heavily on these three subsistence crops to feed their families. In some ways the cultivation techniques have changed little since the prehispanic period, with important exceptions—notably the use of animal-driven plows and steel tools (introduced in the Spanish colonial era) and chemical fertilizers (introduced in the late 1960s). In other ways, however, agriculture and diet have changed tremendously. Specifically, two Old World crops—sugarcane and coffee—have been added to the agricultural and culinary repertoires, though both are cultivated and processed using centuries-old technologies. Talean agriculture has been invented and reinvented: it is a blend of New and Old World crops, knowledges, and techniques that farmers have creatively adapted to local conditions. The vast majority of villagers work the land on privately owned plots of less than three hectares; indeed, farming is carried out by more than three-fourths of the "economically active population," according to recent census figures (INEGI 1991:315). Knowledge about farming and food is passed down from parents to their children verbally and materially as children listen to and emulate their superiors. In this way they learn about subsistence agriculture.

Initial observations in the field led me to think about the sophistication of Zapotec farming. Shortly after arriving in Talea, I witnessed government consultants, mainly agronomists, struggling vainly to understand the complexities of local farming methods and their underlying assumptions. The visitors seemed to have a different way of talking about plants, soils, and the weather—sometimes radically different—than campesino farmers did.[1] Afterward, a few Taleans told me that such experts periodically arrived with good intentions but were handicapped because they did not understand local soils. Furthermore, they said, the consultants had much theory but little practical experience (*mucha teoría, poca práctica*). Campesinos, they implied, know more about farming on Talean soil than agronomists do because they actually do the farming. "The government programs usually don't amount to much," said an informant. "Most campesinos who show up pick up their materials [fertilizer or direct cash payments], sit silently through the meetings, and leave." Over time, it became clear to me that Zapotec farming was immensely successful—not "underdeveloped"—when seen from the perspective of its practitioners; indeed, how else could the region's villages have survived for so long?

The irony of this and similar situations sparked a curiosity about the nature of science. Specifically, I began to ask the following questions. Does Zapotec food production, preservation, and preparation constitute a science—and, if so, how? If Zapotec farming and foodways do constitute a system of local science, how is it similar to modern cosmopolitan science and how is it different? How might the two be related?[2] These questions became the focus of my research. In general, the answers seem to depend on whether one takes the view that science is characterized by universal features or by culturally specific ones—that is, a relativistic view or one that is culture bound. The anthropological debate over "rationality" ("Does the 'Other' think rationally?") might be seen as another statement of the same problem.

For now suffice it to say that Zapotec farmers, to the extent that they employ a "considerable store of knowledge, based on experience and fashioned by reason" (Malinowski 1948 [1925]:26), certainly practice science, as does any society whose members engage in subsistence activities. They hypothesize, they model problems, they experiment, they measure results, and they distribute knowledge among peers and to younger generations. But they typically proceed from markedly different premises—that is, from different conceptual bases—than their counterparts in industrialized societies. Indeed, it is striking that in spite of technological changes in the Sierra—in crops, farming techniques, and food preparation—a distinct kind of logic appears to operate, one which holds together in a coherent fashion but is inexplicable or incorrect according to the dominant paradigms of contemporary international science or theories that model human behavior after an ideal "Economic Man."[3] In this sense, Zapotec farming might be thought of as a set of activities stemming from and informing conceptual foundations that are culturally incommensurable with those predominating in industrialized societies.

In this book I attempt to unravel some of the characteristics setting apart Rincón Zapotec farming and foodways from contemporary cosmopolitan science on the one hand and the ideologies of development planners on the other. I focus on how concrete, everyday agricultural practices flow from the logical calculus, epistemologies, or key concepts underlying Zapotec food production and consumption. I also concentrate on the ways in which Rincón Zapotec agricultural concepts and practices inform each other and the ways in which some modern cosmopolitan sciences inform and are informed by agricultural concepts and practices resembling those of the Zapotec. A central theme is that Rincón Zapotec farming and foodways mixed with and were juxtaposed to agricultural systems from many other parts of the world from the time of contact with the Spanish; consequently, the boundaries separating Zapotec knowledges and practices from those of other parts of the world have overlapped and blurred. Another theme is that crop cultivation may be viewed as a highly successful set of techniques when assessed using the criteria of Zapotec farmers and that, significantly, the fundamental concepts which inform them may bring to light a number of phenomena that lie outside the purview of cosmopolitan sciences. These concepts or categories include household and ecological maintenance, reciprocity, food quality, the personification of nonhuman and supernatural actors, the inevitability and normality of physical work, and hot/cold. Together they form a partial foundation for agricultural practices that are entirely coherent, rational, and quite effective for Rincón Zapotec farmers in the Northern Sierra.

Rationalities, Knowledges, and Sciences

In this book I attempt to build upon an emerging body of work on the anthropology of science and technology by analyzing a particular set of agricultural and dietary knowledges and practices that combine "local" and "cosmopolitan" components; "ancient" and "modern" idea systems and artifacts; and techniques and methods developed in "industrialized" and "nonindustrialized" societies. In short, this is an examination of the cultural implications of a subsistence system that might be described as a case of scientific and technological syncretism.

Increasingly, anthropologists have taken an interest in the anthropology of science and technology, and some have attempted to synthesize earlier work on science, magic, religion, and rationality. Robin Horton and Ruth Finnegan, for example, edited a collection on *Modes of Thought* (1973) a generation ago. Stanley Tambiah (1990) examines *Magic, Science, Religion, and the Scope of Rationality* in historical perspective, emphasizing their trajectory in anthropology. Laura Nader (1996), in an edited volume entitled *Naked Science*, reviews previous studies in the anthropology of science and focuses on configurations of power and the boundaries separating magic, science, and religion. This is perhaps the first major collection of anthropological essays juxtaposing cosmopolitan and local science cultures. Scholars from other disciplines, particularly philosophy and sociology, have also shown an interest in combining anthropologically informed analyses of sciences drawn from a variety of settings (Jasanoff et al. 1995; Harding 1998).

These debates have a long history in modern anthropology. For example, the renowned social evolutionist Edward Tylor (1871) argued that magic in "civilized" European societies existed only as a "survival" from a barbarous past. He divided religion, "the belief in spiritual beings," into a number of stages, the earliest of which coincided with magic in "primitive" societies. For Tylor, science was characteristic of "civilized" peoples and "changed ideas from notions of personalized force to impersonal force... [and] therefore necessarily dissolved animism" (Tambiah 1990:50). He ascribed scientific thought to Europeans and magical and religious thought to their colonial subjects. James Frazer (1911–1915) argued that magic corresponded to "lower cultures," while religion and finally science corresponded to "higher cultures."

French theorists were also active participants in rationality debates. Emile Durkheim and Marcel Mauss (1963 [1903]) hypothesized a continuous development from "primitive" to "modern" thought. Lucien Lévy-Bruhl (1923), in his earlier work at least, rejected the notion that "primitive" thought was a rudimentary form of the "modern" civilized mentality, but instead argued that it was altogether different from modern rational scientific thought and that "it had its own characteristic organization, coherence, and rationality" (Tambiah 1990:86). Claude Lévi-Strauss's view can be contrasted to that of Lévy-Bruhl; in *The Savage Mind* (1966) he describes the difference between "concrete" science and "abstract" science, present in both "primitive" and "modern" societies. For Lévi-Strauss, the complex, particularistic categories deployed in "concrete" thinking are constructed in a piecemeal fashion, like that of the *bricoleur* or "tinkerer" who pragmatically assembles things from odds and ends. This differs from the more general "concepts" constructed when "abstract" science is done.

British structural functionalists engaged the French scholars, particularly Lévy-Bruhl, in different ways. Bronislaw Malinowski (1948 [1925]) promptly dismissed his views, arguing instead that magic, science, and religion coexist in all societies, though each serves a different function. In particular, he stressed the universality of scientific thought by providing examples from Trobriand gardening, fishing, and canoe-building to show how science was used to deal with the natural world (Malinowski 1948 [1925]:26). Trobrianders deployed magic and religion for influencing the sacred supernatural sphere, which served psychological and social functions, respectively. Malinowski showed that the thought processes of Trobrianders and Europeans were similar, for each practiced magic, science, and religion. He demonstrated that Trobrianders had genuine science (not just magic and religion) and that Europeans had magic and religion (not just science).

Malinowski's work was influential and probably inspired Ludwik Fleck's critique of science (1979 [1935]). Fleck, a Polish physician, reviewed changing notions of syphilis over time and succeeded in revealing science's constructed nature, its contextuality, and its contingency. He influenced science studies by way of Thomas Kuhn's *The Structure of Scientific Revolutions* (1962); following this lead, social scientists—including anthropologists—began doing work among modern cosmopolitan scientists (Traweek 1987; Dubinskas 1988; Gusterson 1995).

In *Witchcraft, Oracles, and Magic among the Azande* E. E. Evans-Pritchard (1976 [1937]) focused upon how the Azande answered questions that modern scientists did not even attempt to ask. He found that the Azande deployed oracles and magic to explain random events—for example, a death caused by a collapsing granary. They used divination to expose otherwise inexplicable occurrences as premeditated acts carried out through witchcraft. Evans-Pritchard (1976 [1937]:16) concluded that "[t]he Zande mind is logical and inquiring within the framework of its culture and insists on the coherence of its own idiom." Tambiah (1990:92) summarizes the similarities in the approaches of Malinowski and Evans-Pritchard:

[Both were conscious of] the danger of double selection by which ["primitive" peoples] are described entirely in terms of their mystical beliefs, ignoring much of their empirical behavior in everyday life, and by which Europeans are described entirely in terms of scientific rational-logical thought, when they too do not inhabit the mental universe all the time ... A person can in a certain context behave mystically, and then switch in another context to a practical empirical everyday frame of mind.

The rationality debates have continued in anthropology. Robin Horton (1967), for example, has argued that "theoretical thinking" in modern cosmopolitan science has its equivalent in African thought, but that the latter is inferior since "it is not reflective or critical, is closed rather than open, it is unable to entertain alternative conceptions to its dogma, it is ignorant of the experimental method ... and it resorts to secondary rationalizations to protect its premises, rather than face courageously the possibility of falsification" (Tambiah 1990:91). Horton fails to mention how such characteristics may also describe what Kuhn (1962) called "normal" activity among modern cosmopolitan scientists.

Max Weber's work (1946) has embedded within it a critique of rationality, not unlike Kuhn's critique of "normal science." For Weber, rationality was not a quality in and of itself, but part of a broader process, *rationalization:* the rational reorganization of production for maximizing productive efficiency. He argued that rationalization stemmed from the introduction of a new "capitalist spirit" of entrepreneurial enterprise and the subsequent "disenchantment" of the world. In this sense, Weber's rationality assumed a progressive, unilinear development. Unlike the evolutionists, however, Weber saw rationalization not as a liberating force, but as an all-enveloping ethos linked to the creation of bureaucratic, dehumanizing "iron cages" (Weber 1958:181).

Similarly, Karl Mannheim (1960:508–509) noted that rationality has at least two distinct and apparently contradictory meanings. Mannheim, like Weber, drew links between rationality and industrialization and noted that, although they may represent a particular evolutionary stage, they tend to paralyze, not liberate (Mannheim 1960:512):

Increasing industrialization, to be sure, implies functional rationality, i.e. the organization of the activity of the members of society with reference to objective ends. It does not to the same extent promote "substantial rationality," i.e. the capacity to act intelligently in a given situation on the basis of one's own insight into the interrelations of events... The violent shocks of crises and revolutions have uncovered a tendency which has hitherto been working under the surface, namely the paralysing effect of functional rationalization on the capacity for rational judgement.

The work of the Frankfurt School—particularly that of Herbert Marcuse (1964) and Jürgen Habermas (1971)—might be seen as elaborations of this critique of formal rationality, so prevalent in modern bureaucratic societies.

Unresolved Issues in Studies of Local Science

Within anthropology, there is abundant evidence that local knowledge systems are internally coherent and logical. Studies of local science traditions, following Malinowski, have described, among other things, native astronomy in the Caroline Islands of Micronesia (Goodenough 1953), knowledge about marine life off the coasts of Hong Kong (Anderson 1967), navigation in the Puluwat Atoll of Micronesia (Gladwin 1970), Tzeltal Maya plant classification in Chiapas, Mexico (Berlin 1974), and "ethnoecological" knowledge (Nazarea 1999).

The interdisciplinary field of "traditional ecological knowledge" (TEK) has become something of a growth industry in recent years. Edited TEK collections have become increasingly popular (Brokensha, Warren, and Werner 1980; Freeman and Carbyn 1988; Johannes 1989; Williams and Baines 1993), and marine biologists (Berkes 1977; Johannes 1981), geographers (Richards 1985; Wilken 1987; Nietschmann 1989), entomologists (Altieri 1987), and others have taken an interest in local environmental and resource management strategies. Such projects demonstrate that scientific thinking appears to exist everywhere, even if different societies have particular scientific styles or begin from distinct theoretical premises.

Even so, several issues remain unsolved in the TEK field. To begin with, a number of anthropologists and TEK researchers describe the benefits associated with certain local science practices, but often analyze local knowledge systems in the terms of modern cosmopolitan sciences rather than from the "native's point of view." To the extent that this research demonstrates the effectiveness of local knowledges by translating the results into cosmopolitan science terms, such analyses are laudable. But there are hazards. In the words of anthropologist Colin Scott (1996:71), "if the sharing of knowledge were to be reduced to a skimming-off by Western specialists of indigenous empirical insights, and their mere insertion into existing Western paradigms, then it would be an impoverished and failed exchange that would ultimately contribute to undermining indigenous societies and cultures." Angus Wright (1990:254) agrees: "There are major problems about how to proceed with such an effort because there are fundamental conflicts between the way Western science views the world and the systems of knowledge and practice of traditional agriculturalists." He cites Richard Norgaard (quoted in Wright 1990:254):

In the absence of a consensus about epistemological beliefs, agroecologists have resorted to pragmatism. Western knowledge is not rejected, for the mechanical world view has given us many insights, and conventional agricultural explanations help the agroecologist understand traditional systems as well. At the same time, agroecologists are receptive to the explanations of traditional peoples ... traditional knowledge has survived the test of time—the selective pressures of droughts, downpours, blights, and pest invasions—and usually for more centuries than Western knowledge has survived.

Perhaps part of the problem is the search for a "consensus about epistemological beliefs"; indeed, such beliefs often derive much of their power from their specificity, uniqueness, and geographical exclusivity—in a word, their "localness."

The question of how magic, science, and religion might be related in all cultures is often not addressed in TEK studies or in studies conducted in "high-technology" settings. *Naked Science* (Nader 1996) begins with a discussion of these connections, and several of the essays illuminate linkages between the knowledges and practices of local and cosmopolitan scientists. A branch of cultural materialism has drawn links between symbolic structures and ecosystemic principles (see, for example, Rappaport 1968; Reichel-Dolmatoff 1976), but as Scott (1996:71) notes,

it can appear as if the "totalizing" view of Western science has captured what remained unconscious or invisible to native subjects. The intellectual processes involved in framing practical knowledge within cosmological categories, from the actors' point of view, remain largely obscure. The adaptation of native cosmologies to their material-historical environments can then appear to be . . . the outcome of blind selective forces, rather than the outcome of theoretical work and proactive environmental management on the actors' parts.

Furthermore, there is sometimes a tendency to exaggerate cross-cultural distinctions, as if different systems of thought were not linked together: "The dominant . . . tradition of science is one among many traditions. Historians of science who describe science as a tradition originating from Europe are incorrect and ignorant of the remarkably diverse science traditions internal to Europe itself, traditions that have vied for control over knowledge production" (Nader 1996:8). Nader writes about struggles to create boundaries demarcating magic, science, and religion, how such struggles are embedded in political and economic structures, and how limits surrounding them often imply a civilizing process used to justify the imposition of an allegedly universal science and technology in regions thought to lack development. Only in this way can we understand international development projects such as the Green Revolution (Shiva 1991; Escobar 1995; Scott 1998). Such cases reveal the connections between scientists who may see their work as value-free and the political agendas of some development agents.

The Green Revolution highlights another dimension of some kinds of cosmopolitan science, especially "Big Science" projects like the Manhattan Project: an unshakable faith or belief in the necessity and goodness of the scientific process and the final product. According to Nader (1996:24), "the belief in the omnicompetence of [cosmopolitan] science has been steadily gaining ground throughout this century in this culture, and operating on a core-periphery model in the world." What is needed, she argues, is a relativizing or "lower-casing" of a Big Science that, in Paul Feyerabend's terms (1978), has the ideological grip of a religion.

The Sociology of Local Science: Connecting Concepts to Practice

Where does this leave the anthropological endeavor, which itself emerged as a modern cosmopolitan science in the nineteenth century? Colin Scott (1996:71) suggests that "anthropology is unique in the degree to which it emphasizes the more inclusive cultural contexts of our local teachers and values ways of translating indigenous knowledge that reflect the symbolic and institutional contexts in which the knowledge is generated." His research focuses on the way in which actors model "social-environmental practices" in terms of "mythico-ritual categories."

Specifically, Scott examines the construction of hunting knowledge by the James Bay Cree in Canada, particularly the paradigms of pan-species personhood and communicative reciprocity. He shows how these categories allow Cree hunters to predict the behavior of geese in ways that are only now beginning to be understood by modern biologists. Central to his argument is the idea that all sciences—including cosmopolitan sciences—are constructed upon certain paradigms. For example, in industrialized societies the idea that nature exists apart from culture and humans (and is therefore subject to their domination) undergirds much of science and may be linked symbolically to notions of hierarchy—with serious implications for the way nature is manipulated in practice. There is no reason, argues Scott, that cosmopolitan science concepts should be accepted a priori; indeed, they should be subjected to the same scrutiny as local science concepts.

Similarly, marine biologist Robert Johannes (1981) takes an anthropological look at coral reef fishing in the Palau district of Micronesia from the "native's point of view" with impressive results. He discusses local conceptions of time used by successful fishermen, including lunar and seasonal rhythms that are often seen as superstitious by city dwellers. Johannes also devotes an entire chapter to a discussion of how seabirds are used as instruments by Palauan fishermen and another to the "traditional conservation ethic," the ideological basis for sustainable fishing practices. It includes established norms for reef and lagoon tenure, avoidance of waste, and laws restricting fishing at certain times. Johannes pays special attention to the influence of "modernization" upon local fishing practice and its foundational concepts. In particular, he notes that the conservation ethic has declined in the wake of imported food and the dual processes of industrialization and official state education, which have drawn young Palauans away from fishing and into bureaucratic jobs. Johannes's monograph gives special attention to links between theoretical concepts and practices in subsistence activity.

Scott and Johannes suggest new directions in the analysis of local subsistence systems and address some of the shortcomings that characterize many local science studies. They both analyze their respective subject matter using the conceptual categories of local specialists. Both relativize modern cosmopolitan science by contrasting its assumptions with local assumptions about the environment, human-animal interaction, and the relationship of nature to culture. The results force us to consider the possibility that modern cosmopolitan sciences, like any knowledge system constructed upon a set of given precepts (the theory of relativity, the second law of thermodynamics, the effects of lunar patterns, pan-species personhood, etc.), might inherently filter out or exclude certain possibilities that might have a high degree of predictive value for practical application. Both Scott and Johannes examine the interrelationships between local and cosmopolitan science traditions, how each has influenced the other, and how third systems have had an impact on both. Finally, they take Cree hunting and Palauan coral reef fishing, respectively, seriously enough to probe the relationship between local concepts or theories and subsistence practices. What emerges in each case is a sociology of local science—a serious, sophisticated treatment of locally based knowledge production.

In agriculture such an approach is lacking, though important initial steps have been made by anthropologists. For example, in *Coral Gardens and Their Magic* (1935) Malinowski exhaustively describes Trobriand agriculture and links it to social organization, kinship, and the legal order. Although Malinowski alludes to the influence of European society (noting, for example, that the Trobrianders exported food surpluses for consumption in the colonial plantations of New Guinea), he minimizes outside influences. Trobriand horticulture was essentially a means for Malinowski to demonstrate functionalist theory: garden magic integrated farmers' attitudes and organized their cooperation. One reviewer claimed that *Coral Gardens* challenged ethnocentric notions of "primitive" society: Trobriand agricultural methods were "eminently fitted to their environment and tribal structure . . . there is nothing primitive about them" (Madgwick

1936:141). However, this was attributed to the Trobrianders' indigenous farming, not to their intelligent selection of foreign technologies such as steel axes.

Harold Conklin's analysis (1954) of shifting (swidden) agriculture among the Hanunóo of the Philippines is another ethnography which explicitly reviews local farming knowledge and practice. One of the central points is that swidden cultivation occurs in many forms, not all of which are destructive. In fact, the relation of the Hanunóo to their environment appears to be remarkably harmonious. This was an important point to make, for at that time many assumed that all forms of swidden agriculture were ecologically disastrous. Conklin, by contrast, argues that Hanunóo farming is integrated, self-contained, ritually sanctioned, and ecologically viable. Although a detailed list of cultivated crops is included, they are not discussed in wider contexts. One is left with the impression that Hanunóo agriculture is static, stable, and in near-perfect equilibrium with the environment. Absent from the discussion is a consideration of what kind of an impact chili peppers, maize, cacao, coffee, tobacco, and other crops might have had once they were introduced on the island.

Both *Coral Gardens* and "The Relation of Hanunóo Culture to the Plant World" are characteristic of studies which portray local knowledge systems as organic or "home grown." While providing good descriptive data and functional, linguistic, and ecological analyses, they tend to downplay cross-cultural factors. As in many other ethnographies of science,

> there is a tendency to focus too closely upon relatively isolated "villages" or communities of specialists, whether they are Nobel Prize-winning researchers at the Jonas Salk biological laboratories or "traditional" navigators sailing among the islands of Melanesia . . . Perhaps of greatest concern is the general inability of any approach to affect the self-knowledge of physical and biological scientists. (González, Nader, and Ou 1995:867)

But local knowledges are often a complex combination of ideas, artifacts, and institutions that have traveled rough temporal and cross-cultural trajectories, as world-systems theorists and others have made abundantly clear (Wallerstein 1974; Wolf 1982; Mintz 1985; Weatherford 1988).

Clifford Geertz's monograph on *Agricultural Involution* (1963) in Indonesia takes a different approach. His analysis of two agricultural systems shows how the intensive wet rice subsistence farming of Central and East Java was linked to the extensive swidden and cash-crop farming in much of the rest of the country. Specifically, as the latter system was imposed upon a greater portion of Indonesia's land, the former system became increasingly intensified and undiversified ("involuted"). Geertz's study differs from those of Malinowski and Conklin to the extent that it discusses agriculture in a historical, ecological, and geographical context. But it has shortcomings: as one reviewer noted, Geertz's ideal types ignore the important role of intermediate regions practicing both wet rice farming and cash cropping (Jaspan 1965). Nor does the study focus on the various meanings attached to agriculture by the Indonesians themselves.

Other research on local agricultural systems generally falls within three categories: (1) anthropological work describing agricultural techniques in the terms of local farmers;[4] (2) studies analyzing "agroecological" or local "resource management" techniques in terms of cosmopolitan science (Altieri 1987; Wilken 1987);[5] and (3) work reviewing local subsistence strategies in relation to development programs (Chambers 1983; Richards 1985).[6]

These studies successfully demonstrate the effectiveness of local agricultural techniques but rarely focus on how these knowledges and practices have incorporated elements from the outside world and vice versa — in short, they often rely heavily on a separation between the local and the global without considering how ideas and artifacts are borrowed cross-culturally. In this book I attempt to fill part of the void by analyzing the historical trajectory of food plants in a global context, the techniques and knowledges associated with them, and their meanings in a Rincón Zapotec village.

Fundamental Concepts Underlying
Rincón Zapotec Farming and Foodways

This study analyzes some of the conceptual bases of Rincón Zapotec farming and foodways and their relationship to actual farming practices. Some are encoded in myths and rituals (as in the case of Cree hunters); others are embedded in institutions (as in the case of Palauan fishermen); still others are expressed more explicitly in informants' explanations of farming and food practices. In a study of myth and ritual among the Valley Zapotec, Fadwa El Guindi (1986:3), invoking the work of Lévi-Strauss, notes:

Like myth, ritual has been shown to be enormously rich in symbolism and productive as a reservoir of cultural conceptualizations ... Through ritual's repetitive statements and activities, concepts that are embedded in the collective unconscious of culture-bearers become activated, revealing to the ethnographer and analyst not only their content but their structure as well.

Institutions such as reciprocal exchange or "gift-giving," to the extent that they are "total social phenomena" (Mauss 1954 [1924]:1), may also be exceptionally revealing, as they find "simultaneous expression ... [in] religious, legal, moral, and economic" spheres of social life. With respect to terminology, explanations, and other utterances, one of the aims "was to discover how people construe their world by the way they talk about it" (El Guindi 1986:23). Ritual, myth, institutions, and language can illuminate a group's shared underlying cognitive structures.

For the Rincón Zapotec, some of the most important concepts include *mantenimiento* (Sp., "maintenance"), reciprocity, the personification of nonhuman and supernatural beings, the normality and inevitability of physical work, food quality, and hot/cold. These six concepts are interrelated and often overlap in ordinary conversation and daily practice; even so, they may be separated for analytical purposes. They do not neatly fit any single typology, but instead might be seen as a mixed bag of interconnected ideas and postulates.[7] Some are genuine theories about the world; others define the goals of local science practice. To avoid ambiguity, I shall refer to the Zapotec ideas as concepts, precepts, and epistemological or conceptual bases.

A number of disclaimers are in order. Not all of the concepts can be mapped mechanically onto all farming and food preparation practices. Some serve as general guides, models, or prescriptions for certain aspects of food production or consumption, in much the same way that Heisenberg's Uncertainty Principle and Bernoulli's Principle dictate practical measures taken in physics experiments and fluid mechanics projects, respectively. They structure action and provide the conceptual tools for real-life activity in the minds of practitioners. Some concepts may be relatively new ideas recently introduced from the outside world. Most of them, however, have probably been deduced over the course of hundreds or even thousands of years.[8]

Nor would I suggest that these terms are used by all Talean farmers. Like fundamental concepts in cosmopolitan science, key agricultural concepts in Talea are generally consensual, but there are those at the fringes who, for various reasons, do not share the more uniform views of other campesinos—these are farmers who, in Kuhnian terms, might not adhere to "normal science." Perhaps the most obvious example is the case of farmers who have abandoned subsistence cropping completely in order to cultivate coffee. They tend to take a radically different approach to farming than their fellow villagers—most notably, bookkeeping becomes more important than householding.⁹ Whether this outlook will eventually gain acceptance and replace one based on the notion of household maintenance, remain at the margins of agricultural theory, or disappear entirely in Talea is still an open question.¹⁰

The genesis of these ideas is a complex question. As we shall see, it is difficult enough to follow the trajectories of artifacts and crops over time, and it is an even greater challenge to track ideas. It is tempting to refer to the following concepts as indigenous, but this obscures the fact that much of what presently encompasses Rincón Zapotec farming and foodways, in both theoretical and practical terms, can probably be traced to the Spanish colonial or postcolonial eras. Because the boundaries separating local and global science traditions blur and overlap, the question of origins is often difficult or impossible to answer. Consequently, such an analysis lies outside the scope of this book.

Mantenimiento (Maintenance)

Mantenimiento, a Spanish term literally meaning "maintenance," is a concept that Talean campesinos use to guide the entire gamut of activities related to food production and consumption. In fact, one of the most significant things about Zapotec farming and foodways is that, when talking about their work and that of their families, many campesinos do not make a sharp distinction between agriculture, food preparation, and consumption. *Mantenimiento* glosses these distinct moments in the life cycles of crops by making reference to their common aim: household maintenance over the course of human lifetimes. The goals or objectives of the activities are thus more important than the setting in which they occur (in the fields versus in the home/kitchen).

This concept of household maintenance, which shares a remarkable congruence with the ancient Greek notion of *oikos* (from which the term "economy" is derived), has a number of practical ramifications for farmers. It implies that a balance must be struck between producing sufficient food so that household members have enough to eat and at the same time limiting production (by allowing land to lay fallow) so that farmland is not overtaxed. The two may be contradictory, but each is necessary to maintain the household. In the case of the former, short-term maintenance is assured by coaxing a sufficient quantity of food from the earth; in the case of the latter, long-term maintenance is guaranteed by preserving land for future generations. *Mantenimiento* spills over into other domains as well. For example, draft animals and pack animals are "maintained" so that they can assist in the larger role of household maintenance. Children are "maintained" so that they can contribute to household maintenance through farm and domestic work.

Mantenimiento can be drawn into sharper focus by juxtaposing it to growth models that dominate the language of development planners. In growth models the objective of productive processes (including agricultural production) is not equilibrium or maintenance but rather economic growth, generally realized through profit maximization and expressed in terms of "Gross Domestic Product," "average yield," and "productivity." Furthermore, agriculture and nutrition are conceived of as separate categories corresponding to distinct sites.

Reciprocity

Reciprocity is most apparent in the Zapotec institution of *gozona*, a mutual aid arrangement. It is a good example of what Marcel Mauss referred to as *The Gift* (1954 [1924]:1), in which "contracts are fulfilled and exchanges are made by means of gifts. In theory such gifts are voluntary but in fact they are given and repaid under obligation . . . [they are] economic prestations between the component sections . . . [of] 'archaic' societies." *Gozona* touches many aspects of village life, but in its essential form it involves lending goods or services to be repaid at a future date. In farming this typically involves helping kin, neighbors, or friends with several days of agricultural work which are then returned upon request. In food preparation it implies food exchanges and the pooling of labor to cook food, particularly for feasts associated with life-cycle events.

Reciprocal arrangements prescribe relationships with kin, neighbors, and other campesinos, but in reality they extend even further. *Gozona* may be carried out between two villages (for example, when musical groups are exchanged for fiestas) or between families in neighboring villages (for example, when one hosts another during a fiesta). Furthermore, reciprocity in the form of successful farming and good health is expected from the earth, non-Christian spirits, Catholic saints, and the deceased in exchange for sacrifices, prayers, and other ritual offerings. Reciprocity might be related to *mantenimiento* to the extent that it involves the maintenance of social relations with supernatural actors who can help the household successfully sustain itself.

We might contrast such reciprocal exchanges with labor contracts and food purchases which are the norm in many industrialized societies. For example, at life-cycle events in the United States (such as weddings and funerals) buildings must typically be rented, cemetery plots purchased, food catered, morticians and musicians contracted, etc. Reciprocal arrangements have, in a sense, been replaced by cash transactions. Hence Mauss's observation (1954 [1924]:74): "It is only our Western societies that quite recently turned man into an economic animal... *Homo economicus* is not behind us, but before... For a long time man was something quite different; and it is not so long now since he became a machine—a calculating machine."

Personification of Nonhuman and Supernatural Actors

The personification of nonhuman and supernatural actors is another concept employed by the Zapotec, not only in the Rincón but in other regions as well (Nader 1969; Marcus 1983). It is most clearly expressed in ritual sacrifices and myths. The earth, for example, is assumed to be a willful being who feels pain when her surface is cut open by a plow or scorched with fire or when her fruits are plucked. Farmers customarily reciprocate by offering sacrifices directly to the earth or her emissaries: stone idols buried in the ground, mountain spirits in the forest, Catholic saints with jurisdiction over farming, or all three. The earth can inflict punishment upon offensive humans by causing accidents, holding human souls hostage, or withholding harvests. Individual crops such as maize are also seen as willful actors, as are deceased humans and certain animals such as snakes and bulls.

Enchanted beings are a part of the local mythology. Perhaps the best-known is the *matelacihua* or *mal aire* (Sp., "evil air"),[11] an enchantress usually described as a beautiful woman with fair hair and skin. She appears in the evenings on footpaths between the village and farms and mesmerizes men. According to Taleans, those enchanted by the *matelacihua* wake up minutes or even hours later, miles away from the place of the encounter. The *bhni guí'a* (Zap., "man of the mountain") is another supernatural being; he dwells in the forest above the village, appearing as a well-dressed man in city clothes. He offers wealth in the form of enchanted money in exchange for one's soul. The *bhni guí'a* seems to represent the excesses of a cash economy; one could interpret his curse as a symbol of the consequences that the pursuit of cash might have for householding, reciprocity patterns, and so on.

For the Rincón Zapotec, even one's own body may express willfulness by "requesting" certain foods—"hot" foods if it feels "cold" or "cold" foods if it feels "hot." Furthermore, it may crave certain items (often without the individual being aware of it), giving startlingly explicit indicators: an inflamed thumb or big toe may signal a craving for spicy pork sausage; a yellowish abscess on the lip, a craving for pork skin; lumps around the neck or lower jaw, a craving for eggs; parted skin behind the ear, a craving for fish.

These personified entities contrast sharply with conventional concepts among development agents and many modern cosmopolitan scientists, who view intentionality and willfulness as the exclusive domain of humans. Earth, water, forests, and, more generally, nature are considered to be inanimate objects, while for developers animals and plants are usually treated as instruments, capital, or commodities. Michael Adas (1989:213) notes that during the Enlightenment, "through scientific discoveries, the long-hidden secrets of the 'world machine' were being revealed, while a great proliferation of technological innovations was making it possible for Europeans to tap vast reservoirs of natural energy previously not even known to humans"—the earth emerged out of the Enlightenment not as a living being, but as a mechanical object. Thus the dominant post-Enlightenment view of "nature" stands quite apart from that of the Rincón Zapotec.

The Normality and Inevitability of Physical Work

Most Rincón Zapotec assume that physical work is a normal and inevitable part of life for all but a select few. In the words of an informant, "There's no rest until they lay the leaves on your belly" (a reference to the broad leaves of a plant used to cover earthen funeral mounds). Michael Kearney (1972:45) observed a similar, though more fatalistic and tragic, outlook on the part of Sierra Zapotec in the village of Ixtepeji. From a young age, Taleans learn how to do simple farming and food-processing tasks, and parents take special precautions to ensure that their children do not grow up to be lazy people. They train their children to carry out chores responsibly; to ensure that their infants will collect firewood efficiently later in life, some farmers place tiny amulets around their necks, formed from a cocoon resembling a miniature log. One of the most striking things about the fieldwork experience was the sheer physical strength of many villagers—men, women, and children. It was not unusual for a campesino to carry 130 percent of his or her own body weight using a tumpline. Many were able to stoically endure pain or physical discomfort for long periods.

Critiques of those who escape manual work are harsh. Taleans often describe the inhabitants of Villa Alta, a neighboring village that has served as the political and administrative seat of the region since its founding by the Spanish in the 1520s, as parasitic, lazy people who make an unscrupulous living by collecting legal fees from hard-working country folk. Politicians and, in some cases, schoolteachers are criticized along similar lines. During fiestas, villagers ritualize such criticism by lampooning government bureaucrats such as the *agente del ministerio público* (Sp., a locally stationed district attorney representing the federal government), secretaries, and police in colorful Conquest plays. The officials are portrayed as overpaid people who dress in city attire and sit uselessly before typewriters, sipping sodas.

Such a view differs from a recurrent theme among many modern economists, who view physical work as distasteful and hold out the possibility of the liberation of humans from the drudgery of manual labor, expressed by theorists of both the left and the right. Karl Marx predicted that technology and automation would eventually render manual labor obsolete and ultimately lead to socialist revolution. Neoclassical economists, in a similar vein, currently tend to describe countries with many white-collar workers and "service sector" employees as more developed than countries with many people devoted to "primary sector" (agriculture, mining, logging, etc.) or "secondary sector" (blue-collar) work. Sidney Mintz (1985:35) has suggested that such a view may have predated the industrial era among New World colonists: in sixteenth-century New Spain many colonists viewed physical work as ignoble, using the Spanish term *deshonor del trabajo* to articulate this view. In short, many economists have attached a negative value and low status to physical work, and automation itself—in agriculture, in the industrial sector, and in offices—may represent an effort to minimize it as much as possible. Thus in many parts of the industrialized world manual labor is often viewed as neither normal nor inevitable.

Food Quality

For Taleans, the food quality of crops is often more important than the quantity harvested. This means that, for many, there is a threshold below which crops become unfit for human consumption. This is frequently articulated by Taleans in conversation and in their maxims: *Somos humildes pero delicados* (We are humble [poor] but delicate [picky]), I was told by a number of informants. Others noted that "[e]ven though we only have our little beans, they are legitimate, and we can eat them with pleasure because we know they're clean." In practical terms, this is manifested by a distinct preference for locally grown maize, beans, sugarcane products, coffee, poultry, beef, and other foods over imported items. "Knowing where it comes from" is critical in assessing foods, and many people purchase imported maize and beans only in cases of extreme privation. This implicitly forms part of a broader civilizational scheme opposed to urban models: indeed, viewed from the countryside, urban folk appear animal-like, since they eat anything, at any hour, without knowing whether it has been irrigated with sewage or drenched in pesticides—an unexpected situation in which city dwellers are classified as "brutes" or "barbarians." Food forms part of a broader scheme in which a high value—and an underlying civilizational assessment—is attached to those substances most important for the survival of humans: high-quality food, pure water, and clean air.

Among many bureaucrats in industrialized societies, quality may sometimes be a consideration, but in distinct ways. For example, in the United States, from the perspective of multinational food corporations and government bureaucrats, "food quality" often refers to the external appearance of fruits and vegetables or ritual inspections of meat and poultry processing plants by U.S. Department of Agriculture (USDA) officials. More subtle conceptions of quality such as taste, texture, or the presence of pesticides are not always taken seriously (indeed, the ability to taste quality may be effectively impaired if consumers have no sense of how foods may vary). The requirements of a long, economically rational food chain may often take precedence over other criteria (though in Chapter 8 we shall see how this may be changing as organic foods gain popularity). This may explain the justifications for the injection of hormones in cattle, the use of pesticides and herbicides on farms, and other cases in which quality may be sacrificed in order to realize greater quantities of food.[12] Across much of the United States, the quantity of food (including its year-round availability) has often taken precedence over its quality.

The Hot/Cold Concept

Nearly all Taleans describe food, drink, plants, medicine, soil, and other substances as "hot" and "cold"; to the North American observer, the deployment of the categories seems ubiquitous. Humoral theory (or the hot/cold concept) has been extensively documented across Latin America. It is described by George Foster (1994:2–3) as a system in which

> all foods, all herbal and other remedies, and many other substances as well (such as iron), have a metaphoric "quality" . . . a humoral value of "hot" or "cold" (and occasionally "temperate") that serves primarily to distinguish classes of foods and remedies . . . [Illnesses are] due to hot and cold insults (sometimes thermal, sometimes metaphoric) that upset the bodily temperature equilibrium that is believed to spell health. A hot insult produces a hot illness, while a cold insult produces a

cold illness. Therapies... conform to what has been known since the time of Hippocrates as the "principle of opposites": a Cold remedy for a hot illness, and a Hot remedy for a cold illness.

In Talea, as in many parts of Mesoamerica, "other substances" include soil, water, certain minerals, human or animal fluids (for example, snake venom or bulls' blood), and air. Personality characteristics are often described in these terms (e.g., people from hotter regions are described as "hot-blooded").

Classificatory schemes are subject to inter- and intravillage differences, and Foster notes that even the same informant may give contradictory data. Some foods are consistently classified as hot (chiles and garlic), however, while others are consistently classified as cold, particularly plants with a high water content (for example, avocados, raw sugarcane, and cactus leaves) or game animals hunted in the ("cold") forest. Many others are classified according to criteria that are more elusive: oranges are hot, but limes are cold; beef is hot, but pork is cold; mezcal (distilled from the blue agave [*Agave* sp.]) is hot, but *aguardiente* (distilled from sugarcane juice) is cold; Coca-Cola is hot, but Alka-Seltzer and beer are cold. When I asked how hot and cold are distinguished in the latter cases, several informants told me that cold items make one ill if taken at night, while hot items do not. It may be that the effects of food items, rather than any of their intrinsic properties, serve as a guide for classification.[13]

In recent years anthropologists have "hotly" debated the origins of the Latin American humoral system. Foster (1994) claims that Spaniards—who inherited the system from the ancient Greeks—introduced the system in Latin America. Others argue that humoral theory evolved independently in the Americas (Butt Colson and de Armellada 1976; Ortiz de Montellano 1990). It now seems to have been established beyond a reasonable doubt that the American and Old World systems evolved independently before fusing in the decades following the Spanish Conquest.[14]

The Science of Zapotec Farming[15]

In a recent anthropological exploration of science, Sandra Harding (1998: 308–314) asks, "Could there be other sciences, culturally distinctive ones, that also 'work' and thus are universal in this sense?" Though the conventional view in industrialized society holds that only the "cosmopolitan" sciences deserve such a designation, a reconsideration of Harding's question might be fruitful. We might ask: Do the knowledges and practices of Zapotec farming and foodways—or any other system of local knowledge—constitute a science? If they are so thoroughly infused with quasi-religious and other problematic premises and assumptions (viewed from the perspective of contemporary international science), how can we describe Zapotec farming and foodways as a science in their own right?

Looking at the science concept in historical perspective is helpful. The word "science" came into the English language in the fourteenth century as a term for knowledge in the general sense. This continued until the nineteenth century, when it began to refer exclusively to the physical and experimental sciences. Science became associated with experiment rather than experience, and the terms "scientific," "scientific method," and "scientific truth" were used to describe the successful methods of physics, chemistry, and biology (Williams 1985:277–279).

My position is that science, in its most essential form, is a practical quest for truths about the world—a dynamic search for effective "knowledge, based on experience and fashioned by reason" (Malinowski [1948] 1925). In order to differentiate this definition of science from others, we might contrast it with a range of competing meanings:

1. Science as a collection of certain truths that have been verified (for example, Newtonian physics).
2. Science as high technology (as portrayed, for example, in the mass-media views of space travel, cloning, etc.).
3. Science as a social enterprise (for example, activities associated with the U.S. "Big Science" establishment).
4. Science as "what scientists do" (often meaning modern cosmopolitan scientists). This includes their errors, miscalculations, life problems, worldviews, etc.
5. Science as a form of inquiry that makes use of a particular method (the scientific method).
6. Science as the achievement of Superior Western Man. This is, at its root, a form of racism posing as science and long predates modern cosmopolitan science. Note, for example, the Spaniards' use of the term *gente de razón* (people of reason) to describe Europeans, as opposed to *gente de costumbre* (people of custom) or Indians. This view of science continues to the present day.

A great deal of what is written about "science" today mixes up these and other definitions, and in practice they are often confused by the mass media, government and non-government organization (NGO) bureaucrats, and development agents, as well as by the lay public. But by proposing a working definition of science that emphasizes the practical process by which people seek to discover truths about the world around them, I hope to create an area that is broad enough to include the activities of such diverse groups as Zapotec farmers, Chinese acupuncturists, nuclear physicists, agroecologists, and many others.

A critical part of my formulation is the notion of science as *practice*, as a practical search for knowledge to understand certain aspects of the world in which actors, while constrained by certain structures (conceptual, social, and material), can and do transform them over time, through practice (Farquhar 1994; Nyerges 1996). That is, people engaging in scientific activity begin with given frameworks, conduct practical experiments, analyze results, and modify given frameworks or invent new ones when faced with too many anomalies (Kuhn 1962). From this perspective, the search for knowledge takes the form of a dynamic accommodation that people make between experience and received wisdom.

Objectives and Themes of This Book

Here I have intentionally presented the Zapotec concepts in preliminary form so that they may be more fully developed later in the book. They take on more meaning when they guide daily activities (that is, when they are put into practice), and illuminating the way in which conceptual categories inform farming practices is an important goal. Ultimately, I shall use these concepts to examine the possibility that certain methods prescribed by the Zapotec concepts might long have eluded cosmopolitan scientists. This is closely related to one of the book's underlying themes: the legitimacy of local science. As we look at how the concepts are actually used to inform the fashioning of implements and maize, sugarcane, and coffee cultivation, it will become clear that Zapotec farming is remarkably successful when we assume the perspectives of its practitioners.

Another theme is the interconnectedness of Zapotec agricultural science and cosmopolitan science. Specifically, this book illustrates that the boundaries demarcating Zapotec farming and cosmopolitan science may be artificial because of the constant diffusion of crops, implements, and techniques across continents—what Alfred Crosby (1972) refers to as "the Columbian exchange." This theme illustrates the complexity of farming systems worldwide following centuries-old exchanges of plants, techniques, and knowledges. In short, local agricultural sciences have become "cosmopolitan" even as "cosmopolitan" sciences have become "localized" because of the multidirectional movement of crops and technologies.

Globalization is another theme. Analyses of local knowledge systems, particularly in the field of agriculture, could potentially enrich globalization studies by exposing possible connections between international political and economic policies and specific regions or communities. I have situated Zapotec farming within a global context to show how international institutions—as well as more general globalization processes—form a critical part of the reality, history, and local science of the Rincón Zapotec. I do this by focusing on three "world" crops grown in Talea, with distinct histories and complex trajectories over space and time: maize, a New World crop that diffused across Europe and Asia rapidly after 1492; sugarcane, an Old World crop thought to be native to New Guinea and introduced in the Americas shortly after the Conquest; and coffee, an Old World crop discovered in Ethiopia, popularized in the Middle East, and brought to Mexico (by way of Europe) relatively recently. Conventional notions of Eurocentric diffusionism—that the "West" developed the "rest"—are highly problematic if we trace the paths of crops in the era of empire (Kloppenburg 1988; Blaut 1993). Globalization also enters into my analysis of imposed development—from obligatory mining, cotton, and cochineal production organized by the Spanish in the colonial era, to the massive Papaloapan River basin development project of the mid-twentieth century (modeled after the Tennessee Valley Authority), to the U.S.-led Green Revolution and Mexican factory farming.

A number of issues in the concluding chapter also concern the theme of contemporary globalization. Specifically, I discuss the interrelated problems of global food security, environmental deterioration, the rapid diffusion of genetically modified crops, and the loss of genetic diversity—issues affecting small farmers across the world in the wake of the North American Free Trade Agreement (NAFTA), the General Agreement on Tariffs and Trade (GATT), and other "free" trade regimes that have radically extended the reach of multinational agribusiness firms (Nash 1992, 1994).

I suggest that as grain yields stagnate, as world population continues increasing, and as usable farmland around the globe decreases, possible solutions to the problems of twenty-first-century agriculture might be found among the Zapotec. They are extraordinarily efficient consumers of food who derive the vast majority of their nutrition from maize while consuming few meat products. They enjoy a high level of food security because nearly all their food is grown locally, using little fertilizer and virtually no pesticide. In essence, I argue that subsistence and semisubsistence agricultural systems are becoming more important because of, not in spite of, global integration. Unlike conventional factory farming, local farming systems often use tried and true techniques for controlling pests without chemicals, for managing droughts without irrigation, for feeding people without poisoning them, and for preserving high levels of biodiversity in and around farms. Cutting-edge scientific work by agroecologists, botanists, mycologists, entomologists, and others researching local ecological knowledge is testimony to an interest in producing nutritious food in a sustainable fashion—an activity that the Zapotec have successfully engaged in for more than 5,000 years.

This study should contribute to at least three growing fields. Literature on local knowledge systems is expanding rapidly, but appears to be marginalized within the anthropology of science and technology. Part of the problem may be the low status attached to local science studies in the United States; some of the best collections have been published by Australian, Swiss, and Canadian institutions (Williams and Baines 1993; Johannes 1989; Freeman and Carbyn 1988). With respect to local agricultural systems, the works of Robert Chambers (1983), Paul Richards (1985), and Gene Wilken (1987) are especially noteworthy. In Latin America a small but growing number of researchers are building a core body of work on local farming systems, resource management, and sustainability (Hernández Xolocotzl 1977; Toledo et al. 1985; Altieri 1987; Trujillo Arriaga 1987; Tyrtania 1992; Toledo 1995).

As an anthropological critique of development, this book should supplement previous work, including analyses of the depoliticizing effects of development projects in Lesotho (Ferguson 1990); of development as discourse (Escobar 1995); of development as an enterprise that often ignores local knowledges (Richards 1985); and of development as the product of high-modernist, authoritarian state planning (Scott 1998). In the field of Mexican rural development this book should complement analyses of recent land-tenure reforms (DeWalt and Rees 1994), the effects of and possibilities for rural development programs among the Maya (Faust 1998), and the impact of Latin American development on indigenous peoples and the environment (Wright 1990; Nigh and Rodríguez 1995).

Finally, this study updates the still modest ethnographic record of the Northern Sierra. Apart from a handful of ethnographies based on research conducted in the Rincón and surrounding villages (Nader 1964, 1990; Parnell 1989; Tyrtania 1992; Hirabayashi 1993), relatively few monographs on the Zapotec of the Northern Sierra have been published (Schmieder 1930; de la Fuente 1949; Mendieta y Núñez 1949; Pérez García 1956; Berg 1968; Kearney 1972; Young 1976; Ríos Morales 1994).

Methodological Approaches

Anthropologists enjoy a host of methodological approaches, and it is important that the fieldworker select analytical units and methods appropriate for specific research questions. Because my study is about local science in a global context, I focus on a particular place—a village community.

Community studies have a long tradition in Latin American anthropology, from the work of Robert Redfield (1930), Elsie Clews Parsons (1936), George Foster (1948), and Julio de la Fuente (1949) to the more recent work of such anthropologists as Lynn Stephen (1991), Frank Cancian (1992), and Jeffrey Cohen (1999). For most of the twentieth century, however, many ethnographers did not explore links connecting villages, regions, nations, and the world system. After the 1960s Mexican anthropologists in particular critiqued these mostly functionalist ethnographies (some of which were used to guide projects in applied anthropology) on the grounds that they negated the need for structural change in Mexico by limiting the scope of analysis to village affairs rather than national or global politics (Bonfil Batalla 1962; Warman 1970; Stavenhagen 1971). More recently, some U.S. anthropologists have attacked community-based ethnographies as inadequate for capturing the complexity of the multi-sited world in which we live. In this book I attempt to combine both local and global approaches by maintaining a strong focus on one village while exploring global events through history and prehistory. In other words, I have tried to view global changes from a local vantage point.

In terms of field methods, the most valuable was "learning by working," a variant of participant observation in which the researcher accompanies informants to their work sites and toils with them in order to learn through firsthand experience. This method has been suggested mainly by people in the international development field looking for creative research methods for solving practical problems (Chambers 1983; Richards 1985). In practice, it has been carried out in only a surprisingly small number of cases (Hatch 1974; Devitt 1977; Richards 1979; Johannes 1981). To collect the data for this book, approximately thirteen months (out of a total of eighteen) were spent doing fieldwork—literally in the fields—an average of five days each week on farms and in forests and ranch houses more than an hour's hike from the village. During this time, I learned to use and maintain machetes, axes, hoes, plows, coffee depulpers, sugarcane mills, and other implements; to select and cut firewood; to cook beans, stews, and other foods; in short, to do most of the things that a young Talean campesino might be called upon to do in the course of the agricultural season. My education was put in the hands of a man described by several people as Talea's most dedicated campesino, highly recommended for his profound knowledge of plants and animals, his unremitting patience, his household's economic situation (he had no sons old enough to accompany him to the fields), and, not least of all, his thoughtfulness and sobriety. He and the farmers located near his ranch house became informants, neighbors, and close friends.

In spite of initial hardships, this approach offered advantages that would have been difficult, if not impossible, to realize in any other way. As farmers patiently taught me how to work without unduly punishing myself, they inculcated me with the frameworks that guided their own agricultural practices and elaborately explained the logic underlying their actions. They were occasionally bewildered at my inability to grasp some of their most basic, taken-for-granted concepts; but over time, as I unraveled the structures underlying everyday conversations, the assumptions became much clearer. Through their spoken words, they also fleshed out intricate links between their work and their lives, between agriculture and personal history. Perhaps just as importantly, I think that learning by working gave me a level of rapport—what the campesinos call *confianza* (Sp., literally "confidence")—that cannot easily be described. Suffice it to say that working alongside others in the most fundamental set of activities that constitute subsistence farming, day in and day out, quickly earns one a kind of respect that may be impossible to duplicate by other means.

Learning by working was supplemented with participant observation in the village. I gained a great deal of insight by participating in Talea's band and orchestra (where I played the trumpet) through most of the fieldwork period. Because one of the two musical groups was typically asked to play at fiestas, funerals, weddings, and other functions, I was automatically invited to many critical life-cycle events and village celebrations by virtue of being a musician; many Taleans greatly appreciated my services, judging from their favorable comments and encouragement. (The tradition of village bands is strong throughout much of Oaxaca. Most rural communities, even those with fewer than 300 inhabitants, have at least one musical group and take great pride in it.) It would be difficult to overestimate the significance of music in the Sierra; indeed, musical groups are important organizations with political implications at the village level (Nader 1990), and participation in the groups has customarily been less a pastime or diversion than a serious obligation. The musicians themselves, who are seen as role models by younger villagers, became an important pool of informants, since the vast majority were campesinos ranging in age from twelve to about seventy-five. The camaraderie among musicians is strong, perhaps because of the collective nature of making music and the shared sacrifice that long hours of practice and performance entail, and I gained many insights through close rapport with group members. Because membership in the musical groups drove home the obligatory and burdensome nature of reciprocity, it was profoundly educational. I learned that although musicians take pride in their musical abilities, they must meet rigorous demands and high expectations: participation inevitably means long, sleepless nights spent playing in the cold; the occasional neglect of farm work, animals, and families; and, in general, fatigue. For me, being a musician meant striking a physically exhausting (and sometimes impossible) balance between musical responsibilities, agricultural work, and note-taking. Nevertheless, to the extent that I was drawn into networks of reciprocity and obligation—and expected to participate in organizational matters (official meetings regarding appointment of group officials, negotiating sessions with municipal officials, festival committees, and church representatives)—music was an invaluable source of education about village politics and small-scale democracy.

I conducted a village census with the help of four research assistants, which included questions about agricultural production, household self-provisioning, and migration. I also interviewed members of campesinos' families, merchants' families, and other villagers. In addition, I consulted physicians, teachers, and other government employees in formal and informal settings as well as representatives of civil and religious organizations with a strong presence in the region, such as Catholic missionary groups and NGOs. I spoke with others interested in the region, including biologists, anthropologists, and government officials in Oaxaca City. Finally, aerial photographs provided the basis for an analysis of village land-use patterns.

Organization of This Book

To understand Rincón Zapotec farming and foodways, we must know something about the historical contexts in which they developed. In Chapter 2, I outline the geography and history of the region. First I focus on the Rincón and Talea, particularly its geographic features and its fields and forests. I continue by reconstructing Sierra society on the eve of the Conquest and then by describing colonial-era political, economic, and social structures. The chapter includes a discussion of farming practices in the wake of post-Conquest plagues, pestilence, and tribute payments to Spanish administrators as well as more recent changes in village history. It concludes with a description of the village's economic, cultural, religious, and political landscapes, including an analysis of its regional, national, and "transnational" contexts. This clears the way for a detailed look at Zapotec farming.

In Chapter 3, I review weights, measures, farming implements, and other artifacts essential to successful food production. For campesinos who do not resist the idea that physical work is an inevitable part of life, the importance of custom-made tools should not be underestimated. Crafting implements requires complex calculations—all of which are retained mentally—and a degree of control over bodily coordination to ensure accuracy in the construction process. The chapter concludes by discussing the creative uses of scrap materials, the role of the campesino as an artisan who produces tools for his or her own use, and the implications of changes which threaten the survival of Zapotec farming.

In Chapter 4, I analyze maize, a New World crop dating far back to prehispanic times, with special reference to how the crop fits into distinctly Zapotec notions of living matter. The chapter begins with a review of several legends, myths, and rituals in which the crop and its bearer, the earth, are personified. From these accounts, certain key concepts come to light, including the personification of nonhuman and supernatural actors and the importance of reciprocity. This review is followed by a brief social history of the crop, particularly the role it played in the demographic expansion of the Old World and other "agricultural revolutions."

In Chapter 5, I take a closer look at maize farming practices and food preparation in Talea. This is followed by a discussion of distinctive methods (intercropping, the construction of earth mounds, and techniques based on lunar rhythms) that are of special interest because they have only recently been recognized as effective practices by some cosmopolitan scientists. The chapter concludes by looking at the issue of food quality as a dominant concept among the Zapotec.

Another important staple food in Talea is sugarcane, the topic of Chapter 6. As an Old World crop it differs from maize in important respects, but observers are correct in noting that "most species introduced [by the Spanish] have been totally integrated [into indigenous schemes] and are conceived of as 'creoles' " (Katz 1992:766). After a brief historical profile of the crop, I review farming techniques and the role that reciprocity plays in sugarcane production. Throughout, I focus on two questions. Why is it that sugarcane production has remained a smallholder activity in Talea? And why do Taleans prefer locally produced *panela* (unrefined brown sugar) over white sugar imported from outside the region, even when the latter is much cheaper? I conclude with a discussion of the symbolic aspects of sugar.

Coffee is a more recent crop. Since its introduction in the Sierra approximately 100 years ago, it has become integrated into Zapotec farming and social life. One of the main points of Chapter 7 is that coffee has become a part of the very identity of Taleans and that it has supplemented, not replaced, subsistence farming. Coffee forms part of a long-term strategy for household maintenance for the vast majority of campesinos. It is generally grown on parcels of land smaller than four hectares; unlike other parts of Mesoamerica, the Rincón has no large plantations (greater than ten hectares). At the same time, coffee has presented a number of contradictions because the crop is produced primarily for distant overseas markets, sometimes with severe price fluctuations. As a result, most campesinos have been unwilling to give up subsistence crops. The section ends with an account of the shifting political economic realities of coffee in the region.

In the final chapter, I examine the broader implications of Zapotec farming and foodways and the status of Zapotec agriculture vis-à-vis cosmopolitan science. This opens up a long-standing set of issues in the anthropology of science and technology. What constitutes science, and who is qualified to carry it out? In what ways is cosmopolitan science more successful than local science and vice versa? Is cosmopolitan science subject to ideological blinders that filter out new ways of constructing knowledge? These questions are addressed by examining the rise of new cosmopolitan sciences such as ecology, agroecology, alternative medicine, and others that have been informed by local science traditions. Specifically, I examine certain elements in campesino food production and consumption that were once considered superstitious, illogical, or prescientific by "normal scientists" but are now being taken more seriously by innovative cosmopolitan scientists. (By no means are all cosmopolitan scientists cut from the same cloth.) In short, the recent rapid rise of cosmopolitan science concepts that closely resemble those of some local science systems—including sustainability, ecology, alternative medicine, organic farming, and others—is probably part of a corrective trend in modern cosmopolitan sciences. Finally, I look at Zapotec farming and foodways in the context of international development initiatives and transnational firms and conclude with a discussion of the boundaries of science and technology.

Throughout, I follow three threads that constitute the fabric of Talean farming: campesino conceptions of living matter or what we might call "nature"; science (specialized knowledge about farming that campesinos bring to bear on their work); and technology (implements and techniques used in crop cultivation). Viewing these intertwined elements as part of a dynamic agricultural system (with disjunctures and contradictions as well as continuities) is a central part of my story and might serve as a framework for integrating Zapotec farming and foodways.

CHAPTER 2

LOCATING TALEA

Geography, History, and Cultural Contexts

Contextualizing Talea poses a formidable task because the village, though relatively remote in some respects (it is a five-hour trip from Oaxaca City by bus), sits squarely in the world of international migration, mass communication, and global trade. The economic reliance of most Talean households on maize and beans has not impeded their participation in "modernity." Investigating how Talea fits into bigger pictures—geographic, economic, cultural, ecological, and political—is a vital part of understanding its place in the world. These contexts link the local, the regional, the national, and the "transnational." The criteria used to describe Talea and its links to the outside world are flexible, and the notions held by a government functionary or an anthropologist do not always coincide with the villagers' conceptions. Even among Taleans there is disagreement about these concepts.

The Rincón

Talea lies in a region that today is commonly referred to as the Rincón of the Sierra Juárez.[1] At its southern end, the Sierra rises from the Valley of Oaxaca to a peak of more than 3,000 meters in the Mixe region at Mount Zempoaltéptl. From there, the mountains descend toward the coastal plains of Veracruz into the Papaloapan river basin, called Los Bajos by local people. Thus the Rincón forms part of the area between the Valley of Oaxaca and the coastal plains of the Gulf of Mexico.

Figure 2.1. Map of the Sierra Juárez and the Rincón.

1 Betaza
2 Choapam
3 Guelatao
4 Ixtlán
5 Santa Gertrudis
6 Solaga
7 Tabaa
8 Talea de Castro
9 Tanetze de Zaragoza
10 Totontepec Mixes
11 Villa Alta
12 Yaeé
13 Yagavila
14 Yalálag
15 Zoogocho

Two rivers flow northward into the Papaloapan basin from the Rincón, the Río Cajonos and the Río del Rincón. They flow toward Veracruz through the Rincón's only aperture, for on its other three sides the region is hemmed in by mountain chains (Nader 1990). Abundant humidity from the Gulf of Mexico—located 150 kilometers away—reaches the Rincón through this opening, often remaining trapped ("cornered") by the high peaks surrounding the region. This provides the Rincón with more rainfall than any other part of the Sierra; not surprisingly, it has been described as one of the most fertile parts of the state (Nader 1990:18; Tyrtania 1992:22). The difference between the lush pine and oak forests of the Rincón and the dry, baked lands of the Cajonos towns is especially striking. As one travels from the villages of the Rincón Alto (upper Rincón) to those of the Rincón Bajo (lower Rincón, in the direction of Veracruz), the climate is progressively hotter and more humid.

The history of the term "Rincón" is unclear, but it appears to be of relatively recent origin. The first reference I found dates back over fifty years (de la Fuente 1962), and the term appears to have been in broader use by the end of the 1950s (Nader 1964).[2] It has a certain derogatory connotation in Spanish and perhaps was coined by outsiders. Given the period in which it came into currency (perhaps in the 1920s or 1930s), it seems reasonable to guess that the term may have been popularized by government functionaries, schoolteachers, or missionaries. It is interesting to note that current political boundaries do not coincide with most definitions of the Rincón, for the region consists of communities pertaining to several districts, including those of Ixtlán and Villa Alta.

Nader (1964:199) noted that "there are probably as many boundaries [marking the limits of the Rincón] as there are inhabitants." This is as true today as it was then. There is no consensus on what constitutes the Rincón, and over the years it has been defined as a particular ecological and geographical zone, as an ethnolinguistic zone inhabited by Nexitzo Zapotec speakers, and as a base for political mobilization. Most definitions demarcate an area including approximately thirty to thirty-five villages that extend from San Juan Juquila Vijanos and Porvenir at the southern end to La Chachalaca, Las Palmas, Yagalaxi, and Lachixila in the north.

Village Geography

Though Talea is a relatively large village by Sierra standards (with approximately 2,500 inhabitants), it is a face-to-face society. Taleans share a set of common geographic spaces that form the stage for both everyday events and extraordinary dramas. Whether these are physical features (such as a hill or a spring) or architectural structures (such as a church or a road), they form meaningful settings for social interaction.

As in communities across Latin America, Talea's center is the nucleus of political, religious, and economic activity embodied in the municipal palace, the church, and the market buildings, respectively. The village itself sits on a patch of level ground, a kind of patio interrupting the slope of

Figure 2.2. Panoramic view of Talea in 1999. The name of the village is apparently derived from the Rincón Zapotec word *ta'a-lh'a*, literally "slope-patio," which is probably a reference to the village's geography. *(Photo by Gabriela Zamorano)*

the mountainside. The Zapotec name for the village, Ralh'a, means "close to the patio." The word "Talea" is apparently a simplified Spanish pronunciation of the Zapotec terms *ta'a-lh'a*, literally "slope-patio."

The Catholic church, a massive stone structure that took decades to complete, was initiated at the end of the 1800s. According to older informants, the building was modeled after a painting of an Italian church and was constructed entirely with local materials.[3] Heavy stones, some larger than a cubic meter, are said to have been hoisted into place using a wooden crane powered by oxen, and some were lifted more than ten meters.

After the church was completed in the 1920s, construction of the municipal palace began. This stately stone structure—the largest of its kind

Figure 2.3. The municipal palace is the center of village government. Taleans constructed the massive building in the first half of the twentieth century out of stone, brick, and a locally produced cement made of limestone and sand. *(Photo by Gabriela Zamorano)*

in the Rincón—was built over two or three decades with labor provided through *tequios* (Sp. from Nahuatl, obligatory communal work days) and by defendants in lawsuits who were unable to pay their fines in cash (Nader 1990:29). Both church and municipal palace are proudly cited by many villagers as evidence of the ingenuity, industriousness, and aesthetic appreciation of Taleans.

The market buildings are less than 100 years old. The Centenario, an adobe market building located alongside the municipal palace, was constructed in 1910, one of a dozen built with the assistance of the Mexican government to celebrate the centennial of Miguel Hidalgo's Grito de

Dolores (the cry that marked the beginning of Mexico's War of Independence from Spain). Though a number of merchants still use this space to display their wares on Mondays (the weekly market day), more of them use the much larger Salón de Usos Múltiples (Hall of Multiple Uses), a brick and cement structure that was constructed in the 1970s in the space where a second market building had previously stood.

Although important prehispanic Mesoamerican centers such as Monte Albán were centralized (Flannery and Marcus 1983:102-104, 131-136), the Sierra was probably composed of small and rather dispersed hamlets prior to the Spanish arrival. This implies that in the Rincón village centralization may have been a product of the Conquest. It is interesting to note that the church, municipal palace, and plaza form the heart of the vast majority of Latin American communities and many cities as well, and they are typically much more developed in the New World than in Spain (Foster 1960:71-95). Some have argued that only in the context of a "New World" could central plazas and geometric village layouts become a reality for utopian-minded colonial planners (Wolf 1959:165-166; Foster 1960:94-95).

The relative importance of the Latin American plaza is open to another interpretation, however. In light of evidence that many Mesoamerican communities were either newly founded (as in the case of Talea) or else profoundly restructured by the Spanish in the early decades of the Conquest (Wolf 1959:214-216), it is probable that the new centralized villages gave the conquerors an efficient way of monitoring their new subjects. Given the fact that many people were forced into "congregations" to facilitate political and religious administration, the plazas probably gave the Spaniards a central location from which to oversee entire villages. From this perspective, the Spanish played the defensive role of police rather than being progressive social architects.[4]

The centrality of the church and municipal palace in Mesoamerican villages is striking physical testimony to the importance of the civil-religious hierarchy or cargo system. The use of space illustrates the relationship between social structure and architecture. If the civil-religious hierarchy formed the backbone of Talea's social structure, the church and the municipal palace probably served a similar function from an architectural point of view.[5]

Immediately in front of the municipal palace is the basketball court, where many civic ceremonies are held: graduation ceremonies; "sociocultural programs" to celebrate Independence Day, Mother's Day, and fiesta days; fireworks displays, musical concerts, and intervillage basketball tournaments on festival days; and receptions for visiting government functionaries. During the rest of the year, the basketball court is one of the most important informal meeting places for young people.[6] It serves as a secure public space away from home where teenage boys and girls can interact, especially in the late afternoons and evenings when junior high students are done with classes and campesino youths have returned from the fields. To understand why so many villages across the Sierra rank basketball courts high on their list of public works projects (along with electricity, roads, health clinics, and telephone service), one must first understand the important social role they play in village life.

Talea's other outdoor stage is the atrium located alongside the church. Large receptions and fund-raisers sponsored by the Catholic church are held here. The atrium also is the beginning point of religious processions, in which large wooden images of various saints are carried around the outside perimeter of the church building, accompanied by brass band music, prayers, and incense. It is customary for Conquest plays to be performed in the atrium during festival days, for the space lies two or three meters below the level of the road above, forming an open-air auditorium.

Two perpendicular axes radiate from the central plaza, dividing the village into four sections named after the cardinal points. They are used for assigning communal work projects. Each section elects a chief who is responsible for keeping the streets of his section clean and safe. Many Mesoamerican communities had such sections (called *barrios* in Spanish or *calpulli* in Nahuatl) in the prehispanic period (Wolf 1959:220), but they probably did not exist in the Sierra until the colonial era (Chance 1989).

Talea's houses conform to the pattern found in many Latin American communities: wealthy villagers tend to live close to the center, while *humildes* ("humble" or poorer people) live on the fringes. Consequently, the adobe houses near the center are often two or even three stories high and have many rooms. There are a lot more cinder-block buildings near the center as well. Cinder block began to be used about twenty years ago and has caught on quickly. As one moves further away from the center, the houses are typically two-, three-, or perhaps four-room adobe structures. The most distant houses are located upon a hill at the extreme northern edge of town, referred to as "the colony of the poor" because a small chapel there honors La Virgen de los Pobres, one of the Virgin Mary's manifestations. When newcomers are granted citizenship in Talea, they may request small homestead plots in this area, which has grown quickly over the last generation.

Figure 2.4. Scene from a Talean fiesta. This area forms part of the church's atrium, which serves as a public open-air auditorium. Clowns and acrobats are a principal form of entertainment at fiestas, as are *danzas* and musical events. *(Photo by Gabriela Zamorano)*

Unlike those in most Latin American communities, Talea's streets do not follow a square grid plan, but instead follow the contours of the mountainside. In fact, this is typical of nearly all Sierra and Rincón villages except Ixtlán and Villa Alta (Nader 1964). The streets are narrow, and the most important ones are paved with cobblestones. Only the two largest streets are paved with concrete.

Talea's infrastructure has been radically transformed in the last thirty years. Potable water, electricity, and garbage collection have been mixed blessings. Today all but a handful of houses have running water and electricity. Unfortunately, the town's sewage winds up in the Río de la Hacienda, below the village, and there is no longer wildlife in this river. Electricity generated by the dam constructed near the city of Tuxtepec (this was a vital part of the Papaloapan Commission's development project) is transmitted to the state capital, Oaxaca City, and the Sierra is fed along these lines from a substation located at Guelatao, between the Rincón and the capital. Talea's garbage is collected by the village dump truck once or twice a week, which takes the trash—a growing percentage of which is nonbiodegradable—to the edge of the village's land, at the border with neighboring San Juan Juquila Vijanos, where it is dumped just below the road in a smelly pile of garbage. It is clear that septic tanks, solar-powered technologies, and recycling programs have been overlooked in favor of inappropriate technologies that are taking a toll on Talea's natural environment.

There are many other important public spaces in Talea. *La curva* (Sp., the curve) refers to a place where the road makes a hairpin turn at the northeast end of the village. Just beyond is *la loma* (Sp., the hill), alongside of which is located *el panteón* (Sp., the cemetery). On the hill is *el campo aéreo* (Sp., the airstrip), a long, level area cleared approximately thirty-five years ago so that visitors could reach the village by plane. *Los tres chorros* (Sp., the three springs) is a place where three streams of spring water emerge from the earth. Older informants say that this was the site of intense social interaction for women before the advent of potable water, for they would gather to wash clothes and to collect water. Today it plays a less important role in village life. The large volume of water provided by the streams is said to have been the key reason why Talea's ancestors relocated away from Sudo', an area on the mountainside that faces the current site of the village (Nader 1964).[7]

Perhaps no other site has as much daily movement as the elementary school. Twelve teachers—most of whom are from other parts of the state of Oaxaca—are responsible for teaching nearly 400 students. The junior high school is significantly smaller (with approximately 50 students), and many of those attending are from other Rincón villages.

Talea's architecture is being rapidly transformed. I have already mentioned the importance of cinder-block structures. In addition, a number of other projects were in different stages of completion when I left the field in mid-1997: new priests' quarters, new offices for state and federal employees, a new drainage system, and a fountain and bandstand for the atrium of the church. Not all of these projects were conceived by resident Taleans. For example, the new bureaucratic offices were the brainchild of a Talean native son who left more than thirty years ago, a relatively influential member of the Partido Revolucionario Institucional (PRI) political party. He acquired materials for the project, and much of the labor was provided by Talean men through communal work.

Farms, Fields, and Forests: Talea and Its Land

One of the clearest divisions Talean campesinos make is that between forests and cultivated areas. Forested areas are limited and compose perhaps 10–15 percent of the village's land area of 2,875 hectares.

The largest forests are those which lie above the village, a small area below the village which forms the boundary with San Juan Juquila, and a thin strip of forest on the other side of the Río de la Hacienda (most of which was lost in a 1991 land conflict with neighboring San Juan Tabaa). Also, behind the mountain on the way to Villa Alta there are so many abandoned fields that a sizable area is reverting to forest.⁸ Nearly all of the forest above the village is communal land where those who collect fire-

Figure 2.5. Aerial photo of Talea and its lands, December 1995.

wood may gather pine, oak, and other woods for burning in their homes.[9] Trees are occasionally cut down with chainsaws on private plots of land for reflooring or reroofing homes, but felling trees for commercial purposes is prohibited. Some informants claim that this policy is paying off; they note that deforested areas above the village are rebounding rapidly. The forest is talked about as an area teeming with vitality, as a living being, and is home to a wide range of animals including deer, javelinas, squirrels, and many others. One kind of supernatural entity, the *bhni guí'a* (Zap., "forest man" or "man of the mountain"), is associated exclusively with the forest.

Cultivated fields cover most of Talea's land, particularly coffee groves, milpas (Sp., maize fields), *frijolares* (Sp., bean fields), and sugarcane fields. They are mostly held as private property, either with official land titles or as *bienes ocultos* (Sp., literally "occult" or "hidden" lands, so called because no formal land title exists). *Bienes ocultos* were popular in the past as a way of evading government tax collectors, but among villagers they have been respected as private property. *Mojoneras* (Sp., large stones that mark the boundaries between plots of land) continue to serve as property divisions, and this knowledge is passed on from generation to generation. In addition to forests, farms, and the village there is some pasture land, and several abandoned farms are being reabsorbed into the forest. Livestock are allowed to graze freely in most places except cultivated fields and the stubbled maize fields of other farmers.

Each parcel of land has a specific place name. The name is often (but not always) in Zapotec and may refer to a geographic feature (for example, Lachirgúna, Zap., "muddy plain"), to a person's last name (for example, Franco, a nineteenth-century Spanish mine owner), to some historical artifact (for example, the rubble at Sudo'), or to other characteristics.

The Sierra in Historical Context: First Contact

The Northern Sierra was probably inhabited by hundreds of thousands of people before the Conquest (Chance 1989:13). The Serranos had a significant degree of contact with the inhabitants of the coastal lowlands to the east and the valley Zapotec and Mixtec to the west and probably traded extensively with them for many years. Once the Spanish arrived, however, the region was quickly drawn into world economic systems, and patterns of commodity production and consumption that were established during the colonial period had wide repercussions. Specifically, they helped form the blend of commodity production and subsistence farming that characterizes many Sierra communities today.

The earliest Spanish forays into the Northern Sierra appear to have been motivated by the search for riches. In an account of Hernán Cortés's first visit to Tenochtitlán in 1521, friar Bernal Díaz (1963:265) described how Cortés was impressed with the quantity of gold he saw there and asked Moctezuma to reveal the location of the mines. Moctezuma said that, among other places, "the country of the Chinantecs and the Zapotecs" was a good source for precious metals. It seems likely that he was simply attempting to rid himself of the conquistadores by sending them to one of the most remote and inaccessible parts of Mexico. During the next five years, the Spanish initiated a series of failed attempts to conquer the Sierra, struggling against extreme weather, difficult roads, unfamiliar territory, and the Mixe, Chinantec, and Zapotec, who were experienced in warfare. The Spanish launched attacks from both the Valley of Oaxaca and Veracruz until 1526, when approximately 100 soldiers founded the first Spanish settlement in the region, Villa Alta. The initial site was located near the zone of conflict between the Mixe and the Zapotec, who were at "war with each other as well as with the Spanish" (Chance 1989:16-19).

The early years of Conquest were particularly brutal for the Serranos, who suffered terroristic abuses at the hands of the Spanish. Entire towns—Tiltepec, Yagavila, and Yaeé, to name but a few—were attacked, resulting in hundreds of deaths. Villagers were mutilated, hanged, burned, or literally thrown to the dogs if they refused to search for gold and silver mines. Others had their limbs amputated or were branded and taken as slaves. A series of *alcaldes mayores* (Sp., governors of New Spain's "Indian Republics") treated the Serranos brutally and harassed them throughout the colonial period (Chance 1989:18-19).

These events, which occurred to varying degrees across New Spain, provide a historical backdrop to the later colonial period. What makes the Sierra exceptional is that, comparatively speaking, it was traumatically affected by the Conquest. One reason was that the relative isolation of the area made it easier for Spanish officials to terrorize their new subjects without fear of being monitored by higher-ranking authorities. But just as importantly the Zapotec, Mixe, and Chinantec—unlike many peoples in New Spain—had probably never experienced colonial domination: "Both fear of the conquerors and resistance to their demands were considerably greater than in other parts of Oaxaca, which had long since become accustomed to Aztec dominion. This was the first time that these Serranos found their freedom placed in jeopardy, and they did all within their power to oust the foreigners" (Chance 1989:19).

The Political Economy of the Northern Sierra I: Colonial Society

The process of amassing wealth using the labor and land of indigenous peoples was made possible in New Spain through colonial structures, hierarchies, and institutions that ensured the delivery of tribute to officials and guaranteed them virtual trading monopolies. This section reviews the Sierra's political economy in the colonial period.

Alcaldes Mayores, Repartimientos, and Tribute

The *alcaldes mayores*, the most powerful colonial officials in New Spain's "Indian Republics," were officially responsible for administering justice and collecting tribute payments in jurisdictions called districts. By the 1660s the Villa Alta district—where Talea was located—was ranked at the top of the "first-class" districts of New Spain because it allowed the *alcalde mayor* to make more money than in any other district through illicit trading practices called *repartimientos de efectos*.

The *repartimiento* system was quite simple. The candidate for *alcalde mayor* established a business contact, typically a merchant in Antequera (Oaxaca City) or Mexico City, to pay a cash advance to the crown. Then the *alcalde mayor* would handle his sponsor's trading activities as a monopoly in the district to which he was appointed. The system has been described as "a system of forced production and consumption":

Alcaldes mayores would forcibly "sell" commodities to the Indians in their districts at artificially high prices. Cattle, mules, oxen, wheat, tobacco, sugar, cotton, fish, and even corn were often distributed in this way. Alternatively, and especially in Villa Alta, repartimientos often consisted of advances of cash or raw materials, such as cotton, to Indian communities and individual households that in turn were obligated to use the money to produce cochineal or other crops or weave the cotton into cloth. Then on a specified date, the alcalde mayor would purchase the harvest or finished textiles at below market prices. (Chance 1989:103; see also Hamnett 1971:13-14)

Profits were split between the merchant, the *alcalde mayor,* and his legal lieutenant, who carried out actual trading operations. Cochineal, cotton *mantas* (Sp., bolts of cloth), and other products were typically purchased from villagers at one-half to three-fourths of the market value, while goods were sold to them at prices that could reach more than twice the market value.

Sometimes Serranos worked as commercial agents for the *alcaldes mayores* or Antequera merchants, taking their patrons' merchandise from Antequera into the Sierra and selling it at high prices then purchasing cochineal at low prices. Later in the colonial period, others participated in commodities trading as independent merchants. It was not unusual for such Serranos to learn Spanish and even to adopt Spanish dress (Chance 1989:98, 112–115).

Early *alcaldes mayores* frequently rewarded their soldiers with grants for collecting tribute (*encomiendas*) or with deputy appointments (*corregimientos*). Encomiendas were grants awarded to ex-conquistadores who were given the right to collect tribute from one or more villages. During the sixteenth century, more than twenty Rincón villages were subject to these payments, including Talea. Villagers who refused to pay the grantees (*encomenderos*) might expect severe corporal punishment (Chance 1989:23–26). Deputies (*corregidores*) were appointed by Villa Alta's *alcaldes mayores*, ostensibly to monitor the villages of the district. In practice, they functioned as tribute collectors; indeed, their incomes were derived in this way. *Corregidores* were first appointed in the Villa Alta district in the 1520s or 1530s and peaked during the mid-1700s, when approximately twenty existed throughout the Sierra.

Tribute payments were never easy to collect, partly because Serranos were frequently saddled with a triple tribute—to *alcaldes mayores*, to *encomenderos*, and to *corregidores*—and often resisted paying tribute by fleeing at collection time. Also, even though Spanish officials were required to live in Villa Alta, most resided in Antequera, making it difficult to collect tribute. Finally, because of smallpox, measles, and typhus epidemics in the sixteenth century, the Zapotec, Mixe, and Chinantec often had trouble feeding themselves, much less providing for the Spaniards at Villa Alta (Chance 1989:28–29).[10] These methods of exacting tribute did not last long after the end of the sixteenth century, for the Spanish viceroy revoked the *alcaldes mayores*' privilege of awarding *encomiendas* and *corregimientos* (Hamnett 1971).

The Political Economy of the Northern Sierra II: Commodities

Before the Spaniards arrived in the Sierra, villagers produced cotton products and cochineal dye for local use. Cotton, cochineal, and precious metals were drawn into new and geographically far-flung economic systems in the early years of the Conquest, however. For Serranos, these economic facts were accompanied by fundamental political and social transformations.

From the time of the Conquest, commodities have been drawn out of the Sierra in a number of successive currents that have not replaced subsistence farming—which remains at the core of Sierra village economies today—but instead have coexisted alongside it. In the case of cotton and cochineal, tribute payments were collected from villages, but production was distributed among campesino households. Unlike the situation in other parts of Mesoamerica, where full-time landless wage workers became dependent on owners of large commercial sugarcane plantations, in the Sierra subsistence householding continued to survive even as commodities were produced for foreign markets. Commodity production supplemented subsistence production.

Cotton

Cotton was grown locally and woven on backstrap looms in the prehispanic period. During the colonial era, it was in high demand across New Spain because of its durability and low price. According to one historian, "indigenous [cotton] production from Oaxaca, in particular the town of Villa Alta, reached markets as far and diverse as Mexico City, Puebla, and the silver-mining communities of Taxco, Guanajuato, and Zacatecas" (Hamnett 1971:3). It was grown in small quantities in the region before the Conquest, but afterward much of it was grown by Zapotec people under the direction of Spanish *encomenderos* and *corregidores*. The cotton was then sold or distributed for village women to weave under exploitative conditions:

In Yaguila [Yagila], for example, there would be two repartimientos of cotton per month, in which eight pounds of cotton would be issued to each family, deducting the cost from the price of each finished cotton cloth of five yards (*varas*) in length and one in width. In general, in Villa Alta, one mantle was to be finished in twenty days. This mantle would bring a market price of 16 reales. However, the Indian weavers would receive a price of only 8 reales from the Alcalde Mayor. Besides this abuse, the Indian population, which manufactured its own clothing, would be required to receive imported clothing issued in the repartimiento. For such clothing, it was obliged to pay in the products of the region. In such a way, the activities of the Alcaldes Mayores forced the Indians to be both producers and consumers of commodities from which the justices and their financial backers derived a large measure of their wealth. (Hamnett 1971:13-14)

Cotton was destined primarily for markets within New Spain. In Villa Alta the cotton industry was probably more important than the cochineal industry during the second half of the colonial period. The district of Villa Alta, in fact, was forced to specialize in weaving more than any other in Oaxaca; it was estimated that by the late 1700s approximately 50,000-60,000 bolts of cloth were being produced there annually (Chance 1989:107). Few, if any, pueblos in the district were exempt from weaving; consequently, women spent long hours at their looms while their husbands and children grew cotton, food crops, or cochineal.

Cochineal

One of New Spain's most lucrative commodities was cochineal, a scarlet dyestuff extracted from a cactus-feeding insect (*Dactylopius coccus*). During the Conquest, the Spanish coerced Serranos into producing the dye on an expanded scale for international markets. Throughout the 1500s the market for cochineal was modest, but by the mid-1600s the dyestuff ranked second only to silver among New Spain's exports. It was far and away Oaxaca's most important export. Throughout the colonial period, production was confined almost entirely to the diocese of Oaxaca, with most of the dyestuff produced in the Villa Alta district and the neighboring districts of Ixtepeji and Nejapa (Chance 1989:105-107), where production techniques were most refined and widespread.

The coerced production of cochineal dye was so prevalent in Oaxaca that an eighteenth-century bishop, Fray Angel de Maldonado, was concerned that villagers would suffer food shortages because they were neglecting their subsistence crops in order to cultivate the insects. He estimated that over half of Oaxaca's agricultural population was involved in cochineal production and that "the Indians had left the greater part of their fields uncultivated for products of alimentation" (Hamnett 1971:14-15). The work was done by men, women, and children, while cotton was woven almost entirely by women. With the invention of chemical dyes in the 1850s, the cochineal industry declined significantly.

The cochineal industry did not exist outside of the larger political-economic realities of New Spain. As we have seen, the cochineal trade was monopolized by Villa Alta's *alcaldes mayores*, who worked in partnership with merchants from Antequera or Mexico City.

Mining

If one considers the Sierra as a whole, silver and gold mining never completely dominated the region as in other parts of Mexico. However, mining did have a large impact on a small number of villages during the colonial period and for some time after.

As in the case of cochineal and cotton, gold and silver were commodities during the prehispanic period, extracted from shallow placer mines or else panned from rivers. Early Spanish expeditions were motivated by the quest for gold and silver, but the colonizers were able to extract little from the region during the initial decades of the Conquest. It was not until about 1580 that the Spanish discovered larger amounts of precious metal from a site near the Mixe town of Totontepec near Mount Zempoaltéptl.

In the late 1770s the Santa Gertrudis mine, a 45-minute walk from the present site of Talea, began to be exploited. This mine and several others nearby were acquired by a powerful Antequera merchant named Juan Francisco de Echarrí. During the 1780s, approximately seventy-six workers were employed at the mine and ten more worked on an adjacent agricultural hacienda where foods were grown for the miners. In these early years of mining, workers were drafted through *repartimientos de labor* provided by nearly all of the pueblos of the Rincón (including Talea) and by the Cajonos region (Chance 1989). Under this system, *alcaldes mayores* sent citations to each village, demanding a specified number of workers each week (typically 2–4 percent of the population). It is unclear whether the mines near Talea functioned during the War of Independence from Spain, but they would be exploited on a large scale once again in the 1800s.

Perhaps the most remarkable aspect of the colonial period was that, in spite of severe economic pressures, the people of the Sierra never lost their sufficiency base:

> In Oaxaca most Indian households became integrated into expanding markets while at the same time retaining considerable control over the means of production. In both the Valley and the Sierra, Indian communities succeeded in holding on to most of their traditional lands and many of their traditional subsistence activities . . . [production of tribute products] did not totally replace subsistence activities, but came to coexist alongside them. (Chance 1989:121)

Later we shall see how this pattern repeated itself in the twentieth century, when many Rincón villages began to balance a system of coffee production for world markets with subsistence farming of maize and beans. The preservation of the *mantenimiento* ethic—household maintenance—was never entirely lost by the Rincón Zapotec.

The Political Economy of the Northern Sierra III: The "Indian Republics"

There were many Mesoamerican communities in existence prior to the arrival of the Spanish, but it appears that in the Sierra the pattern of head town (*cabecera*) and subject hamlets (*sujetos*) was not as developed as in some other parts of Mesoamerica, particularly among the Mixe (Beals 1945; Chance 1989). Apparently, only when the Spanish arrived did the pattern of small, relatively independent and highly concentrated villages intensify.

Evidence indicates that many people in the Sierra were forcibly resettled with the aim of facilitating administration and religious indoctrination. The relocations were a New World version of a practice employed in the Christianization of the Spanish peasantry (Ricard 1966 [1933]:136). The forced concentrations, called *congregaciones* (Sp., literally "congregations"), were devastating. They began in the Sierra near the end of the sixteenth century with the relocation of many Rinconeros to Tanetze. People from Yaeé, Lalopa, and Otatitlán were forced to move there, and many if not most of those affected died in the process.

The system of municipal governments imposed by the Spanish in the colonial period was based on Old World models and was probably "alien to the indigenous cultures of the Sierra" (Chance 1989:132). The Spanish were reluctant to get involved in village affairs in the Sierra until 1550, when a colonial judge arrived in the region with an order from the viceroy to appoint twenty-four *alguaciles* (constables) from the local population. In 1556 full municipal governments were appointed in Sierra villages for the first time. The following offices were established: *gobernador* (mayor), *regidor* (councilman), *alcalde* (judge), *mayor* (police chief), *escribano* (scribe), and *fiscal* (a servant of the church, responsible to the local priest).[11] Ideally, young male villagers entered this "civil-religious hierarchy" at the lowest levels and every few years reached progressively higher levels. Although a number of additional positions were later added to the system, for the most part it endured unchanged throughout the colonial period.

It is likely that municipal governments were established to help *alcaldes mayores* efficiently administer the *repartimientos*. In fact, *repartimientos* were made with the assistance of village officials, who often stood to gain from the arrangement. The authorities frequently served as middlemen between their constituents and the *alcaldes mayores* and gained power and perquisites during their terms of office.

In the years immediately following independence from Spain, municipal governments probably became more democratic,[12] and it is likely that village governments, once used hegemonically by Spanish administrators seeking efficient tribute collection, were increasingly used counterhegemonically by Serranos. Eric Wolf (1959:148, 214) has called this a "retreat from utopia" in which "a tug of war between conquerors and conquered" resulted in the appropriation of certain European institutions by indigenous peoples in the colonial and postcolonial periods. Ironically, these small-scale democracies—based on a Spanish model—failed miserably in the region's only Spanish outpost, Villa Alta, yet thrived in indigenous villages for centuries and still form the structural core of communities today.

The people of the Sierra did not take abusive treatment passively. When legal complaints were ineffective, uprisings, riots, and rebellions were a regular feature in the region, just as in other parts of New Spain (Barabas 1985). For example, within ten years of initial contact with the Spanish, the village of Tiltepec rose up against them and reportedly killed seven. Large attacks on Villa Alta were launched in 1550 and again in 1570 by the Mixe. Abuses by *alcaldes mayores*—particularly the excessive burdens created by the *repartimiento* system—motivated at least nine rebellions during the colonial period, including uprisings at Lachirioag, Yojobi, Lachixila, San Francisco Cajonos, Yalálag, and Yagavila (Chance 1989:110).

Others caused riots as well. Abusive clergymen sparked at least four uprisings during the colonial period,[13] while mine owners and their foremen provoked indigenous workers toiling at Santa Gertrudis in 1782. The following year, villagers from Tabaa, Yojobi, and Solaga filed a suit against a mine owner and, when it was apparent that they would lose the contest, began to riot (Chance 1989:95-96).[14]

The Introduction of Old World Plants and Animals

One aspect of the encounter between Spaniards and Native Americans that is seldom discussed is the process by which Old World crops were adopted by farmers in the decades following the Conquest. Across most parts of New Spain, Catholic missionaries introduced the new seeds, animals, and technologies, and campesinos often experimented with them enthusiastically (Ricard 1966 [1933]:142-145).

Although there is not much material on the introduction of new crops in the Northern Sierra, François Chevalier (1963:50) notes that

> the sites of the first garrison towns were often useless for agriculture. San Ildefonso de los Zapotecas (Villa Alta), for example, was located in so mountainous and forbidding a region that it could be reached only on foot. The [Spanish] inhabitants themselves said that no business or profit was possible.

This passage is interesting because it reveals the significance the Spanish placed upon commercial agriculture. What was frustrating for the Spaniards was not the quality of the Villa Alta region's soils, but rather its remoteness and the difficult terrain, which made cash cropping impossible. The region's agricultural "uselessness" thus referred to commercial farming, not to subsistence farming.

Yet subsistence farming seems to have thrived among the region's Zapotec campesinos, and the available evidence indicates that the Serranos enthusiastically adopted new food plants and animals. Although apparently "Spanish crops did not invade native lands on any appreciable scale until the seventeenth century" (Chevalier 1963:189), by the late 1770s Spanish officials reported that Sierra and Rincón Zapotec people had adopted many kinds of food plants and animals: citrus, fig, and pomegranate trees grew in gardens and fields in San Miguel Amatlán; villagers in Tabaa enjoyed more than five varieties of bananas, peaches, and many other fruits and vegetables; and the people of Yagavila cared for chickens and pigs (Esparza 1994).

María de los Angeles Romero (1990) notes that in Oaxaca the Dominicans were especially active in encouraging (and sometimes coercing) farmers to cultivate Old World crops, using collective labor for the maintenance of their Catholic churches. By the middle of the sixteenth century wheat was grown in the Sierra villages of Ixtlán and Capulalpan, and new kinds of fruits and vegetables spread rapidly across Oaxaca. Romero argues that the rapid adoption of these plants by the Zapotec and Mixtec is a good example of how people subjected to the ravages of the Conquest and subsequent epidemics still attempted to improve their lives actively and intelligently by enriching their culinary and medicinal repertoires.

From another perspective, it also seems likely that the sudden arrival of so many new food plants and animals might have appeared to be an event of miraculous and staggering importance for Mesoamerican peoples, who often, as in the case of the Zapotec, viewed the earth, forests, plants, and animals as living, willful beings who directly influenced the lives of humans (see Chapter 4). Thus, from the point of view of a Zapotec farmer in the colonial era, the sixteenth and seventeenth centuries represented not only the destruction of millions of human lives, but the discovery and adoption of hundreds of useful and wonderful new plant and animal species.

The Religious Conquest of the Sierra

Prehispanic Zapotec religion has been outlined by a number of scholars working in different parts of Oaxaca (Parsons 1936; de la Fuente 1949; Weitlaner and de Cicco 1958; Weitlaner 1965; Nader 1969; Alcina Franch 1972; Marcus 1983). Joyce Marcus, in a description of Valley Zapotec religion on the eve of the Conquest, notes that the Zapotec recognized a supreme being, but that more attention was focused on the supernatural forces it created, including lightning, sun, earthquakes, fire, and clouds. There was also a strong emphasis on ancestor worship, the ritual sacrifice of animals and sometimes humans, blood offerings made by piercing body parts, ritual intoxication, and the use of hallucinogenic substances for divination. Like Catholicism, Zapotec religion also featured a trained hierarchical priesthood, confession, feasting, a complex pantheon of supernatural beings, possession and worship of wooden and stone images, fasting, penance, and a ritual calendar (Marcus 1983:345–351).

Colonial documents indicate that Fray Bartolomé de Olmedo was probably the first Catholic priest to visit the Sierra and that he baptized some 500 people before his death in 1524. Over the next three decades, only a handful of priests (Dominicans and secular clergy) were stationed in Villa Alta; but, beginning in 1558, the Dominicans established themselves firmly in the Sierra by fortifying their presence there (Ricard 1966 [1933]:72). The outpost served as the base for Catholic missionary activity throughout the first half of the colonial period. In the early years most of the friars dedicated themselves to converting the Zapotec, Mixe, and Chinantec to Christianity, while one friar remained in Villa Alta, teaching at a school attended by the children of Spanish residents and Zapotec elites.

Robert Ricard explains how the Augustinians, Dominicans, and Franciscans effectively transformed the area between the Isthmus of Tehuantepec and what is today northern Mexico. He argues convincingly that the era between 1523 and 1572

> is the period in which the clash of civilizations . . . occurred in the sharpest form, a period in which native American elements and imported Spanish traits are some-

Figure 2.6. Talean *danzantes* reenact the wars between the Spanish and the Moors. This drama, known as the Danza de los Negritos, was probably introduced in the Rincón by Dominican friars in the sixteenth century. *Danzas* draw many spectators during village fiestas. *(Photo by Gabriela Zamorano)*

times fused and sometimes juxtaposed, together giving Mexico its present personality. The sixteenth century contained in embryo the subsequent evolution of the country; it was to leave its strong imprint upon the following centuries. (Ricard 1966 [1933]:3-4)

Ricard examines the various institutions that were established by the religious orders, the process by which this was made possible, and the way in which conversion was undertaken by the friars in the decades following the Conquest. The study shows how customs and institutions which today are seen as characteristic of "indigenous" communities—for example,

Conquest plays, fiestas, music, and, more generally, "folk Catholicism" —are in fact complex arrangements between Mesoamerican peoples and Spanish clergy. Many Spanish, insisting on the adoption of Catholicism, eventually accepted what in some instances might be seen as loose interpretations of Catholic doctrine and ritual activity by Native Americans. In some cases, prehispanic cosmologies were left intact; in others, even prehispanic rituals survived, with Catholic rituals and deities simply added to ancient repertoires. Ricard's discussion of "native resistance" problematizes the notion of "syncretism" by showing that what happened in the sixteenth century was not a simple blending of two traditions, but a much more nuanced, uneven, and often incomplete combination of many traditions.

The Northern Sierra is a case in point. At a cosmological level, prehispanic religions were powerfully affected by the Conquest; the Zapotec supreme being, Betao, diminished in importance. In contrast, supernatural beings (for example, "mountain spirits") influencing more local spheres of power continued to flourish well into the twentieth century (and, as we shall see, to the present day), often alongside Catholic saints (de la Fuente 1949). Chapter 4 analyzes the complementary role of the "mountain spirits" and the Catholic saints with respect to agriculture in more detail. Suffice it to say that even though idolatry on the part of the people of the Sierra was punished severely, prehispanic religious elements and practices endured much longer in the region than in other parts of Mesoamerica.

The Exceptional Case of Talea

Talea has much in common with other Rincón villages, but in many respects it has an exceptional history. Perhaps the most significant aspect is evidence suggesting that it was not founded until several years after the arrival of the Spanish. A document in the municipal archives relates how the village was founded in the sixteenth century by a group of Spaniards.[15] According to the account, a party of village elders from various Rincón villages went to Mexico City in 1524 to request that Cortés send representatives of the Catholic religion to the area, perhaps hoping to circumvent an attack. The invitation was answered the following year when Fray de Olmedo came with a number of other Spaniards to baptize and preach Catholicism. On this trip, they founded Talea on the border between Yatoni and Juquila. The account would explain why surrounding villages—

most of which probably predate the Conquest by hundreds of years—describe Talea as a "new town" (Nader 1964).

Archival research casts doubt on the legitimacy of the document, however. According to John Chance (1989:79), the Talea document is one of a group of five which detail accounts taken from painted *lienzos* (Sp., cloth paintings) after 1550.[16] In spite of the fact that all five documents are dated in the 1520s, they mention two Spaniards who did not arrive in the Sierra until the 1550s.[17] It has been suggested that "the falsification of dates on all these documents was probably done intentionally to enhance their 'authenticity'" (Chance 1989:194). Even so, Chance admits that it is "very probable" that Talea was founded before 1548.

Another discrepancy—and perhaps one with broader implications—involves the circumstances of Olmedo's visit. According to the surviving document in Talea, Olmedo and his party came as guests invited by the elders of the Rincón villages, but an important historical account identifies Olmedo as a participant in one of the initial military expeditions into the Rincón (Gay 1950 [1881]:1:380, 416, 435). It is plausible that Olmedo was part of Rodrigo Rangel's military expedition of February 1524, an offensive in which Cortés had authorized the taking of slaves. The Rincón Zapotec and Mixe effectively defended themselves against the Spanish and forced back the Rangel expedition, but not before Olmedo had baptized approximately 500 people (Chance 1989:17). In light of this evidence, the question essentially boils down to this: did the Taleans hospitably accept Catholicism from missionaries they personally invited, or was the new religion forced upon them by missionaries who were backed by military muscle? The answer is not clear.

If it is true that Zapotec elders sought out the intervention of the Spanish, this would represent an exceptional (and almost implausible) version of the usual pattern in New Spain, where the valiant defense of prehispanic religions was common. Ricard (1966 [1933]:264) notes that "the great force of inertia" was the most common obstacle to Spanish evangelization efforts: "they were faithfully attached to their ancient gods by their convictions and habits, and could make only a kind of sly resistance to the action of the missionaries by inertia and dissimulation." Nowhere in Ricard's study is there an indication that Mesoamerican groups might actively have sought out religious conversion or voluntarily solicited Spanish intervention in religious affairs.

The question takes on more meaning in the contemporary context, when one considers the place of Talea vis-à-vis the other villages of the

Rincón. At first glance it is easy to pass Talea off as a mestizo (literally "half-breed" or non-"Indian") or even Europeanized town; the story of its founding might lead to the hasty conclusion that, like Villa Alta, Talea is European (or at least more so than its neighbors) and always has been. But an examination of the village's colonial history reveals that Taleans were subjected to the same mechanisms of colonial control, tribute extraction, and violence as other villages. Unlike Villa Alta, the village was populated primarily by Zapotec speakers—*indios* from the perspective of the Spanish—who were subjected to exploitation by Spanish officials and forced to provide labor for the mines. In short, Talea was in nearly every sense an "Indian" village.

This probably did not begin to change until approximately 150 years ago, when the new owners of the Santa Gertrudis mine began expanding production. This led to a rapid population increase that drew outsiders from other Sierra villages, the Valley of Oaxaca, and other regions and eventually led to Talea's development as a commercial center that supplied the mines with provisions (Nader 1964). It seems likely that with the influx of so many outsiders, including relatively acculturated (that is, Hispanicized) miners from the Sierra Juárez and the Valley, Talea began to assume the place of a mestizo center from the point of view of people living in the neighboring villages of Juquila, Yatoni, Yaeé, Tanetze, and Tabaa (here I mean mestizo in the sense described by Rodolfo Stavenhagen [1965]: as an ethnic category associated with centers of economic domination). The tension between *lo mestizo* and *lo indígena* has probably been more acute in Talea than in other Rincón villages and, as we shall see later, continues to survive today.

Talea and the Rincón in the Twentieth Century

From Mining to Commerce and Coffee

Talea's mining industry gathered new strength when Miguel Castro, an Antequeran entrepreneur, reopened the Santa Gertrudis mines in the 1840s. Talea and Santa Gertrudis grew rapidly in the following years; although agricultural land area was limited, the wages earned by the miners in the latter half of the 1800s allowed them to purchase food from growers in adjacent villages. These growers probably sold their maize and beans to the miners at the weekly market, which at the turn of the century was located at Santa Gertrudis. Talea benefited from the boom years by supplying the miners with manufactured items: it is estimated that at the turn of the twentieth century the village had approximately 1,000 people, up to 200 of whom were employed in nonagricultural activities, including blacksmithing, carpentry, charcoal making, candlemaking, baking, and weaving (Nader 1990).

Because of the political and economic instability leading up to the Mexican Revolution, the mines closed in about 1905 and the miners were forced to relocate. Many chose to establish themselves in Talea as campesino farmers, but others turned to commerce; shortly after the arrival of the ex-miners, the village became the most important commercial center in the Rincón. The weekly market moved from Santa Gertrudis to Talea, and the ex-miners, perhaps pressured by the village's limited land, began to manufacture and sell items that were needed by people in other villages (Nader 1990).

Coffee was probably introduced in Talea for the same reason: a shortage of land. For most of the century, it has been more profitable to grow coffee than to grow maize; because Talea had little land relative to its large population, cash cropping was an appealing option. Like the cotton, cochineal, and gold and silver mining industries in previous centuries, coffee linked the region in general and Talea in particular to the world economy. Indeed, the impact of coffee on the Rincón was a profound one with many repercussions (see Chapter 7).

During the mid-1940s, a group of twenty to thirty young men left Talea to work in the United States under the auspices of the Bracero Program, in which Mexican laborers were hired to meet wartime production shortages. The event is indelibly etched in the minds of older Taleans: the departure of the young men was a traumatic event. In the words of an informant: "Fathers, mothers, wives didn't know if their loved ones would ever come back. Many saw their departure as something that would maybe last forever." The Braceros traveled as far as Idaho, Montana, and Washington state—working on railroads and, above all, farms (where they picked sugar beets, apples, corn, hops, and other crops)—and returned to Talea with cash. According to Nader (1990:24), "These Taleans returned wearing shoes and colored trousers and shirts, and gradually, by conscious persuasion, they convinced other citizens of the town to change to this new dress."

Roads and the Decline of Craft Industries

Coffee was probably the single most important factor leading to the construction of a road to Talea in 1958—the first road linking the Rincón to

Oaxaca City. Nader (1964:213) notes that "when Taleans began producing large quantities of a crop which had to be shipped to the valley of Oaxaca to be sold and distributed they committed themselves to the problem of how to improve transportation and lower shipping costs." Much of the organizational work was carried out by a wealthy Talean merchant and coffee buyer named Agustín García, who was active in seeking government support for the project and rallied other Taleans around the idea.[18] Some ex-Braceros, who during the 1950s and 1960s typically supported progressive causes and were often at the forefront of changes, were among those leading the effort to construct the road. It was eventually completed with the assistance of the Papaloapan Commission, one of Mexico's most ambitious development projects, aimed at improving conditions in the Papaloapan River basin (Poleman 1964).[19]

It would be difficult to overstate the significance of this event in Rincón history. "When the road arrived, *everything* changed," informants told me time and again. The changes brought on by the road were indeed profound ones. In the 1950s Taleans had many small industries that produced matches, soda pop, fireworks, soap, shoes and boots, huaraches (sandals), hats, rifles, machetes, and other tools.[20] Carpenters, blacksmiths, weavers, and tailors did a brisk business, especially during the Monday market, but many of these businesses had disappeared by the late 1960s. According to some sources, the change had to do with the introduction of cheaper, more durable, and more prestigious goods manufactured by companies in large Mexican cities or overseas (Berg 1976). Many products were produced cheaply in Mexico as a result of import substitution policies implemented in the 1940s and 1950s, which promoted nascent industries by levying high taxes on imported goods and subsidizing domestic firms. The new roads made it easier for trucks to bring these items to weekly markets in the region, while making it difficult for local artisans to compete with the industrial goods. Eventually many items produced by artisans in the Sierra nearly disappeared: reed raincoats (Sp. *capisayos*) were replaced by *nailos* (Sp., "nylons" or plastic sheets), leather huaraches by plastic shoes, clay vessels by plastic containers, and cotton fabric by polyester.

A number of informants also blame the federal Secretariat of Treasury and Public Credit (SHCP), the Mexican government's revenue agency, for the decline of local craft industries. During this period, SHCP officials began to enforce laws requiring small businesses in even the remotest parts of the country to pay fees. This, according to a number of Talean merchants, reduced thin profit margins and drove some out of business.

Other Recent Changes

Since the completion of the road, other technological changes have been rapid and continuous. An airstrip was constructed by the villagers in the early 1960s and used periodically until daily bus service was established in the 1980s. (By the late 1990s a bus trip to Oaxaca took only five hours. All but the last forty kilometers of road are paved.)

Electricity and potable water were introduced by 1970 in several Rincón villages as part of the Papaloapan project. A government health clinic and a junior high school were opened in Talea in the 1970s, and government-subsidized food stores (Compañía Nacional de Subsistencias Populares [CONASUPO] outlets) began selling staple foods in the 1960s. In the late 1980s and early 1990s a private high school was established, the only one in the region. More than 100 TV satellite dishes, many of them fabricated in Talea, are precariously perched upon roofs made of tile and corrugated tin. Telephones were installed in January 1994, and there were more than twenty-five private lines in service by mid-1997.[21] Even personal computers have recently made a debut in the village.

Some of the most important devices affecting life in the region are simple implements used daily. Many now have depulpers (machines for removing the pulp from coffee beans), powered by either electricity or gasoline. Practically every Rincón village has an electric mill for grinding maize into dough, and Talea has five. Many households also have electric blenders and gas stoves. All of these devices undoubtedly cut down on labor time, though cash is required to purchase and maintain them.

At least one technology has affected the entire Sierra since its inception: radio. The station XEGLO has broadcast indigenous-language programs in Sierra and Rincón Zapotec, Chinantec, and Mixe since the early 1990s. The station is based in Guelatao (three hours away from Talea by bus) and is funded by the federal government's National Indigenous Institute (INI). Villagers point out that it is the only station that broadcasts *sones* and *jarabes*, the lively and addictive regional music introduced in the latter half of the nineteenth century. It is also an important source for regional and national news.

Along with these changes have come a host of institutional offices established by organizations (governmental and otherwise) in the latter half of the twentieth century. An independent credit union based in northern Mexico opened a branch office in Talea approximately twenty-five years ago and has been successful in spite of severe currency devaluations and economic crises. It has more than 400 members from Rincón villages. Since the collapse of the state-run National Coffee Institute (INMECAFE) in 1989, a number of independent coffee cooperatives have been organized in the Rincón (see Chapter 7). Several government offices have been established in Talea in the last three decades, including offices for the Mexican Postal Service, the state-owned Federal Electricity Commission (CFE), the Oaxaca state department of education (IEEPO), the Civil Registry, and most recently the *agente del ministerio público*, a lawyer who typically serves as a federal investigator, judge, and jury for serious crimes. Protestant missionaries have been increasingly active in the region since at least the post–World War II years. One group of mostly foreign-born Catholic missionaries (including, in the last decade, a German, an Irishman, a Canadian, an East Indian, a Filipino, and an Australian) belonging to the Society of the Divine Word has also had a notable influence in the Rincón since the 1970s.

Attitudes toward Technology and Change

Many Taleanos expressed a fascination with "modern" devices and technologies during the time that I was in the field. Nader (1964:213) noted that "Taleans desire modern facilities, whatever they may be, and modern medicine." It was not unusual for Taleans to tell me the minutest details concerning the introduction of a specific technology. Sometimes there was a comparison, either implied or explicit, made between Talea and neighboring villages that are considered by some to be technologically "backward." The technological hero tale would thus become a lesson on Talea's superiority vis-à-vis other villages. These villagers told me of men such as Agustín García, who had harnessed electrical power a decade before the CFE felt an obligation to do so, and Adán Mejía, who had designed a wooden water mill for extracting the juice from sugarcane. A small number told me that they would like to see as much technology as possible introduced into Talea, as quickly as possible.

Yet by the time I completed my fieldwork in 1997, I had detected a certain disenchantment on the part of at least some villagers with the effects of these changes on everyday life. One reflective man, an older merchant, told me that the road had done more to damage village life than anything else in recent history. He explained that the road had made it easier for undesirable elements to come into town and for the brightest Taleans to leave. A merchant speaking in a film produced by Laura Nader (1981) noted that technology was destroying the "creativity" or "know-how" of young people. Another informant told me that a great irony had occurred in Talea since the arrival of "modern things": the young men and women of the village had become physically lazy and less imaginative: "They have everything given to them, and so they don't have to struggle the way our ancestors did." Others complained to me privately about the recent practice of chlorinating the water supply, instituted by the health clinic as part of a national anticholera campaign. Frequently, the water is overchlorinated and pours out of the faucets white. Still others have observed noticeable behavioral differences between Talea's first TV generation and older generations.

I took these observations with a grain of salt—particularly since those critiquing technology were often the same people who most aggressively sought it. It may have been a case of the informant telling me what he or she thought I wanted to hear. And, of course, it is entirely possible to want "modernization" or "progress" while recognizing its costs, which may be seen as annoying but inevitable (as is often the case in the United States) or as avoidable but imposed by outsiders. In general, it did seem that forty years after the completion of the road a more tempered and reflective view of new technology was beginning to emerge (albeit on a limited and generally nonactivist scale) among at least some Taleans.

Contemporary Economic, Cultural, and Political Contexts

Talea and Rincón Markets

Talea is a market town—the biggest Rincón market is held there on Mondays; though Yaeé and Tanetze have recently tried reviving their markets (on Saturday and Sunday, respectively), these attempts have not been immediate successes.[22] As informants were quick to point out, one still must go to Talea to find a good blacksmith or welder, to pay the light bill, to get a birth certificate or marriage license, and to get a junior high school education. Indeed, at least part of some Taleans' enthusiasm for the government's bureaucratic offices has to do with the fact that they attract people who might not otherwise attend the declining Monday market.

Archaeological evidence indicates that markets existed among the Zapotec of the Valley of Oaxaca long before the Conquest (Beals 1975; Malinowski and de la Fuente 1982 [1940]; Flannery and Marcus 1983). Oaxacan artifacts found at Teotihuacán (north of Mexico City) imply that ancient Zapotec trading networks were indeed extensive.

There are no references in the literature to markets in the Sierra until the colonial period. This does not mean that interregional trade did not exist between the Sierra and the Valley, however. Glottochronological data indicate that from 500 to 1000 A.D. the Sierra Zapotec dialect began to diverge from Valley Zapotec, which "might reflect lessened contact... following the decline of Monte Albán" (Flannery and Marcus 1983:7). The implication is that before this period there was a high degree of contact (at least enough for the language to have remained fairly uniform), and it seems reasonable to assume that trade was an important component of the relationship between the Valley and Sierra Zapotec. The latter may well have served as intermediaries conducting trade between the Valley and the gulf coastal lowlands (Atlantic shells, for example, were crafted by the Valley Zapotec).

During the colonial period, the right to host markets was a point of contention between Rincón villages — particularly Tanetze and Yaeé. This dispute, and a similar one between Yagila and Yagavila, extended through much of the 1700s (Chance 1989:119). Neither Talea nor Santa Gertrudis appears in colonial documents as a contender for weekly markets; it is probably safe to assume that not until the latter half of the 1800s (when the mines near Talea were at high production levels) was a market established at Santa Gertrudis. At the turn of the twentieth century it was moved to Talea.

Ethnic Landscapes

Talea may be contextualized in terms of ethnic or sociolinguistic groupings. The villagers themselves often divide the people of the Sierra into five categories, each of which is referred to by a Rincón Zapotec term. They refer to themselves as the *bhni shídza* (people of the Rincón), while others are commonly referred to as one of four groups: *bhni li'hj* ("fence" people), *bhni zhreb* ("dry" people), *bhni mish* (Mixe people), or *bhni ga* ("nine" people). These groupings correspond roughly to five groups cited by John Chance (1989:3-10): the Nexitzo (or Rincón) Zapotec, the Sierra Zapotec, the Cajonos Zapotec, the Mixe, and the Chinantec, respectively.[23]

Some of these terms describe an aspect of the lands in which the people reside. For example, the *bhni li'hj*, or Sierra Zapotec, apparently got their name from their habit of enclosing fields with "fences" or rows of *zompancle* trees (*Erythrina* sp.); this is still common in a number of villages. The *bhni zhreb* (Cajonos Zapotec) got their name from the dry, scrubby land they inhabit — remarkably different from the more lush and tropical

Rincón. No informants were able to tell me how the Chinantec might have gotten the name "nine" people.

The boundaries separating these groups are not always clear; they have changed over time and vary according to who is asked. There is apparent disagreement surrounding those villages located in what Leonardo Tyrtania (1992) and others refer to as the Rincón Bajo, the relatively low-lying area in the direction of the Gulf Coast. De la Fuente's ethnography of Choapan (1947:144) includes these villages (e.g., Yagalaxi, Lachixila) on the southern (Zapotec) side of the boundary dividing the Nexitzo Zapotec from the Chinantec. But some Taleans I spoke with seemed to place these villages in the *bhni ga* (Chinantec) category rather than the *bhni shídza* (Rincón Zapotec) category. Apart from this, Taleans consistently report that *bhni zhreb*—though much closer geographically (Tabaa and Solaga each lie only three hours' walk from Talea)—are much more difficult to understand than *bhni shídza* of the Rincón Bajo, who may live as far as ten or twelve hours' walking distance away.

Many Taleans make reference to Jarochos (people from Veracruz), Istmeños (from the Isthmus of Tehuantepec), *gente del Valle* (from the Valley of Oaxaca), and Costeños (from the Pacific coast of Oaxaca). Some Taleans ascribe certain characteristics to these people, invoking the hot/cold concept in the cultural construction of hypothetical regional personality traits. Jarochos and Costeños are generally regarded as violent and hot-blooded and tend to get into fights with machetes on sweltering nights, I was told by informants. On several occasions this was explained to me as being causally related to their hot, humid homelands. The Sierra Zapotec (e.g., from Guelatao, Ixtepeji, and Lachatao), by contrast, were described to me as "brutes" but also as "clever" (Sp. *listos*)—the implication was that they were aggressive or violent, but in a cool, calculating way—they are, after all, from an area at a higher altitude than Talea and therefore colder. (The important role of the Sierra Zapotec in the Mexican Revolution earlier in the century was presented as evidence of their craftiness.) The Rincón Zapotec from the other side of the mountain—especially those from Zoogochi, Teotlaxco, and to a lesser extent Yagavila—were sometimes derisively referred to as *delicados* (Sp., literally "delicate" or thin-skinned—that is, not able to take a joke).

The ethnic landscape has certainly shifted over time, though there are conflicting data about how identities have been transformed in the Sierra. Although Martha Rees (1996:111) reports that "historical evidence shows that modern mountain Zapotecs identified as Mixe in the sixteenth century," I have been unable to confirm this even after consulting her data source (Carmagnani 1988). Indeed, other scholars arrive at the opposite conclusion, positing that Mixe and Zapotec identities were probably stronger and more distinct at the time of the Spanish arrival than at any time since, for the archival record indicates that the Mixe were engaged in a bloody war with both the Rincón and Bixonos Zapotec at the time of the Conquest (Chance 1989:14). In spite of some apparent cultural similarities between the Mixe and mountain Zapotec (particularly recent introductions such as philharmonic band music, dancing styles, and certain foods; see Beals 1945; Kuroda 1984; Lipp 1991), the groups distinguish each other very clearly.[24] It appears that anthropologists themselves had something to do with the recent self-awareness of the mountain Zapotec as indigenous subjects (Nader 1990:16).

What seems unequivocal is that in northern Oaxaca the strongest ethnic allegiances are those related to community. Regional alliances among the Zapotec of the Northern Sierra are often tenuous, and intercommunity conflicts flare up from time to time. The differences between the situation in the Sierra and other parts of southern Mexico are made dramatically clear in varying notions of indigenous peoples' autonomy. In recent conferences and workshops, the Maya and the Isthmus Zapotec expressed a preference for regional autonomous zones linking together many villages, while the Zapotec and Mixe from the Northern Sierra argued that autonomy is likely to work only at the community level (Hernández Castillo 1997).

Talea is considered by at least one anthropologist to be a mestizo (non-"Indian") town (Tyrtania 1992:28). Based on the criteria outlined in a classic paper on "Classes, Colonialism, and Acculturation" (Stavenhagen 1965), this analysis is valid, for Talea is an important market town that is a point of "surplus extraction" for Zapotec and non-Zapotec intermediaries buying from and selling to people from more remote communities. It has served as a commercial center between Oaxaca City and the Sierra for most of the twentieth century and also has more of the superficial trappings of modernity—automobiles, TVs, satellite dishes, and electrical appliances—than other Rincón villages. Nevertheless, the material conditions of most Taleans are remarkably similar to those of people living in these other villages. They farm for a living on small parcels of land, generally live in one-, two-, or three-room adobe houses, and have a largely autonomous system of self-governance. They also continue to speak Zapotec. (Although it is not commonly spoken in conversations between young people, the language is understood by many of them and is used when speaking to older people. In many cases children who do not speak Zapotec publicly begin using it after they reach adulthood.) What is more, if asked, many Taleans describe their village as "indigenous" (with an increasing sense of pride, especially among younger people). Rather than becoming embroiled in the long-running debate over what is "Indian" and what is mestizo, it is perhaps more important to analyze how the categories themselves may be transformed over time (Frye 1996) and deployed hegemonically (Friedlander 1975) or counterhegemonically (Campbell et al. 1993) in different situations.

National, International, and "Transnational" Landscapes

Migration has played an important part in Talean life in the past century. The Braceros were among the first to have traveled far from Oaxaca in the 1940s, and their exploits are remembered vividly. At about the same time, Taleans began to migrate to Mexico City and had already begun moving to Oaxaca City seeking work (Hirabayashi 1993:69–71). Once they established themselves, it was easier for relatives and friends to follow them.

Tepatitlán, Jalisco (near Guadalajara), probably has more than thirty Taleans living and working there. This is an especially intriguing case, because migration from Talea to Tepatitlán only happened in the last ten years when a missionary priest who is a native of "Tepa" facilitated the relocation of the Taleans. Another destination that seems to be growing in popularity is rural Washington state, where migrants have worked mostly in agriculture, picking apples and pears.

Tijuana became a popular destination in the 1960s and 1970s and has since become the major stepping stone for the hundreds of Taleans living in the Los Angeles area. Many are concentrated in Orange County, particularly in Anaheim and Santa Ana, but others live in Santa Monica or central Los Angeles. A small number have been to the East Coast, as in the case of one young man who met a group of Mixtec with New York City connections while working for the Mexican postal service in Oaxaca City. He worked with them for more than five years busing tables at New York's Hard Rock Café. Migration tends to have a snowball effect.

A small number of Taleans have had the opportunity to travel even farther than the United States, owing to their contacts with foreigners. Two had visited Canada, and another had been to West Germany for a number of weeks with a retired Catholic missionary priest who is a native of Germany. The distance traveled by Taleans has undoubtedly expanded over the years, but this has not always resulted in a loss of identity. In Los Angeles, Mexico City, and Oaxaca City, Zapotec migrant associations thrive and are called upon at times to help with expenses for fiestas or other obligations.[25] Lane Hirabayashi (1993) conducted a study of mountain Zapotec migrant associations in Mexico City and found that the Taleans had an especially extensive network. Although village-centered organizations exist in the United States, some have observed the formation of pan-ethnic indigenous alliances that link together ethnolinguistic groups that have historically been at odds, such as the Mixtec and Zapotec, who have formed unions in the agricultural fields of California (Nagengast and Kearney 1990). To a certain degree, a kind of segmental regionalism can be observed, in which villagers identify themselves as Talean when in the Rincón, as Serrano when in Oaxaca City, as Oaxacan when in Mexico City, and as either Zapotec or Mexican when in California. Identity is never static, and as the Rinconeros move around the world their ethnic affiliations are transformed to meet new situations.

The Political Scenario

Little has been written about the Sierra in relation to contemporary Mexican politics. Part of this may be related to the fact that, since the Mexican Revolution of 1910 at least, the Sierra has been a much more tranquil place than other parts of the state such as the Isthmus of Tehuantepec, where political upheavals began nearly twenty years ago, or the Southern Sierra, where a guerrilla army appeared in 1996 (Campbell et al. 1993; Gatsiopoulos 1997).[26] Talea in particular has been described as a place where a powerful "harmony ideology" exists, resulting on the one hand in a remarkably pacific approach to dealing with conflict within the village and on the other hand in a general reluctance to allow state or federal authorities to interfere with local affairs (Nader 1990).

A number of points are worth mentioning in this regard. Although nearly all municipal elections in the Sierra are conducted without the involvement of political parties, it has become standard practice for elected officials to register as party members of the Institutional Revolutionary Party (PRI). Even when village officials strongly denounced the PRI and its practices in private conversation or at a village assembly, officially they always signed as party members and would occasionally attend official PRI events. The advantages appear to be manifold. There is a perception that it is easier to arrange an appointment with PRI officials in Oaxaca City (to request building materials or other needed items) if one is a card-carrying

PRI member. Villagers also strongly desire to keep their towns from dividing along political lines, as has occurred in Yalálag, Betaza, and other villages (Martínez Luna 1995a), where factions formed in years past have led to protracted violent conflicts and probably to higher homicide rates (Nader 1990:220).

It seems that in Talea, as in the Rincón more generally, the PRI has been extraordinarily successful in maintaining power for most of the twentieth century at all levels of government. State governors, federal senators, and federal deputies (representatives of the lower house of the Mexican national congress) have consistently been PRI members. In the Sierra it is unclear whether their success is due to (1) overwhelming support for the party's candidates; (2) an absence of opposition party candidates to run against; (3) electoral fraud, either in Talea itself or further "downstream" in Ixtlán de Juárez or in Oaxaca City; or (4) some combination of these elements.

In recent years it seems that widespread support for the PRI's candidates might be a plausible factor in Talea because a number of relatively high-ranking PRI officials have taken a special interest in the village. Hirabayashi (1993) reports that one influential politician helped Taleans secure various projects at the state level in the 1970s before his untimely death. More recently, a native son of Talea was elected on the PRI ticket as the Sierra's federal deputy (representative to the lower house of Congress) from 1990 to 1994 and before this was responsible for distributing some of Oaxaca's Solidaridad public works funds. A second native of Talea held a position as federal deputy on the PRI ticket for another electoral district in Oaxaca. To what degree these officials have actually helped Talea is an open question. But there is clearly a perception among Talea's neighbors that the village is a PRI stronghold that has unjustly received more than its share of government handouts.

If this was once true, it appears that it is no longer the case. Within weeks after having left the field, I was told by informants that in the July 6, 1997, elections the opposition candidate (of the Party of the Democratic Revolution or PRD) for federal deputy received slightly more votes than the PRI candidate in village polls. (Although the PRI candidate lost in Talea, he received enough votes across the region to win the office.) This is apparently the first time that such a thing has happened in recent decades. Part of it may have been due to the transparency of elections (supervised by the newly independent Federal Electoral Institute or IFE), including the supervision of voting booths by the Taleans themselves.

Even if we assume that the PRI won fairly in previous elections in the Sierra, there are both external and internal factors that almost certainly tipped the scales against the PRI in 1997. First of all, in the last decade a drastic political reconfiguration has occurred in the region, with opposition party candidates now campaigning across the Sierra and openly criticizing the PRI. On a different level, the villagers' access to the mass media, most notably television and radio, has increased their awareness of opposition parties, while the widely publicized assassination of presidential candidate Luis Donaldo Colosio and scandals involving ex-president Carlos Salinas de Gortari and his brother Raúl have led people in Talea, as in other parts of the country, to question the legitimacy of the PRI. So have their experiences in Tijuana, Mexico City, and Oaxaca City, where indigenous people typically arrive as members of an urban underclass besieged by crime, corruption, and economic crisis. Rumors about state and regional misuse of government funds and authority circulate as well.

But perhaps the events that had the most impact on the village of Talea—and probably on the region as a whole—began less than three months before the July 1997 elections, when a number of armed guerrillas (apparently troops of the Popular Revolutionary Army or EPR) swept through Talea and San Pedro Yolox on April 16 and 17, respectively, distributing propaganda and even staging a press conference (Ramales 1997; Ruiz Arrazola 1997a).[27] Though the EPR carried out its mission quickly and peacefully, the Mexican army responded within hours, and more than fifty troops occupied Talea for the next three days.

Perhaps in response to these events, the Mexican military authorities installed an outpost in the Sierra, on communal lands pertaining to Ixtlán de Juárez. According to the Mexican daily newspaper *La Jornada*, Ixtlán's officials were not asked for permission, though the constitutionality of such actions (not to mention the intimidating and arbitrary searches conducted at highway checkpoints) is questionable (Ruiz Arrazola 1997b). These acts may be seen as part of a much broader militarization that is occurring in southern Mexico, a process that has had a frightening and intimidating effect, particularly on rural people but also on missionaries, prospective tourists, NGO workers, and researchers (Corro 1996). It is possible that Serranos, living under conditions of increasing repression and vigilance, might resent the unnerving military policies of the PRI-controlled federal and state governments and choose to vote against the party.

Nader (1990:10) noted that the Mexican government would stay out of the Rincón "only as long as the Mexican state (like the Spanish Crown previously) continues to regard local rule as in its best interest. As long as the Mexican state can continue to regulate the economy (labor, resources, consumption) and as long as local disorder does not threaten the state, local village law will continue in its present manner." Are the limits of local rule and order being reached in the Sierra, with upsets at the voting booths and visits from guerrilla insurgents? It is still too early to predict the outcome. But what seems certain is that the Rinconeros in general and Taleans in particular are becoming rapidly sensitized to political issues; local rule appears to be increasingly threatened even as the Zapatistas in Chiapas try to come to an agreement with the federal government guaranteeing regional autonomous zones.

CHAPTER 3

THE CRAFT OF THE CAMPESINO

Measures, Implements, and Artifacts

Talean campesinos make use of many implements, methods, and measures, some dating to the prehispanic period, others (with Old World roots) to the colonial and postcolonial eras. Here I examine some of the most important in order to set the stage for an analysis of maize, sugarcane, and coffee cultivation. Specifically, I attempt to explode the myth of the campesino as unskilled laborer by demonstrating that the calculation and mental operations necessary to conduct even the simplest farming tasks require a great store of knowledge. In many cases, the campesinos themselves become precision instruments with senses and skills sharpened over the course of a lifetime. Like the Micronesian navigators described by Ward Goodenough (1996), Talean campesinos are often more "sensible" than factory farmers: that is, they rely more on sensory information and their own bodies in the absence of high-tech instruments and mechanized equipment. Rincón Zapotec material culture is quite different from material culture in industrialized society and appears to work toward different ends.

Zapotec Farming before and after the Conquest

The Impact of the Conquest on Rincón and Sierra Zapotec Farming

To understand the success of plow agriculture, iron implements, and new crops and techniques, one must understand the historical context in which technological change occurred. It is not enough to assume that the Spanish technologies were superior to those of the Native Americans or that their incorporation represented an evolutionary advance for Mesoamerican societies. Instead it is likely that Sierra agriculture was radically transformed during this period as the result of colonial policies and demographic catastrophes. Serranos had to resolve the problem of how to dedicate more land to growing cash crops (cotton and cochineal) while setting aside more time for obligatory labor service during a period marked by fatal smallpox, measles, and typhus epidemics. Though farmers were generally able to maintain a subsistence base of maize and beans and retain ownership of their lands, they might not have been able to do so without agricultural implements introduced by the Spanish. The situation probably changed drastically from one in which there was more or less a balance between the population and the land's capacity to feed it to one in which there was a labor shortage and a land surplus—all within a matter of decades. The new technologies changed the fabric of Zapotec farming.

It is probable that slash-and-burn cultivation was practiced throughout most of the Rincón in the prehispanic era, using few implements other than wooden digging sticks. There are a number of clues that might lead us to believe this was the case. Much of the Rincón consisted (and continues to consist) of vast forests, and slash-and-burn is ideally suited to areas of abundant woodland. Furthermore, slash-and-burn would have made it much easier for subsistence farmers in the Sierra—who had no metal tools and probably did not use neolithic hoes—to work soft forest lands rather than the tough lands covered by grasses and rhizomes associated with fallow cultivation (Wolf 1959:61–62).[1]

Spanish tribute policies had a profound impact on Sierra agriculture and affected both land use and labor patterns. Much of the tribute had to be paid in cochineal dye, and for farmers in the Villa Alta district this often meant converting land to cactus plantations to meet the cochineal quota. The demand for labor was another element that weighed heavily on the people of the Sierra. Cultivating cochineal bugs and processing scarlet dye was a time-consuming task, as was an important activity carried out by women, cotton weaving. Other tributary crops included maize and beans and undoubtedly required households to increase agricultural production. Still other forms of tribute were institutional. The *repartimiento de labor*, as we saw earlier, obligated villagers to provide service for colonial officials; consequently, less time was available for farming. It is likely that the imposition of the cargo system in the early colonial period also impeded the farmers' ability to concentrate fully on subsistence cultivation.

One sobering colonial phenomenon tremendously eased land pressure while exacerbating labor shortages in the region. Epidemics suffered by the villagers, along with a low-intensity war waged against them by a number of notorious Spanish *alcaldes mayores*, led to high death rates in the first hundred years of the colonial period. This "depopulation" (which might more accurately be called genocide) led to a precipitous population drop in the Sierra, from nearly 96,000 in 1548 to just over 20,000 in 1622—a decrease of almost 80 percent—according to one estimate (Chance 1989:48-63). It is likely that this led to an overall abundance of land and a severe labor scarcity.

Solving the Post-Conquest Labor Shortage: The Adoption of Spanish Implements

There are at least two possible scenarios for Sierra farming in the colonial era. On the one hand, it could be argued that the diffusion of plows and draft animals occurred shortly after contact. For example, in some parts of New Spain the viceroy organized "joint ventures," as one Spaniard noted in 1554:

> The Indians supply land and labor for weeding and harvesting; the Spaniards contribute oxen, plows, carts, and other implements, in addition to their skill. The latter take one-third of the harvest and the former two-thirds . . . The natives also gain experience in plowing with oxen, sowing, and tilling as we do in Spain. (quoted in Chevalier 1963:194)

Chevalier (1963:60) argues that the plow was used across much of New Spain by the end of the sixteenth century and notes that more than 12,000 plowshares were imported from Spain in 1597 alone.

In the Sierra the adoption of Spanish technologies like the plow might also have been closely linked to depopulation: plowing is advantageous where labor is scarce (Wolf 1959:196-198), and the deaths of four-fifths of the Serranos surely would have led to a labor shortage.[2] This was almost certainly exacerbated by onerous tribute payments, cargos, and the Spanish practice of prodding villagers into purchasing oxen and mules on credit. These circumstances might explain the rapid incorporation of other labor-saving devices such as steel axes, machetes, hoes, and beasts of burden, all of which would have been useful in a society decimated by plagues, epidemics, and wars and may have been forced upon people by Spanish merchants anyway.

There is one problem with such an analysis, however: the same depopulation that made labor scarce might also have reduced overall land pressure, even after taking into account increased cultivation of cotton and cochineal. And people have little incentive to switch from slash-and-burn to plow-based fallow systems where land is abundant and labor is scarce (Harlan 1975).

On the other hand, a case could be made for late colonial or even postcolonial adoption of plow-based fallow agriculture. Within the first hundred years after the Conquest, it is plausible that agricultural systems in nearly all villages moved toward a system that more closely resembled slash-and-burn since land pressure may have been much lower (even taking increased cotton and cochineal production into account) and labor was scarce. Following this line of reasoning, it would seem likely that not until the 1800s did population densities become great enough in Talea—as a result of highly productive gold and silver mines—to have led to the gradual conversion of a slash-and-burn system to a plow-based fallow system allowing more people to be supported on less land.[3] If one accepts this model, it seems likely that even if villagers quickly accepted axes, machetes, hoes, and other such implements the switch to plow agriculture would still have been delayed. This model is supported by contemporary evidence, for with the exceptions of Talea and Tanetze—the Rincón's most populous villages and important coffee producers—today nearly all Rincón villages continue to rely heavily on slash-and-burn farming.

The earliest explicit evidence I found for plow agriculture in the Northern Sierra is a 1778 report in which a Spanish official, Francisco Antonio Núñez, notes that the people of San Pedro Teococuilco in the district of Ixtlán "cultivate and plow the lands with bulls, [and] transport their wheat and maize on mules" (quoted in Esparza 1994:331–332). As in the case of food crops, it seems likely that Dominican friars introduced these animals in the region.

In sum, plow and draft animals, beasts of burden, steel and iron tools such as axes, machetes, and hoes, and new crops radically transformed Sierra farming. Regardless of when they were incorporated, they were intelligent responses to problems in an era marked by plague, tribute, and war.

Contemporary Units of Measure in Talea

Across Mesoamerica (and, more generally, Latin America) a host of length, volume, area, and weight measures appear to have been adopted during the Spanish colonial period, though it is likely that some may have been preceded by equivalent measures in the prehispanic era. Here I focus on those that are most frequently used in Talea for contemporary farming, which happen to predate the standard metric and English systems used in industrial societies today.

Length

In Talea units of length are measured with reference to the dimensions of the human body. The relevant measures are the *dedo, geme, cuarto, codo, vara, metro,* and *brazada* (all Sp.; see Figure 3.1). By multiplying individual measures, or adding various measures, campesinos can create an almost infinite number of calculations.[4]

It is not entirely clear whether these units are based on Spanish colonial measures, prehispanic measures, or a combination of the two. Foster (1960:108) notes that the *vara* was used in colonial Spain, where it varied between 0.8 and 0.9 meters, though the standard typically used was the *vara de Castilla,* 0.836 meters. Brian Hamnett (1971:viii) provides a

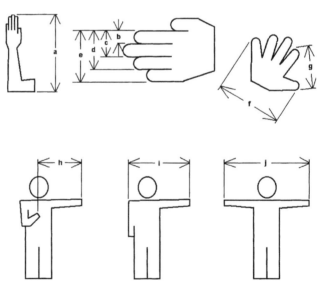

Figure 3.1. Measurements used by Talean campesinos. These include: (a) the *codo;* (b) one *dedo;* (c) two *dedos;* (d) three *dedos;* (e) four *dedos;* (f) one *cuarto;* (g) one *geme;* (h) one *metro;* (i) one *vara;* (j) one *brazada.*

similar description of this unit, used to measure textiles. Oscar Schmieder (1930:28-29), however, notes that prehispanic documents show that the Aztec used a measure corresponding roughly to the Spanish *vara*. This might indicate that both prehispanic and Spanish societies used such a measure. In Talea the *vara* is described in bodily terms: it is the distance from the center of the chest to the tip of the extended arm.

Elsie Clews Parsons (1936:201) lists a host of length measures used in the Valley Zapotec town of Mitla, most of which correspond closely to the Talean system: *cuarta* ("from thumb to little finger"), *geme* ("tip of thumb to tip of index finger"), *dedo*,[5] *cordo* ("the measure from the tip of the middle finger to the elbow"), *braza* ("both arms outstretched"), and *metro* ("from top of right shoulder to top of middle finger of left hand"). According to Parsons, Bernardino de Sahagún reported that the *geme* had a prehispanic equivalent, which is supported by the fact that in the Rincón the Zapotec word *hueni* is used to describe the measure. (The *codo* has a Zapotec name as well, *tu dxit*, literally "one bone.")

De la Fuente (1949:136) claims that in Yalálag one *braza* is the length of a single arm, measured from the shoulder to the end of the hand. This differs from the *braza* in Mitla and *brazada* in Talea, which are equivalent to the distance between the outstretched arms.

A number of things quickly become apparent in analyzing these systems of measurement. There is a great degree of variation, even if we confine our geographic area to the state of Oaxaca. Measures that share the same name do not always represent the same quantity even in neighboring villages. Furthermore, due to a lack of documentary evidence, it is difficult to tell whether the standards that existed in the prehispanic period are congruent with the Spanish colonial system. Rather than attempt to sort out the units historically and geographically, I shall focus on the use of these measures in Talea today, where they are utilized primarily by campesinos for crafting implements.

Among Talean campesinos, the *codo* is used to measure the length of threshing poles to knock maize kernels from their ears. The *vara* is the prescribed length for hoe handles and *ganchos* (Sp., "hooks") that campesinos use in conjunction with machetes to clear away underbrush. It is also the standard length of firewood; cutting the wood in equal lengths makes it much easier to transport using tumplines (Sp. *mecapales*) or beasts of burden. Campesinos use the *metro* for axe and pick handles and the *brazada* for digging sticks. Other implements — notably the plow — require complex combinations of these measures.

Women, particularly older women and those from surrounding villages, rely upon these measures when purchasing clothes from Talean merchants. They can quickly determine whether trousers will fit their children by measuring, for example, *tu cuárt más tu huéni* (Zap., "one *cuarto* plus one *geme*") across the waistband. If a male member of the family is present, a rule of thumb is that a pair of trousers will fit him well if his modified *codo* (with the fist closed rather than open) fits snugly inside the waistband.

For a person schooled in the United States, the use of bodily measures may appear to be arbitrary and problematic. Indeed, I initially thought that these units were not "standards" at all, since they obviously varied according to the physical sizes and shapes of individuals. A tall man's *brazada*, for example, might differ significantly from that of a short man, and both would likely be greater than that of a woman. The measures at first appeared to be quaint but inaccurate guides. My assessment was soon transformed once I realized that bodily measurements have clear advantages over metric and English systems. For one, the Talean system is completely portable: the measures can (and indeed must!) be carried at all times, unlike tape measures and meter sticks. But perhaps more importantly, because the measures vary from individual to individual, they have important implications for the production of implements. Specifically, the measures ensure that the campesino's implements will be custom-made (that is, suited to fit individual size) because they are based on the person's own body. Herein lies the ingenuity of Talean measures.

For engineers in industrialized societies, variable measures may appear to be ill-conceived because standardization is valued. It is no accident that the metric and English systems increased in popularity after the industrial era began, in the first half of the 1800s, for interchangeability of parts is a prerequisite of mass production (Noble 1977:69-83). Talean measures do not conform to these imperatives.

Corporally based measures are not deployed for industrial uses, but for use by artisans. For campesinos, interchangeability is irrelevant because their products do not wind up on an assembly line. They are for one's own use. The criterion of customization — creating a implement with the "right fit" — is much more relevant. It takes only a few minutes of toiling on the farm to realize that an implement poorly fitted to one's body can be inconvenient and even injurious. In a sense, ergonomic designs (which comfortably fit the human body) are generally achieved by Sierra farmers, who are present at all stages of the implement's life cycle, from its design to its construction and finally to its use. This is not surprising, considering that most do not resist the notion that vigorous manual labor is a normal part of everyday life. Ergonomics simply becomes a means of making a necessary part of life safer and easier.

In U.S. society, the problem of ergonomics was only made a priority fifty years ago by the military during World War II (and later by industry). Complex technologies such as fighter planes, tanks, and submarines were among the first tailored to the dimensions of military personnel, whose optimal performance was critical. Dials and gauges were redesigned to be read more clearly; switches, buttons, triggers, and throttles were arranged in a more logical fashion; and seats were equipped with lumbar and leg supports. More than a century after the "Industrial Revolution" began, ergonomic design finally got its start. Perhaps this had to do with the fact that product designers were often far removed from users. It is a striking contrast to the Rincón Zapotec, who solved many ergonomics problems long ago in the course of their daily work.

Volume

Across much of rural Latin America the *almud* (Sp.) and the *fanega* (Sp.) are used quite consistently as measures of dry volume, especially for measuring maize and beans. However, the exact quantities constituting each measure vary. Foster (1960:109) notes that in rural Spain the *fanega* is equivalent to approximately 55 liters and is subdivided into twelve units called *celemines*, which are further divided into four *cuartillos*. Hamnett (1971:viii) reports that in colonial Spain maize, beans, and wheat were measured in *almudes* equivalent to 7.568 liters, which were divided into four *cuartillos*, and that twelve *almudes* were equivalent to a *fanega*. Parsons (1936:413) describes the *almud* in Mitla as a quantity measured with a "wooden box a foot square and about half a foot deep" (approximately 14 liters)—though this is much larger than any other *almud* cited in the ethnographic record and may not be accurate.

According to Carlos Alba and Jesús Cristerna (1949a:453–454), the *almud* used by the Zapotec has "almost always" been equivalent to five liters, though in a few villages a four-liter *almud* was apparently used in the 1940s. Malinowski and de la Fuente (1982 [1940]:177) agree. They describe the *fanega* as equivalent to twenty-four or twenty-five *almudes*. Talean informants describe their own *almud*, called *tu yag* ("one tree" in Rincón Zapotec, probably a reference to the wooden box), as a five-liter measure, though when I measured the dimensions of their boxes the quantity was approximately 5.9 liters.[6] For maize, this is equivalent to approximately 4.2 kg.[7] Other boxes are used to measure half-*almudes*. Talean carpenters construct the square boxes with templates that they have used for generations. Maize or beans are poured into the boxes, then the top surface is leveled off to ensure precise measurement. The Talean *fanega* equals twenty-four *almudes*, approximately 132 liters.

What benefits might the *almud* offer? Malinowski and de la Fuente (1982 [1940]:177), in a thoughtful analysis, describe its advantages:

> The poor Indian and peasant prefer the *almud*, not because they are "conservative" or "dislike innovations," but because this measure enters into all their domestic calculations in a manner which has been standardized for centuries, and they are accustomed to calculate with it. Thus, they know how many tortillas can be produced from one *almud*, or how many cups or bowls of *atole* (maize drink); in short, how many *almudes* per week their budgets require ... Moreover, according to our best informants, grain can be seen more easily in the [wooden] *almud*, which is a shallow measure. This gives a greater sense of security to the buyer than the tall, narrow litre measure. For all these reasons, buyers and sellers are even more opposed to measuring their produce in metric units.

In addition, campesinos are able to "see" an *almud* of maize much more readily than they can a kilogram. Selling maize and beans by volume, in other words, is a way for campesinos to avoid having weight scales tipped against them by unscrupulous merchants. For all of these reasons the *almud* is likely to remain the standard grain measure in Talea for the foreseeable future.

Area

The *almud* is also favored because it is integrated with another important measure: area. In much of Oaxaca the *almud* is used as a measure of both volume and land area, and the two are connected in terms of maize seed. In Talea the *almud* is equivalent to the land area needed to plant a single *almud* (5.9 liters) of maize seed in a field. The Taleans claim that four *almudes* are equivalent to approximately one hectare (2.47 acres).

The (maize) *almud* is also used to measure land that is not planted with maize; consequently, coffee groves may be described in terms of *almudes*. In fact, it appears that hectares are used only when reporting land area to state or federal government officials. In these cases, hectares are calculated by simply converting from *almudes* using a 4:1 ratio (4 *almudes* = 1 hectare).

Weight and Other Measures

Coffee beans are measured by weight, using a unit called the *arroba* (25 pounds or 11.5 kg). According to Hamnett (1971:viii), this unit was used in the Spanish colonial period to weigh cochineal.

In Talea's Monday market, many items are measured in kilograms. For example, tomatoes, potatoes, live chickens, fresh beef, fish, chiles, white sugar, and many other foods are weighed by merchants on scales or else packaged in plastic bags containing ¼, ½ or 1 kg amounts.

Other unusual measures are the "third" (Sp. *tercio*), used to measure firewood, the "bundle" (Sp. *manojo*), used for wild greens, and the *pancle* (Sp.), used to measure unrefined sugar.

Measures of Power: The Politics of Standardization

Historian David Noble, in his book *America by Design* (1977), argues that the standardization of weights and measures played an integral part in the consolidation of corporate power in the United States at the turn of the twentieth century. Standardizing measures and machinery was a costly process, and smaller manufacturers and artisans were often unable to shoulder the expenses associated with such conversions. Changes in the way things were measured became intimately associated with shifts in U.S. manufacturing.

In a broad sense, the adoption of some standard measures in the Americas—particularly those related to volume and weight—can be analyzed in terms of the politics of Spanish colonialism. *Arrobas, fanegas, cargas,* and other Spanish units were used to quantify tribute payments. As an illustration we might consider the case of an ex-conquistador who was granted the right to collect tribute from a Zapotec mountain village (Chance 1989:24, 108):

Half the following goods, due every eighty days, went to the encomendero . . . and the other half to the crown: fifty pesos' worth of gold dust, six *cargas* [approximately 1,350 pounds] of *cacao*, thirty-nine cotton *mantillas* (shawls), twelve *arrobas* of honey, and fifty turkeys. In addition, one hundred *fanegas* of corn were due at harvest time each year . . .

According to Hamnett (1971:viii), *arrobas* were also used to collect cochineal dye. Furthermore, *varas* and English pounds were apparently used by

Spanish merchants who forcibly sold goods to Serranos (Chance 1989:99–100).

Needless to say, many of these measures were adopted by people who, over time, appropriated them and used them for their own purposes. With the eventual shift to metric units in many metropolitan countries during the nineteenth century, Spanish colonial measures began to seem outdated. As Mexican city dwellers began adopting meters, liters, and hectares, the prehispanic and Spanish colonial measures based on corporal dimensions, *almudes,* and *fanegas* probably came to be viewed as "primitive." They effectively served the needs of many Oaxacans, however, and continue to do so in relatively isolated places like the Rincón.

Agricultural Implements

For Talean campesinos, the measurements described above serve as guides for the design and manufacture of farming implements. In this section I review three of the most important—hoes, axes, and plows (see Table 3.1 for a more complete listing of agricultural implements), illustrating how Talean farmers fashion the implements using centuries-old measuring techniques and a profound practical knowledge of materials and construction techniques.

Conscientious farmers are keenly aware of trees and branches which might potentially serve as implements. During long hikes to the fields, informants would often stop and call my attention to a particular tree or branch suitably shaped for a plow, axe handle, or hoe handle. It is not an exaggeration to say that these campesinos had a highly developed sixth sense which allowed them to scan trees visually while walking along pathways. Furthermore, they were often able to detect animals and useful plants long before I was remotely aware of anything remarkable.

Hoes

The short-handled hoe (Sp. *coa*) is among the most important implements used by Talean campesinos. Before the Spanish arrived, hoes were fashioned from organic materials (wood, bone, or antlers) in some parts of the Americas (Weatherwax 1954:61–62).

In 1997 Talean campesinos could purchase steel hoe blades for N$30 (approximately US$2.50 or a day's wage). The campesino fits the blade tightly to the handle (*garabato* [Sp.] or *bládzu* [Zap.]) by wedging a leather tongue between the steel and the wood. The length of the handle should mea-

Table 3.1. Agricultural Implements Used by Talean Campesinos

IMPLEMENT	ORIGIN	MEASUREMENT	SOURCE
plow	Old World	various	all local materials except blade
yoke	Old World	various	all local materials
plow beam	Old World	various	all local materials
hoe	Old, New World	4 *cuartos*	blade imported from factories
digging stick	New World	1 *brazada*	blade imported from factories
hook (*gancho*)	New World	1 *vara*	all local materials
machete	Old World	12–32 inches	imported from factories
axe	Old, New World	1 *vara*	blade imported from factories
pick	Old World	1 *metro*	head imported from factories
threshing pole	New World	4 *cuartos*	all local materials
shovel	Old World	1 *metro*	blade, handle imported from factories
coffee depulper	New World	N/A	imported from factories
sugar mill	Old World	N/A	imported from factories

sure approximately one *vara*; if it is too long or too short, back or arm injuries are likely. The distance between the edge of the blade and the handle should measure one *cuarto*; a shorter distance is likely to result in scraped knuckles.

The ingenuity of the hoe handle is remarkable. Campesinos select branches that are slightly curved (unlike those of mass-produced hoes). The curved handle has a number of advantages: curved branches are typically much easier to find than straight ones, and the arch formed by the curve is able to withstand more stress than a straight rod.

But the most important effect of the curved implement is ergonomic. Campesinos hinted at this by noting that the curve "gives more of an advantage." What they meant by this was not immediately obvious but became clear once I tried using a right-angle hoe. When used for clearing

Figure 3.2. Campesinos use short-handled hoes to weed their fields. The hoe's curved handle gives the farmer a clear ergonomic advantage; he is able to maintain both wrists in a straight position and thereby minimize strain. *(Photo by Laura Nader)*

away rocky soil or weeding a milpa (maize field), the right-angle hoe puts a tremendous strain on the tendons connecting the thumb to the wrist. The curved hoe, by contrast, prevents this problem by straightening the wrist of the forward hand. Only by comparing the differences in an actual work situation can one appreciate the advantage of the curved handle.

Campesinos produce hoe handles using only a machete and insist that they should be cut during the full moon to prevent termite infestation (see Chapter 5). Strong, light woods are best, such as wood from orange trees. A tough handle will last three or four years and sometimes even longer; hoe blades may be spent in as little as two years if used on particularly rocky milpas.

Axes

Axes may be among the oldest implements used in the Sierra. During the prehispanic period, the Zapotec and Mixe probably used stone axes to fell trees for slash-and-burn farming. The Spanish introduced steel axes. Informants report that until the 1950s the only steel axe blades available in the region were of a type that had to be wedged into the wooden handle.

Improved blades (with an eyelet) later diffused into the region, which were much easier to fit to the handle and generally more durable.

Axe handles are often made of oak and are whittled (using a machete) so that the cross section forms an oval. It has a shallow curve to it; I was told that swinging the axe is made easier with a curved handle, though this does not seem to be as obvious as in the case of the short-handled hoe. It should have a length of one *metro* and at its thickest part should measure about three *dedos*. Campesinos hold the steel head fast to the handle by inserting a leather strip between the wood and the blade. An axe blade can last twenty or thirty years. Talean campesinos claim that trees and branches for structural beams and implements should be cut during the full moon to prevent termite infestation.

Since Taleans do not practice slash-and-burn farming, they use axes primarily for chopping pine and oak trees for firewood and for clearing the area around the perimeter of milpas so that sunlight reaches sprouting plants. Within the last fifteen years many campesinos have replaced axes with chainsaws for felling thick trees.

Plows

Plows are undoubtedly the most complex implements crafted by campesinos. Not all are able to manufacture them; perhaps half of Talea's campesinos have this skill. Plows are crafted using only a machete, an iron chisel, an axe, and a wooden mallet.

The campesino fashions the plow beam (Sp. *timón*) from a straight trunk more than two *brazadas* in length and with a diameter of at least four *dedos*. Oak wood is ideally used (though some prefer red oak [*Quercus* sp.], Sp. *encino colorado*). A wood with the Zapotec name *yebágu* (unidentified species) is also suitable. After the campesino fells the tree (again during the full moon phase; see Chapter 5), he carries it to a sunny spot near the ranch house to dry. This takes two to three weeks. Then he peels away the bark and smoothes the surface of the trunk, using glancing machete blows to even out knotholes and other irregularities.

Next the campesino angles the back end (that is, the wider of the two ends) of the beam little by little to approximately 75 degrees with a sharp machete. He shaves the sides of this end with the machete in order to give the beveled back end of the beam a rectangular profile.

Using the chisel, the campesino makes a rectangular hole near the back end of the beam. The hole's back edge should measure one *cuarto* from the edge of the beveled face and should be just over four *dedos* long and a little

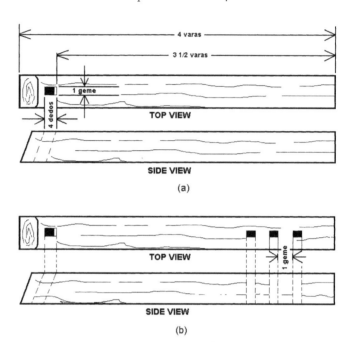

Figure 3.3. Construction of the *timón* (plow beam). This includes measurements for (a) the *telera* (connecting beam) hole and (b) *clavillo* holes.

more than two *dedos* wide. Furthermore, the hole should taper backward, parallel to the angle of the beam's back face (see Figure 3.3a). This hole will receive the *telera* (Sp.), a short connecting beam holding together the plow beam and the plow.

Eleven *cuartos* from the front edge of the *telera* hole the campesino makes a mark with the machete. He also makes marks at distances of twelve and thirteen *cuartos*. These form the center points for three holes that will receive the *clavillo* (Sp.), a thin pin that joins the yoke to the beam. These holes are not angled but perpendicular to the surface of the beam (going down) and should be bored out with the chisel — but only after the beam is secure so that the holes are aligned with the long *telera* hole at the back end. This can be done by driving a stake into the hole so that the back end of the beam is pinned securely to the ground. The *clavillo* holes should each measure three *dedos* long by approximately two wide (see Figure 3.3b). If a *clavillo* is available, the holes can be tested and enlarged slightly if necessary. The *clavillo* is fashioned out of a strong wood—the "heart" of a *rahuás* (Zap.) tree is ideal—and measures approximately two *cuartos* long, with a diameter of just less than two *dedos*.

The campesino uses a thin branch to measure a *vara*, measuring a distance of four *varas* from the top of the back edge of the beam to a point near the front. He makes a mark with the machete at this distance and cuts off the remainder of the trunk. He then finishes the beam by shaving its surface with the machete, removing any remaining bark, knotholes, or other blemishes.

The *telera* is a wide, short connecting beam used to join the plow and plow beam. It is fashioned from oak. The campesino begins by cutting a blank from a log which measures five *cuadros* long and one *geme* in diameter. He then chops the blank into two semicircular halves using the axe (see Figure 3.4), each good for two *teleras*.

The next step is laborious. The campesino smoothes the face of the wood block with a machete by shaving off thin (perhaps 0.5–1.0 mm) slivers of wood. He then swings the machete at an angle nearly parallel to the face

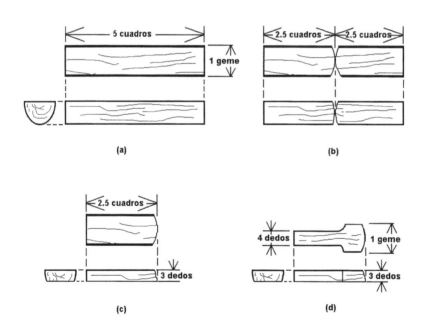

Figure 3.4. Four stages in the construction of the *telera* (connecting beam), which joins the plow beam to the plow.

of the wooden block. Once the process is complete, the face of the block is left smooth and even. The campesino cuts the block in half with a machete; each half measures approximately 2½ *cuartos* long.

The campesino cuts another face, parallel to the face made by the axe in the initial step; he does this with a machete or, if the wood is too tough, with an axe. The thickness of the block should be approximately three *dedos*—slightly more than the *telera* hole cut into the plow beam (see above).

The skinny part of the *telera* will have a width of approximately four *dedos*, so the campesino must lop off the sides of the blocks. The amount to be cut varies, but a good rule of thumb is to measure four *dedos* from the "head," make a mark with the machete, measure two *cuadros* from the first mark, and then shave off all excess with the machete.

Next the campesino trims away the edges until the *telera* takes its final form. This is done after the plow has been constructed, so that the *telera* may be measured by trying to insert it in its hole and narrowing it if necessary. The campesino constantly reduces the width, shaving a little at a time and then attempting to knock it into the plow with the mallet; if it does not fit, it is knocked back out, a little more is shaved off, and it is knocked into place again. The process is repeated as many times as it takes to get a tight fit.

For the plow (Sp. *arado*), oak is the preferred wood. The campesino selects a straight trunk, slightly more than one *geme* in diameter, more than six *cuartos* long, and with a thick branch suitable for a handle at one end. The handle must necessarily be "custom fit": it should rise to the level of the plowman's relaxed arm (see Figure 3.5). If the handle is any lower or any higher, plowing becomes a punishing task.

After the campesino cuts the trunk (during the full moon) and allows it to dry in the sun, he peels its bark away. Then he cuts the leading edge of the plow with a beveled face (approximately 35–40 degrees), which should measure about one *codo* along its long axis.

Next the campesino uses a chisel to carve out a seat (where the back end of the plow beam will rest) where the handle meets the trunk. Its width should be just less than that of the plow beam—four *dedos*. The length of its opening, measured across the trunk, should approach one *geme*. He carves out the seat bit by bit, readjusts it, inserts it into the back end of the plow beam, and readjusts as needed. When the butt of the plow beam can be securely hammered into place with the mallet, the seat is ready.

After the seat has been carved out, a long nail is hammered into the

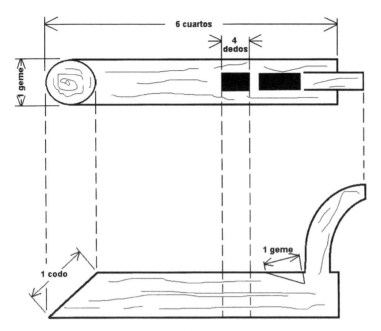

Figure 3.5. Arado (plow) measurements.

trunk, just above the center of the seat's top edge. The campesino ties a strand of thin rope to the nail and draws it across the central axis of the plow to the center of the plow's leading edge. Then, with a pencil, he draws a center line across the leading edge's face and also in the area immediately in front of the seat. The lines are used to center the *telera* hole and the iron cutting edge.

Next the campesino cuts the *telera* hole using a chisel and mallet. He starts its back edge at a distance of 1 *cuarto* from the back edge of the seat. The hole measures 4 *dedos* long by 2½ wide—small enough to give the *telera* a tight fit. When he hammers it into place, it becomes a permanent part of the plow.

Once this is done, all that remains is to attach the iron cutting edge to the beveled face, using long nails made especially for this purpose. Most cutting edges are purchased from merchants who procure them from metalsmiths in the nearby village of Yalina. There is one Talean merchant, however, who sells edges fashioned in Talea out of steel leaf springs from discarded automobiles. This is an ingenious local adaptation of a discarded mass-produced artifact that would simply be seen as junk in many industrialized societies. A plow can last two years or more, depending upon the rockiness of the soil.

The plow may now be joined to the plow beam. The campesino connects the two by hammering a pair of wedges (Sp. *cuñas*) at the joints using the wooden mallet. The distance between the bottom end of the assembled beam and the leading edge of the plow should measure one *codo*. If not, a wedge of a different size may be inserted at the back end.

Making Use of Scrap Material

A Talean campesino once related a wonderful story about an old man who carried two knapsacks rather than one. The first he filled with things that any campesino might carry: matches, a flashlight, a pocketknife, a plastic rain cape, cigarettes, mezcal, a slingshot. The other bag he reserved for useful bits of trash he would find along footpaths. Some days he would arrive at his ranch house with the bag nearly full of chicken wire, bits of rope, cans, bottles, and scraps of cardboard. My companion ended the story by making a broader statement: "Here in the countryside, everything has a use. Smart campesinos don't throw away anything."

Indeed, the creativity of Talean campesinos is perhaps nowhere more apparent than in their uses of scrap material. In a society that still self-provisions itself to a significant degree, bits and pieces of junk are often useful for people with little cash. A complete list would take too long to describe; here I shall review only some of the more interesting uses that some Taleans have for artifacts that are typically discarded in industrialized societies.

I have already mentioned a number of examples. Talean blacksmiths procure truck and automobile leaf springs and make them into cutting edges for plows. Villagers suspend cracked plastic buckets from the roofs of ranch houses to hold dry foods and implements. They may also use the buckets to carry granular chemical fertilizer through maize fields if they are not cracked too badly (see Chapter 5).

There are many examples of long life cycles for products that break or wear out over time. Clay pans, typically used for preparing soups and stews, occasionally crack but are still useful for roasting squash seeds (a popular snack). Some campesinos use worn cassette tape ribbons as a means of preventing rodents from attacking growing maize fields by tying them to wooden sticks (approximately one meter high) inserted around the milpa's perimeter. With a slight breeze, the shiny ribbon flashes brightly in sunlight and, according to campesinos, emits a whistling noise as well. Worn or frayed straw mats (Sp. *petates*) have many uses: to line the bottom of granaries, to hold small amounts of maize for sun-drying, or as extra insulation below newer *petates* used for sleeping.

It is rare to see "disposable" plastic soda bottles disposed of immediately. Instead, villagers use them to hold mezcal, coffee, or fruit juices. Non-returnable glass jars and bottles are often fashioned into kerosene lamps by punching a small hole in the lid and inserting a strip of cotton cloth which serves as a wick. The lamp can last for years and is the primary source of illumination in ranch houses. It burns kerosene very slowly and emits a great deal of light.

Plastic goods have especially long lives. Plastic sheets used as rain capes tear after several years, but campesinos may still use them to line the floor of temporary bins that serve as granaries in the maize harvest, for the plastic prevents ground precipitation from soaking the maize. They also employ fertilizer bags for this purpose, but used bags are guarded with particular care since they are needed to carry coffee beans, maize, and black beans back to the village from the farm. Villagers reuse small plastic bags many times, to carry bread, vegetables, or other foods, cloth diapers and baby bottles, or maize and beans offered to the Catholic saints at the church. When the bags tear, women fashion them into circle templates for making tortillas. Men often use frayed or broken plastic rope to create muzzles for ill-tempered mules or donkeys or weave them into girdles for keeping pack saddles securely mounted on the beasts.

Villagers debut new clothing on fiesta days and Sundays. As the clothes wear out, they are worn on weekdays and then finally on the farm. Clothes that are beyond repair are cut up and used as bandages, as patches for other articles of clothing, as bands for tying girls' and women's ponytails and braids, or as diapers. Most campesinos are able to keep a pair of huaraches together for five or six months. After the sandals are completely spent, they may use the leather to wedge hoe or axe blades in place, as pockets for slingshots, or as replacement wristbands for watches. Incidentally, a great deal of good secondhand clothing—much of it from the United States—is brought to places like Talea by relatives in Mexico City or Tijuana during the fiesta season.

Campesinos place old blankets and burlap bags under pack saddles as cushions, so that their beasts of burden do not get cut by wood or nails. Some use nails to punch discarded sardine cans with holes; these can be used as brushes for combing burrs from mules or horses. Bottlecaps are used as washers, hammered between nails and sheets of corrugated tin to keep the roofing material in place, and as children's toys. Women often use tin cans as flowerpots. Worn files used to sharpen chainsaw blades may be reused for sharpening machetes. Many families use the pages of old school textbooks and newspapers as toilet paper.

There are many other examples. Suffice it to say that campesino families, especially those with little cash, are often resourceful enough to extract "use value" from discarded items. These scraps—in many cases salvaged from the trash heaps of the consumer society—do not appear on the balance sheets of Mexico's Gross Domestic Product, but for many families they serve a critical economic function at little or no cost. The campesino family's "limited purchasing power" is thus compensated for by a seemingly limitless creative power which is tapped to create useful implements out of "junk."

Craftwork and the Campesino

I have used the term "implements" to refer to items crafted by campesinos for doing their work. Why not describe them as "tools" or even "machines"? The problem, eloquently stated by Lewis Mumford in *Technics and Civilization* (1963 [1934]), is that a great deal of confusion surrounds the terms. Sometimes "machine" is used to describe a device that converts energy from one form to another (e.g., electrical energy to kinetic energy or hydraulic energy to rotational energy), while "tool" describes something used to work, cut, chop, or shape materials. The distinction is problematic: plows, for example, might be described as machines (since they convert animal to kinetic energy) but also be described as tools (since they cut earth). Any such distinction, it seems, is ambiguous.

This is complicated by the tendency, by both laypeople and experts, to link the terms to evolutionary schemes. "Savages" and "barbarians" have tools, magic, and religion to help them in the struggle for survival; "civilized" people have machines and science. In this stereotyped view, stone axes, arrowheads, and digging sticks are "tools"; turbine engines, programmable lathes, and nuclear power generators are "machines." Early in the twentieth century many people in industrialized societies referred cheerfully to the modern period as the "Machine Age" (drawing a distinction between it and the "Stone Age," "Bronze Age," etc.).

Mumford (1963 [1934]:12) demonstrated that machines were very much a part of ancient societies and argued that "both [tools and machines] have played an enormous part in the development of the modern environment; and at no stage in history can the two means of adaptation be split apart. Every technological complex includes both: not the least our modern one." Mumford (1963 [1934]:10) attempted to describe some of the differences characterizing the ends of the continuum between machines and tools:

The essential distinction between a machine and a tool lies in the degree of independence in the operation from the skill and motive power of the operator: the tool lends itself to manipulation, the machine to automatic action. The degree of complexity is unimportant: for, using the tool, the human hand and eye perform complicated actions which are the equivalent, in function, of a well developed machine; while, on the other hand, there are highly effective machines, like the drop hammer, which do very simple tasks with the aid of a relatively simple mechanism. The difference between tools and machines lies primarily in the degree of automatism they have reached: the skilled tool-user becomes more accurate and more automatic, in short, more mechanical, as his originally voluntary motions settle down into reflexes, and on the other hand, even in the most completely automatic machine, there must intervene somewhere, at the beginning and at the end of the process . . . the conscious participation of a human agent.

Putting aside the machine/tool distinction for the moment, we might consider Mumford's more profound observations about the human implications of technological complexity. He questions the vulgar notion of the arrested technological development of "primitives" by arguing that those using less-automated implements tend to have remarkably "accurate and more automatic . . . reflexes"—reliant on the "human hand and eye perform[ing] complicated actions"—than those operating highly automated implements which may only require the push of a button.

This point was made clear to me in the first months of my fieldwork when a farmer and I made a short hike from the ranch house to the forest one morning to chop a felled tree into firewood. With two to three well-placed machete blows, he was able to cut rather thick branches from the main trunk. When I attempted to do the same, I failed miserably; my chops would (dangerously) vary by as much as two to three inches from one blow to the next, and my raw blisters and tired arms made accuracy impossible. What took the informant ten seconds to cut took me a full, very punishing minute. I soon learned that the placement of the blow is critical; my companion's finely coordinated eye and arm movements—which gave him an accuracy of less than a single millimeter from one blow to the next—constituted *real* technological complexity. The movement necessary to make a plow is much more complex than swinging a machete, which is among the first tasks learned by young boys and girls when they accompany their mothers to collect firewood.

The proficiency of children in conducting such tasks is not only a corporal education, but also a theoretical one, as they learn to categorize different kinds of wood, soils, plants, animals, and weather patterns, according to logical schemes taught to them by their parents. The beneficial psychological effects of such an education, described in detail by Margaret Mead in *Coming of Age in Samoa* (1928:11–21, 108–138), seem to apply to most campesino families in Talea today.

Craft, Art, and Mass Production

Octavio Paz (1973) takes a different approach in an enlightening essay on handcrafts, in which he contrasts them with mass-produced "technologies" on the one hand and "art" on the other. He begins with a critical analysis of "the religion of art," which has a habit of "consecrating things and imparting a sort of eternity to them," preserved in the temple of the museum. Beauty becomes the hallmark of art, and art, argues Paz, is isolated and made autonomous; it becomes an end in itself. Mass-produced technologies, in contrast, gravitate toward utility. The standards for assessing these technologies are impersonal: efficiency, standardization, and functionality. The artifacts themselves became "precise, obedient, mute, anonymous instruments," associated with art just as the profane is associated with the sacred: "the industrial revolution was the other side of the coin of the artistic revolution" (Paz 1973:19).

Crafts stand in contrast to both high art and mass-produced technologies: they are "beautiful objects, not despite their usefulness but *because* of it"; indeed, they are made not only by human hands, but for them (Paz 1973:17). Craft artifacts are styled regionally, not individually (as in the case of high art) or internationally (as in the case of mass-produced technologies): "Craftsmen have no fatherland: their real roots are in their native village . . . Craftsmen defend us from the artificial uniformity of technology and its geometrical wastelands: by preserving differences, they preserve the fecundity of history" (Paz 1973:23). Significantly, Mumford (1963 [1934]:415–416) also recognized the creative role of craftsmanship as a "safety device and as a means to further insight, discovery, and invention" in an atrophic age of technological standardization and compartmentalization.

Karl Marx and Friedrich Engels's (1960 [1844]) notion of "alienation"

Figure 3.6. Children often learn their first lessons about food collection, farming, and the forest from their mothers. This trip to the forest yielded firewood for the home, decorative flowers for the saints in the church, and *quelites* (edible wild greens). *(Photo by Gabriela Zamorano)*

in capitalist societies describes the separation of workers from the products they create as the result of labor processes. For the worker, alienation "means not only that his labour becomes an object, an *external* existence, but that it exists *outside him*, independently, as something alien to him, and that it becomes a power on its own confronting him; it means that the life which he has conferred on the object confronts him as something hostile and alien" (Marx and Engels 1960 [1844]:498).

Alienation includes the stripping away of the individual worker's potential by compartmentalization through a highly regimented division of labor. For Marx (cited in Mumford 1963 [1934]):

It transforms the worker into a cripple, a monster, by forcing him to develop some highly specialized dexterity at the cost of a world of productive impulses and faculties—much as in Argentina they slaughter a whole beast simply in order to get his hide or tallow. Not merely are the various partial operations allotted to different individuals; but the individual himself is split up, is transformed into the automatic motor of some partial operation.

The dehumanization, degradation, and de-skilling of workers stands in stark contrast to craft. Industrial manufacturing processes began competing with craft traditions, and basic human behavioral patterns changed as they were painfully adapted to the norms and rhythms of industrial work discipline. The consequences of such changes for craftspeople and their families are well documented for the case of England: they suffered severe economic and psychological hardships; not surprisingly they resisted, often violently (Thompson 1963).

The Campesino's Craft

Let us return to Talean campesinos. Many of the artifacts they use in farming could certainly be referred to as hand-crafted items (plows, hoes, axes, gourd containers, etc.). One could argue that these items are inefficiently produced, that the time required to fashion them would be better spent on wage work, or that they are inferior to mass-produced items and should be "modernized" or "automated." Perhaps that is what geographer Oscar Schmieder (1930:44, 142) meant when he referred to the "inefficient" plows of the Zapotec.

But setting aside the efficiency criteria for the moment, we might ask: What advantages do the crafted implements afford the campesinos? In other words, what would the campesinos give up by surrendering these handmade artifacts in favor of mass-produced ones? For one thing, campesinos often have little cash on hand to purchase factory-made goods. In addition, however, handcrafting promotes a custom fit, which may result in ergonomic advantages that are vitally important in conducting physically demanding and repetitive farming work that is viewed by campesinos as a normal part of everyday life. The preference for crafted items may also have to do with the pride that many campesinos take in their work, their knowledge, and their intelligence.

This raises the possibility of a plurality of intelligences. In the case of many campesinos, agricultural knowledge becomes internalized in its practitioner. Sensory perceptions are critical tools to be used in farming. Cloud, wind, moon, and rain patterns are observed; maize kernels and coffee beans are pinched, pressed, or bitten to gauge their ripeness; arms and hands are used to measure lengths. In short, campesino knowledge is literally knowledge incarnate, a kind of intelligence possessed by Melanesian fishermen, Cree goose hunters, and other local specialists. Many cosmopolitan scientists lost these "sixth senses" long ago—the scientific process is not always cumulative. As cosmopolitan scientists gain knowledge about some things, other knowledges disappear or are sometimes reduced to mere possibilities no longer recognized.

So far we have only considered those items that are crafted by campesinos for their own use. But many other artifacts are produced not directly by the user, but by regional craftsmen and craftswomen: clay vessels, products woven from henequen fibers, wool blankets and shawls, fireworks, huaraches, masks for *danzas*, and so forth. These items, sold in local markets, bear a different relationship to their users because they are produced by other individuals and typically purchased (or, much less frequently, bartered for). Still, we might think of them as crafts because they are for local use, are not mass-produced, and are fashioned mostly from local materials.

It would be naive to assume that all crafts have survived. On the contrary, many have disappeared in Talea, particularly since the completion of the road in 1958. Even some agricultural implements have been affected (notably coffee depulpers and sugarcane mills, which were once fashioned from locally procured hardwoods). Chainsaws also increased in popularity in the 1990s, particularly for preparing structural beams and planks for adobe homes, due in large part to government subsidies on the equipment—which many Taleans describe as a way for the PRI to "buy votes." In fact, in Talea chainsaws have completely replaced the older two-handled saws known locally as *bovos* (Sp., referring to a long fish native to the region). Today both *bovos* have all but disappeared: the fish are rare because the rivers are contaminated by sewage draining from the villages above (this represents perhaps the severest environmental consequence of piped water and flush toilets in the region); and the two-handled saws have been made irrelevant as a result of chainsaws. If we take seriously the idea that the Zapotec experiment with new technologies, accepting those that prove to be useful and ultimately rejecting those that are destructive, then it may be that, over time, new alternatives (or old ones reintroduced) will replace recently introduced technologies that have negative environmental effects.

The Ranch House

The ranch house is a site of great economic, cultural, and symbolic significance for campesinos. It serves as bedroom, workshop, and kitchen for most of the year. Many campesinos sleep four or five nights out of every week here, coming home only to pick up fresh tortillas and leave firewood before departing again the following morning.

In Talea the simplest ranch houses are rudimentary structures that are little more than lean-tos patched together with slats of wood and cardboard, while the most elaborate are larger and more luxurious than most homes in the village itself. The latter may have tile roofs, plastered walls, electrical power and appliances (motorized coffee depulpers, CB radios, even TVs), concrete floors, and two or more rooms. Generally, the large ranch houses belong to a handful of farmers who have become relatively wealthy as a result of extensive coffee cultivation. They may have as many as five to eight cultivated hectares.

Most ranch houses, however, are modest structures that fall somewhere between these two extremes. Most consist of a dirt-floor room; many have only three walls. Some have low walls made of adobe or stone, and tile roofs are frequently used. Many farmers have two or three ranch houses of various sizes: one is located at the milpa, another at the *cafetal* (Sp., coffee grove), another at the sugarcane field.

Nearly all campesinos keep a generous store of firewood inside the ranch house, which is often stacked so high that it serves as a wall. Farmers use a corner of the structure as a kitchen and hearth, and clay pots, pans, tins, plates, and cups are stored within arm's reach. So are splints of sappy pine wood (*Pinus michoacana*) used to start fires when coals have died out. In some ranch houses one may find hand grinders and grinding stones for preparing tortillas and griddles for toasting them. Campesinos hang a host of wires and bits of rope with hooks attached from the ranch house's roof beam. From the hooks dangle various containers such as old buckets and gourds used to store foods and hardware: unrefined brown sugar, beans, cloth napkins filled with tortillas, chiles, salt, ground coffee, cigarettes, matches, soap, chisels, squash seeds, slingshots, safety razors, muzzles fashioned from bits of rope or chicken wire, net bags, an occasional plastic bag of powder pesticide, and nylon ropes. Extra pants, shirts, jackets, hats, and huaraches hang from other hooks. (It is important that spare clothes be close at hand in case it rains or one gets drenched in sweat. Leaving damp clothes on as night falls is foolhardy, say the campesinos—it is an invitation to illness.) Suspending all of these items from hooks prevents dogs, mice, and other rodents from consuming them and keeps the morning dew and rainfall from soaking them. Inside the ranch house other hooks hold a wide range of implements: axes, hoes, a hunting rifle, hooks for weeding, other hooks for pulling fruit down from high trees, plastic fertilizer bags, dozens of straw mats, and thick wool or acrylic blankets, carefully stored in fertilizer bags to prevent mice from chewing holes in them.

Some ranch houses have altars complete with votive candles, as Catholic homes in the village do. These are testimony to an earlier era when families spent most of the year in the fields. Today, primarily because of the enforcement of mandatory schooling for elementary school children, this is no longer common in Talea.[8]

Boiled black beans are the standard meal in the ranch house, and farmers season the clay pot of fresh beans with salt, garlic, chile peppers, and the leaves of a fragrant herb called *epazote* (*Chenopodium* sp.). A pot may last three days or more as long as it is boiled at least once daily so that it does not spoil. Other vegetables and plants are often added to the beans: diced squash and cactus leaves, wild greens (Sp. *quelites*), squash flowers (Sp. *flores de calabaza*), the tender leaves of growing squash vines (Sp. *guías*), a small yellow arboreal flower called the *gallito* (Sp.), quartered chayotes, string beans, and green bananas are perhaps the most common. Condiments are added to individual bowls to vary the taste: jalapeño, serrano, bolero, cimarrón, and other chile peppers, onions, chives, and lime juice. Many of these crops are grown in small gardens surrounding the ranch house. Occasionally, game animals are marinated in a chile paste, seasoned with the leaves of a tree called the *aguacatillo* (*Persea* sp.), and steamed to make *barbacoa*. (See Appendix C for a partial listing of edible game animals.)

When wage workers are hired, many campesinos feel obliged to provide other foods as well: pasta, rice, eggs, chorizo (spicy pork sausage), cheese, fish, shrimp, and even chicken may be served to vary the menu and to avoid gaining a reputation as a miserly *patrón* (boss). Most campesinos are exceptional cooks, and many of them eat more of their own cooking than of their wives'.

If a campesino employs wage workers or is in a *gozona* (reciprocal work) arrangement, all usually sleep on the floor stretched out on the same *petates*, each with his own blankets and often shoulder to shoulder. Sleep tends to come early, quickly, and deeply. The rigorous chores are exhausting, and it is common for farmers to sleep solidly for nine or ten hours—something that is often not possible while in the village, according to one informant:

> It's impossible to get to sleep before eleven when I'm at my home in Talea. With the kids doing homework, and the obligations that one has to meet, and the noise from the dogs and the radio, it's tough to get any rest at all. And the little baby wakes us up at four or five in the morning. I love my kids very much, but the truth is that I feel much more tranquil when I'm in the fields—there I work deliciously, eat deliciously, and sleep deliciously, much more than I ever do when I'm in the village.

There are many dangers on the farm, perhaps the most threatening of which are poisonous snakes and black widow spiders. For this reason—but also to enjoy the company of others for dinner and a chat afterward—campesinos often invite neighboring farmers to sleep over.

Campesino farmers are often as sentimental and nostalgic about ranch houses as small-town natives in the United States are about their hometowns, and for some of the same reasons. Frequently, a large portion of a person's childhood was spent on the ranch in the company of parents, siblings, and farm animals. I listened to the accounts of dozens of campesinos who romantically remembered "the old days" with their families on the farms—even if they were suffering from limited lands or disastrous harvests. Others, recalling their years as young men, told of glorious trysts with women they met during the coffee harvest season. Indeed, the immense popularity of such classic songs as "Mi Ranchito" (My Little Ranch House) and "De un Rancho a Otro" (From One Ranch House to Another) is partly due to the resonance such sentimental themes have in a country that was inhabited mostly by rural people as recently as two generations ago. The ranch house lives on as a powerful metaphor in the Mexican consciousness.

Conclusion: The Implications of Change

Times are changing. Even in ranch houses hours away from Talea, the markings of modernity are omnipresent. Some mornings the electric sound of public address systems in Talea and Yatoni can be heard, announcing an upcoming *tequio* (communal work project) or perhaps the recent death of a villager. Cars and trucks can be spotted on the road leading to Talea, visible as colored specks crawling slowly along the dirt road to the village followed by a cloud of dust. And on some clear afternoons a high-flying jet may leave a trail of condensed moisture behind, a white ribbon neatly dividing the heavens into a pair of blue hemispheres. "Is it headed to Mexico City? Is it carrying the *gringos* to their country? Or is it full of *chinos*? They say the Japanese are richer than the *gringos* now—surely they have their own planes," a campesino muses. The signals serve as constant reminders that there is an entire world outside—other peoples, agendas, ways of life, bustling with movement and staggeringly complex.

In the wake of these changes, there is a troubling sense among many older people that somehow younger Taleans are not as skilled at farming as they were a generation ago. Part of the problem, they say, is that it is much easier to earn a living in the village or in the city, working as a *machetero* (loading and unloading trucks), doing odd jobs as a *mozo*, or assisting a more established carpenter, mason, electrician, or merchant. They claim that younger Taleans are increasingly unwilling to do farm work (particularly the more difficult tasks related to maize, bean, and sugarcane cultivation), are reluctant to stay in ranch houses (and then for only a single night at a time), and refuse to work in fields located too far from the village. Some coffee growers are faced with a particularly ironic situation. After sacrificing a great deal of work to successfully cultivate a coffee grove in order to have money to put their children through junior high school, high school, and even college, they are left alone once their sons or daughters migrate out of the village. No one remains to work the land when the campesinos grow old.

The situation is not unlike that described by Robert Johannes (1981:63–75) in the Palau District of Micronesia, where traditional fishermen are witnessing the decline of the "traditional conservation ethic" and environmental knowledge among younger generations. The impact of "foreign cultures," imported foods, formal state education systems, and migration has accelerated the process (Johannes 1981:148):

Today in Oceania knowledge about fishing, as well as farming, hunting, medicine, and navigation, is disappearing because younger members of island cultures are no longer interested in mastering it . . . Why learn to fish well when nine-to-five jobs in air-conditioned offices beckon in the district center and there is an endless supply of fish in cans? Why learn to build a canoe and sail it when fiberglass boats and outboard motors can be bought and operated with little detailed preparation or knowledge? This disdain is reinforced by well-meaning educators, for the exclusion of traditional skills and knowledge from westernized school curricula in many developing countries amounts to a constant, tacit assertion that such things are not worth learning.

The degree to which this process is occurring in Talea is an open question. It is clear that over the last generation many young villagers have migrated to cities, but others from surrounding villages trickle in, and nearly all of them continue to farm using subsistence methods. The threat to farming knowledge seems less severe in the Northern Sierra of Oaxaca than does the threat to fishing knowledge in the Palau District.

Indeed, Tom Barry (1995:2) notes that more people in Mexico are living off the land now than on the eve of the 1910 Revolution (many of them in small subsistence-farming communities), primarily because in absolute numbers the population of rural Mexico has increased rapidly, even if it has decreased as a percentage of the national population. This situation appears to hold true in Talea, as well: approximately 1,000 people lived there at the turn of the century, 80 percent of whom were farmers, while approximately 2,100 people lived there at the time of the 1990 census, approximately 75 percent of whom were supported by farming (INEGI 1991:315). In short, about twice as many people are supported by farming on Talean land today as compared to 100 years ago.

One of the reasons for this is that the mixture of old and new crops and farming techniques has culminated in a profound knowledge about farming that extends far back into the past but also reaches into the present. This is the result of rapid changes that have swept Talea in the last generation, which have given campesinos an opportunity to experiment with a host of technologies from the "outside" world while continuing to use methods developed or adopted in earlier times. In other words, the Talean farmers of today have distinct advantages over their predecessors because they have easier access to and knowledge of twentieth-century technologies, crops, and methods, while at the same time they have an edge over "modern" scientists (e.g., agronomists) because they have inherited highly specialized and effective knowledge systems and technologies that have been developed over thousands of years. Sierra Zapotec farming and food practices have derived strength and resilience from a millennial pattern of eclecticism, pragmatism, and experimentation, a pattern that is perhaps as vital today as it ever has been, due to increased communication and the availability of "outside" technologies. Even so, the erosion of ecologically specific local knowledges about farming, food, medicine, and other areas may well be one of the greatest problems confronting the region in the early twenty-first century.

We have seen how the Taleans (and their ancestors) have constantly tested and experimented with new artifacts. They adopted a panoply of new approaches during the Spanish colonial period (including plows, draft animals, and iron implements) and a system of measures which, though apparently used hegemonically after the Conquest, has afforded certain advantages, including flexibility, simplicity, and perhaps even ergonomic benefits—ends which are not always sought in industrialized settings. The case of material culture in the Sierra illustrates how items exchanged cross-culturally can rapidly become appropriated and normalized. In a similar vein, the campesinos' use of scrap materials (produced in industrial centers) demonstrates how items made in distant regions can be creatively shaped to meet local needs by farmers with a pragmatic sense of curiosity. Many Taleans describe their village as "progressive," and this description seems fitting to the extent that they experiment with and test new artifacts and methods, integrating outside elements into their farming. Many villagers are acutely aware of their talent as innovators and are quick to point out discoveries and inventions that they and their ancestors have made.

What all of this means for Talean farmers is that today there may be more options than at any time in the past and—if globalization processes siphon away too much of the younger generation—perhaps than at any time in the future, since older knowledges may rapidly be lost if they are not transmitted to younger villagers.[9] In Talea, as in the Palau District of Micronesia, "old and new ways... [have] overlapped considerably, producing a degree of understanding of [subsistence activities]... probably never before equalled (and may not soon be equalled again)" (Johannes 1981:15). For the moment, at least, Talean campesinos find themselves benefiting from the richness of a brimming technological crucible.

CHAPTER 4

"MAIZE HAS A SOUL"

Rincón Zapotec Notions
of Living Matter

In the mid-twentieth century North American development agents, in partnership with their "Third World" counterparts, exported a particular version of factory farming—based on the use of mechanized equipment, chemical fertilizers, pesticides and herbicides, and hybrid seeds—to parts of the world deemed to be in need of "development." There was a dual irony to this technological civilizing mission. For one thing, the technologies were deployed to the so-called Third World before the dark clouds of the Dust Bowl—a human and ecological tragedy that has been attributed in large part to factory farming practices—had even settled. But perhaps more ironically, in the Mexican case what were later called Green Revolution technologies were ostensibly exported to increase the "productivity" of farmers who were, in a sense, the heirs of native Mesoamericans who thousands of years ago perfected modern maize—a grain that produces much more food per unit of land than wheat, rice, or barley.

The developers, described as "innocents abroad" in a thoughtful piece by Angus Wright (1984), were in some cases technicians largely unaware of their role in larger political projects, much less the social and ecological realities of rural Mexico. Some assumed that the Mexican countryside was "inefficient" or "underdeveloped," often being ignorant of the highly specialized maize farming techniques that local farmers had been evolving over 5,000 years.[1] Many were also ignorant of maize's centrality in the social, cultural, economic, religious, and psychological lives of the region's peoples. Not surprisingly, the new technologies frequently had disastrous consequences. By the late 1970s it was clear that the Green Revolution had contributed to greater disparities between rich and poor farmers, a greater dependency on chemical fertilizers and pesticides, environmental degradation, and a loss of genetic diversity.

Remarkably, maize farming in Talea has persisted in spite of periodic policy shifts (most notably subsidies to producers and consumers) that have resulted in lower prices for consumers. Even though Talean farmers could have abandoned maize farming completely in favor of lucrative coffee cultivation, they have instead struck a balance between subsistence and cash cropping. To help understand why, here I focus on the Rincón Zapotec view of maize as living matter in the strictest sense — a view that holds maize to be so alive that it displays characteristics that people in industrialized society ascribe exclusively to humans.

Maize and Its Meanings

Maize is especially important for campesino households. It is made into tortillas, which serve as the family's subsistence base; indeed, approximately 75 percent of the calories consumed by Talean campesino families come from maize. Both women and men play a vital role in the life cycle of the crop, and from one perspective maize can be seen as an economic link that helps bind household units together. Maize cultivation is a key strategy for household maintenance.

Maize, like coffee, can be grown according to either an authoritarian logic or a democratic one (Mumford 1964). In other words, maize and coffee may be grown on large-scale plantations spanning thousands of hectares or on tiny plots of a fraction of a hectare without suffering significantly from economies of scale (Nolasco 1985; Strange 1988).

But there are differences in the Rincón. In spite of its adaptability to household maintenance strategies, coffee is exported to extraregional markets almost without exception. By contrast, maize is often planted for immediate consumption by the farmer's family or for sale in regional markets. Furthermore, in Rincón villages maize often serves as a medium for connecting campesino families to their kin, their neighbors, their village, their region, and their deities.

Three Accounts about the Soul of Maize

One frigid January evening a campesino told me a fascinating story as we husked maize in a tiny thatch-roofed ranch house. We had just finished dinner and settled down to chat while working. The story went something like this:

Pablo was a Talean campesino who lived in that time when little coffee was planted in the village. He was a successful farmer, a red-blooded campesino, and nearly every year he harvested enough to feed his family. But one year his mother, a widow, fell ill and was unable to earn enough money to purchase maize at the market. Pablo secretly took a couple of *almudes* of maize from his family's *costal* [straw basket used as granary] to his mother next door. But his wife—who had the ill fame of being a jealous, miserly woman—found out, scolded him, and took the maize from the widow. Pablo was a tranquil man and did not dare contradict his wife. That night Pablo and his neighbors awoke to a terrible sound, like that of a waterfall. It seemed to be coming from a corner of Pablo's house, but it ended before he was able to light a candle. Thinking that perhaps he had been dreaming, he went back to sleep. In the morning, however, he discovered that the *costal* was empty—every last kernel of maize was gone.

After relating this account, my companion paused thoughtfully for a few moments and said in Spanish, "That's why we say *el maíz tiene corazón* [maize has a heart]." He repeated himself, in Rincón Zapotec this time, perhaps sensing my confusion: "We who work in the countryside say *zhua de quie' ladxidáhua*—maize has a heart."

This is, in a sense, biologically accurate. Maize kernels do indeed have a heart (the embryo, nucleus, or germ) from which the plant germinates once it is in the ground.[2] But he meant something more; he was attributing a moral sense to the crop itself—assigning "human" qualities to a plant. The phrase takes on even more meaning when we consider that in Rincón Zapotec "heart" and "soul" are expressed by the same term, *ládxi*. Though my companion said *corazón* (Sp., "heart") there is little doubt that he also meant *alma* (Sp., "soul"). He gave another example:

The villagers of San Juan Tabaa [three hours' walk from Talea] have had a great deal of trouble growing maize since they had the conflict with Yojobi [a smaller village near Tabaa] in the 1950s. Their lands just dried up. You see, they burned most of the village, including a great deal of maize. And the maize, you see, it has a heart, and it remembers. They say that last year campesinos from Tabaa tried plowing the earth near the site of the conflict [with Talea in 1991] but didn't even succeed in breaking ground—it was as if the soil had been made of stone.[3]

Eventually, I learned that *ládxi* is an especially significant concept for the Rincón Zapotec. Asking for the forgiveness or understanding of another is expressed as *ben ládxido zhhn* (Zap., "make your small heart big"). The center section of a tree trunk—which is the strongest and least likely to be infested by termites—is also called *ládxi*. The phrase *búza ládxe* (Zap., "gave from the heart/soul" or "loosened from the heart/soul") is used to express the Spanish term *regalar* (to give a gift).

In a recently published account, a man from Yojobi, a Cajonos Zapotec village, tells the story of how maize inflicted punishment upon him as a child. He was storming angrily away from a milpa, carrying a basket of maize ears and intentionally letting some drop to the ground in his rage, knowing that his mother would pick up after him. Then he mysteriously tripped, and the pain lasted for months—even though the fall had not been a bad one. He learned a valuable lesson: "Later I knew that it wasn't the fall that had injured me, it was the maize, which I had scorned. That's what punished me. That's why I've never despised maize since" (Castellanos 1988:245).

Maize and Responsibility: How Maize Prescribes Reciprocity

How might we analyze these accounts? Strikingly, they each prescribe certain norms or responsibilities associated with maize—and, in a broader sense, with food in general. Pablo's story might be interpreted as a tale about the importance of supporting extended family and kin in times of economic need, perhaps to ensure that they reciprocate. It is also subject to a class analysis; that is, it might be interpreted as a lesson about the responsibility of relatively affluent villagers in lending a hand to the poorest stratum of village society, represented in this case by the elderly widow.

The Tabaa stories might be interpreted in at least three ways. Burning maize—and this might be extended to mean destroying maize in any way (for example, by letting tortilla dough spoil)—is a serious offense that should not go unanswered.[4] Another interpretation is that stealing maize—which is what the citizens of Tabaa did from the Taleans' point of view following the 1991 land conflict (see Chapter 2)—does not go unpunished, whether it is stolen from one's family, a neighbor, or another village. (The alcoholic who takes maize from the family granary in order to purchase liquor is despised by upright campesinos.) Finally, there is an implied notion in both of the Tabaa tales about the way villagers in neighboring communities should act toward one another. In a part of Mexico where land conflicts are endemic (Dennis 1987), maize is a means of linking communities together, either through market transactions or at religious fiestas, where visiting pilgrims requesting lodging are hosted and fed.

The story from Yojobi is perhaps not as obvious as the others, but upon reflection it might be interpreted in the following way: no matter how meager the harvest, no matter how poor the quality of the maize, it should be appreciated and never scorned or used as a medium for expressing anger.

Thus, maize is not only an economic good but a medium through which certain social and moral obligations and responsibilities, particularly reciprocity (toward kin, neighbors, poorer villagers, and people in neighboring villages), must be met. In this way, maize illustrates one of the key components of the fundamental concepts underlying Rincón Zapotec farming.

In all of the stories maize appears as a living being, a plant-person who can help humans meet certain moral obligations and responsibilities toward kin and neighbors. Maize is not a mere crop, but a powerful being living in the midst of humans at all times: present at festivals, baptisms, weddings, and funerals, inside churches, homes, and human bodies, surrounding ranch houses and villages. Maize is a wonderful plant-person with a long memory, a strict moral code, and an unshakable will.

Maize and the Personified Earth: Saints, Spirits, and Agriculture

So far we have considered the responsibilities associated with maize that link villagers together through a web of prescriptions and obligations, but what about other networks? Specifically, how does maize connect humans to the religious and supernatural worlds?[5]

Many campesinos claim that ill fortune may befall those who fail to fulfill religious obligations. A generation ago, one of the village's most dedicated campesinos is said to have tried plowing a milpa during the fiesta of La Virgen de Guadalupe (December 12) only to have an ox run away and break a leg. The animal had to be slaughtered. "That's what I get for working on a fiesta day; I should have known better," he told friends. Restrictions on work also apply to Sundays and many fiesta days (see Table 4.1).

A group of informants told me a similar story that occurred more recently. A campesino purchased a young team of oxen already trained to plow, but they stubbornly refused to work his land. He asked the advice of his father, who recommended that he make an offering to the image of San Isidro Labrador, the patron saint of campesinos. "Ha! I'm the only person who is the master of my animals!" answered the son. The next week he tried again, but to no avail. By this time he was desperate (it was already late in the season) and resigned himself to offer an *almud* of maize and another of beans to the saint's image in the church. Upon returning to the fields, he was successful; the oxen, he later reported, "pulled the plow

Table 4.1. Religious Fiestas Celebrated in Talea, 1996

Dulce Nombre de Jesús	Third Sunday in January
Virgen del Rosario	Wednesday after Dulce Nombre
Domingo de Ramos	Palm Sunday
Viernes Santo	Holy Friday
Sábado de Gloria	Holy Saturday
Domingo de Pascuas	Easter Sunday
Santa Cruz	May 3
San Isidro Labrador	May 15
Santa Rita	May 22
Ascención (*Choa'zá*)	Forty days after Easter Sunday
Espíritu Santo	Fifty days after Easter Sunday
San Juan de los Lagos	June 24
San Miguel Arcángel	September 29
Todos Santos	November 1
Santa Gertrudis	November 16
Santa Cecilia	November 22
La Virgen de Guadalupe	December 12
Fiesta del Niño Jesús	December 24
Parada del Niño Jesús	December 31

by themselves." Since then he has made offerings and consistently attended Masses on San Isidro's day (May 15) and has had good fortune. It is common knowledge among Catholic campesinos that a special offering should be made to the campesino saint after purchasing a new beast of burden or team of oxen.[6]

In addition, campesinos sometimes offer sacrifices for adorning the Catholic church during fiestas. An informant, for example, offered several dozen corn husks which were used to decorate the long colored banners hanging from the roof of the church during the January fiesta. Several months later, his brother-in-law offered six live maize plants—transplanted from his milpa—to decorate the church's statue of San Isidro, who, like Talea's campesinos, uses an armadillo-shell basket, a digging stick, and a team of oxen. One campesino woman regularly makes offerings of flowers,

Figure 4.1. San Isidro, the patron saint of campesinos. Like Talea's farmers, he uses an armadillo-shell seed basket, a planting stick, and a team of oxen and wears huaraches. Most villagers celebrate San Isidro's day (May 15), and many offer the image maize, beans, and *panela*. *(Photo by Gabriela Zamorano)*

often from her kitchen garden, to this saint and others. Significantly, San Isidro's day coincides with the maize planting season.

San Isidro appears to be a syncretic version of the Zapotec god of thunder and rain, Cocijo. (The syncretic connections between the Virgen de Guadalupe and prehispanic fertility deities, most notably the Aztec goddess Tonantzín, have been mentioned by a number of anthropologists; it seems likely that there may be a link to the Zapotec earth deity as well.)

At first sight, non-Catholic agricultural rituals seem to have disappeared in Talea, though archival and ethnographic sources reveal evidence of a rich variety of magical rites performed in the Northern Sierra. Fifty years ago in Yalálag, spirits were said to dwell in the air, on the land, and below its surface and were connected with agriculture, food, animals, and fortune. Like Catholic saints, some of them took a material form, appearing as stone idols (de la Fuente 1949:301-302).

The earth itself was described (and by many still is) as an animate being: it painfully feels the metal blade of a plow, the burning of swidden plots, and the plucking of its fruits. It is therefore entitled to a portion of what it gives, lest it become angry. This is why even today some Serranos first pour a bit of mezcal on the ground before taking a drink. If upset, the earth might capture the soul (Sp. *alma*, Zap. *ládxi*) of a human offender, which leads to "fright" (Sp. *susto*), an illness which can only be cured with folk remedies (de la Fuente 1949:265-267).

To avoid incurring the earth's wrath, elaborate sacrifices were and are made in the Sierra. Mountain peaks, caves, forests, and lagoons were special ritual sites where sacrifices were secretly carried out to avoid the condemnation of priests and colonial officials. Across the Sierra (including the Mixe region; see Beals 1945:147) the magical rites appear to have been strikingly similar. A group of villagers—often led by a ritual specialist—took turkeys, tortillas, candles, and a combination of other offerings (puppies, tamales, ears of corn, cornmeal, eggs, cigarettes) to a site in the forest marked by a stone altar. They slaughtered the animals and poured the blood over the altar and the earth, made a petition for abundant maize, sufficient rain, and good fortune, and prepared food. (It is worth emphasizing that some of these rituals—for example, the use of sacrificial dogs—quite obviously have prehispanic origins.) Part of the food was consumed by the participants and part left at the altar. Upon returning to the village, the group frequently left candles at the Catholic church and sometimes would request a Mass from the priest. Sacrificial rituals were often conducted several times during the year in connection with critical points in the maize cycle. One account, for example, relates how Sierra farmers offered a sacrifice for adequate rain, followed by another for protection from birds, insects, and other pests, then finally by one to protect the milpa from high winds (Zilberman 1966:119-120).

One recent article focuses on the vital role played by women in prehispanic rituals that survived into the colonial era. According to Daniela Traffano (1998), the archival record reveals that women were active participants in rituals related to life-cycle events such as deaths, births, and marriages but also in agricultural rituals. It is significant that today nearly all of the ritual work of collective prayer to Catholic saints is done by women, though men are commonly designated as cantors. These commitments often require attending *novenas*, a series of nine prayer meetings in which the rosary is recited collectively. Apart from prayers to San Isidro and others, I spoke with campesinas who told me that they conduct their own individual private thanksgiving prayers to the earth for successful cultivation in their kitchen gardens. For example, one noted:

Not everybody does this, but I give thanks to her—to the earth. This year, when I cleared away the old plants in my garden and prepared to plant new seeds, I spoke to her, I thanked her for the things she had given us, and I told her that I would take care of her. And I asked her to continue giving us good things. That's my custom. Once I was done, I planted my *chilitos*, chayote, cilantro, and other things.

Individual rituals like this one are a common feature in the ethnographic record. The earth is regarded as a female force by the Rincón Zapotec, which corresponds with evidence from other parts of southern Mexico, particularly Mayan and Nahuatl groups (Guiteras Holmes 1961; Nash 1970; Collier 1975:118; Sandstrom 1991).

In Yalálag, de la Fuente (1949:306-307; my translation) reported a ritual sacrifice conducted by a campesino who made use of a stone idol called the *gwálse* (Zap.) that was buried in the milpa until needed for the ceremony:

The owner of the stone idol and the milpa sacrificed a dog no more than 15 days old or a chicken. Part of its blood was poured into the hole in which the idol normally rested and then the unearthed idol was "fed" by being soaked with the rest of the blood. At the same time, a petition was made for an abundant harvest. The campesino then returned to his home, invited his relatives to eat tamales filled with meat from the sacrificed chicken, and returned to the ranch to give the stone idol a bit of the tamales, since it could become jealous and angry.

The examples cited by de la Fuente among the Cajonos Zapotec and Beals (1945) among the Mixe closely match accounts from the turn of the eighteenth century recorded by Catholic missionaries working in the area (Zilberman 1966; Guillow 1994 [1889]).

Though de la Fuente was able to describe vital elements of the rituals, many of them had not been conducted in decades. He thought that religious syncretism was bringing an end to the rule of the mountain spirits: "The cult of the pagan spirits is diminishing and is increasingly linked to the Catholic deities" (de la Fuente 1949:265).

At first sight, the process appears to have been completed decades ago in Talea, probably because of the heavy influx of relatively acculturated mestizo outsiders in the late 1800s and 1900s. But the animal sacrifices of an earlier era appear to have undergone a symbolic transformation in the guise of a Catholic saint, San Isidro, whose day coincides roughly with the timing of prehispanic rituals in the planting season and to whom food sacrifices are made. It may well be that the saint has partly absorbed the role of the mountain spirits—he often seems to function as a lesser deity with a close connection to an earth that Talean campesinos continue to refer to as a living being.

In the Sierra the rituals continued into the twentieth century relatively unchanged. Ralph Beals (1945) found a vibrant tradition of non-Catholic agricultural rituals and animal sacrifices in the community of Ayutla Mixe in the mid-1930s, and a much more recent ethnography of the Mixe village of San Pablo Chiltepec describes a planting ritual conducted by individual farmers that bears a remarkable resemblance to those described in earlier periods (Lipp 1991:20-23). Farmers in at least one Rincón village recently attempted to reclaim the tradition of animal sacrifices. In the mid-1990s campesinos in Santa Cruz Yagavila, six hours from Talea, tried to recover a ritual in which a turkey was sacrificed, prepared as soup, and then poured into cavities at each of the milpa's four corners with tortillas and other items (Leonardo Tyrtania, personal communication, 1996). Laura Nader (personal communication, 1998) reports that such sacrifices were not unusual in the Rincón in the late 1950s.

There are more recent examples. The spring and summer of 1998 were extraordinarily dry seasons in the Sierra. For the first time in more than ten years, Talean campesinos asked the Catholic priest to conduct Mass at a small lagoon in the forest above the village in order to ensure the arrival of the rains. The site, known as *yhl shna* (Zap., "red milpa"), lies nearly two hours from Talea, near the border with the village of Tanetze. In Ixtlán, between Talea and Oaxaca City, a Mass was held in 1998 for the same purpose. These Catholic ceremonies appear to have a form and function very similar to those of the Zapotec rain ceremonies described by de la Fuente (1964).

Another interesting case is that of the Cruz Verde or Green Cross near San Andrés Yaa, a Cajonos Zapotec village approximately eight hours' walk from Talea. Here, at a site located in the midst of an oak forest, an apparition of the crucified Christ is said to manifest itself during the fiesta days of the Santa Cruz (May 3). The apparition appears on the trunk of a cross-shaped oak tree and is located near a tiny lagoon said to have miraculous powers.[7] A young villager discovered the site in 1951; since then the fiesta has grown much larger, drawing pilgrims from the entire Sierra—Zapotec, Mixe, and Chinantec—to the sacred site in the forest. In 1997 approximately 3,000 to 4,000 people visited the site, offering cash, votive candles, and prayers. The pilgrims' petitions are symbolized by miniature images, including toy oxen, trucks, baby dolls, and houses fashioned from twigs, leaves, and mud.[8] Though the fiesta is officially a Catholic event complete with Masses said by a priest, a number of key elements link the Green Cross celebration to agricultural rituals of the past, including the timing (May 3, during the planting season); the ritual site (in the forest, near a sacred lagoon); a large group of participants; and animal and food sacrifices.[9] More than 100 Taleans have made the pilgrimage annually in the last few years, including campesinos who have petitioned the deity for healthy work animals and success in farming. If the Green Cross celebration is any indication, it seems that the mountain spirits may very well continue to hold importance for Sierra farmers into the foreseeable future.

Whether they are composed of Catholic or prehispanic elements, these rituals may be seen as magical since they attempt to deal with supernatural beings such as San Isidro, stone idols, mountain spirits, and the animate earth, all of whom possess sufficient power to ruin a maize harvest. Malinowski's constellation of magic, science, and religion is especially useful here. The campesinos' arsenal of technical knowledge, as we shall see, assists them in the cultivation of maize. The gifts of turkeys, tamales, and maize, however, are magical because they attempt to minimize the risks associated with subsistence farming, including droughts, heavy rains, blights, and high winds. The magical rites afford a measure of "tranquility for the decision-maker" (Colson 1973)—and therefore function in much the same way as the economist's statistics, elaborate rites performed by professional baseball players, and weapons testing by nuclear scientists in industrialized society (Gmelch 1994; Gusterson 1996:141-142).

On a symbolic level the non-Catholic agricultural rituals link together nature (the forest, *guí'a* in Zapotec) and culture (the cultivated milpa, *yhl* in Zapotec). The ecological connection between forests and farming was not lost on an informant. He explained that forests are a necessary element in successful crop cultivation because, in his words, "the forest pulls [Sp. *jala*] clouds from the sky so that they drop rain on the fields below." (Perhaps this is one of the bases for rituals in which "mountain spirits" dwelling in the forests are consulted for a successful harvest.) Ecological anthropologists have explained the same phenomenon in similar terms:

If in the Sierra Juárez there is a surplus [of soil nutrients] that permits cultivated soils to maintain their natural fertility, this is due to the enormous work done by the forests in moving the hydrologic cycle, a job that is done punctually by nature ... There exists no technology that could replace the forest in this task. (Tyrtania 1992:40)

Thus, the conceptual categories used by the Rincón Zapotec to describe the relationship between forests and farming appear at least partially to map onto categories used by cosmopolitan science specialists.

The Zapotec campesinos' view of the earth as a living being capable of punishing those who unreasonably violate it is strikingly different from the view of many modern "development" projects that see ecosystems as simply means to an end, as media to be used for pecuniary purposes. This has resulted in the unprecedented exploitation of the southern Great Plains that led to the Dust Bowl (Hurt 1981), damage from strip mining (Ponting 1991:218–221), the environmental and public health consequences of nuclear energy and weapons programs, whether in Communist states or market economies (Ball 1986; Fradkin 1989; Feshbach and Friendly 1992), the wholesale destruction of tropical rain forests (Davis 1977), and the ecological and social upheavals caused by large-scale hydroelectric dam projects (Bartolomé and Barabas 1990). These events, which in industrialized society have become normalized to the point of becoming commonplace, would likely horrify the Rincón Zapotec, who maintain a substantive respect for the living earth.[10] Chapter 8 considers how some people in industrialized society are beginning to employ concepts that resemble those of the Rincón Zapotec to understand the earth.

Maize as Social Relations: *Gozona* and the Gift of Maize

Just as sacrificial meals of maize serve as a communion between humans and their deities, reciprocal exchanges called *gozonas* (Sp.) or *gozún* (Zap.) link together villagers. *Gozona* (known in other parts of Oaxaca as *guela-*

guetza) is a mutual aid arrangement used when labor must be pooled to realize a task, whether economic or ceremonial. Nader (1990:42) comments on its prevalence in many aspects of village life:

> It is a prestation of services, of short duration, with implied reciprocity... Taleans may call on kinsmen, friends, and neighbors to build or repair houses and to help in preparing land and in plowing, planting, and harvesting. Taleans volunteer gozona during weddings, funerals, and fiestas. They even carry on gozona between communities; for example, the band and orchestra from Talea have played in fiestas in Villa Alta, Yatzona, Solaga, Yalina, and Yaee ... These communities reciprocate in kind.

Thus *gozonas* customarily include agricultural tasks, weddings, funerals, (adobe) house-raisings, *retechadas* (Sp., reroofing parties), and *padrino* (Sp., religious sponsorship) fiestas but also have begun to include newer events such as *colados* (Sp., cement roofing parties), *quinceañeras* (Sp., female coming-of-age parties), and Alcoholics Anonymous fund-raising dances. Conspicuously absent from *gozona* is cement home construction, for which wage laborers are contracted.

When carried out on farms, *gozona* is a simple affair that involves providing food (except tortillas), coffee, mezcal, and cigarettes to visiting workers. Once the days of work have been given, the host must eventually "pay back" the days upon request. When *gozona* takes place in the village, however, it resembles a fiesta—the *gozona* effectively becomes a stylized work party, not unlike the barn-raisings of the nineteenth-century U.S. Midwest. It consists of several key elements: the invitation, the food gift (occasionally accompanied by money or candles), equipment loans, work, meals, and repayment.

To illustrate how *gozona* works, we might consider a wedding party. Invitations are formally made in person by the families of the prospective bride and groom several days before the event. The day before the wedding (or sometimes two days before) invited families arrive at dawn, with a food gift of some kind—maize or beans or *panela* (Sp., unrefined sugar)—and sometimes with a cash donation. The gift is promptly recorded in a notebook which in effect becomes an IOU record for the hosts.[11] Neighbors and kin lend plates, cups, benches, tables, extension cords, canvas or plastic tarps for covering outdoor areas, and other items. After a shot of mezcal, everyone immediately sets off to work: women take turns converting the food gifts into hundreds or sometimes thousands of tortillas, tamales, and other dishes; men slaughter, dress, and boil the chickens and turkeys and run errands; and children scramble about. Except for the maize, beans, and *panela*

Figure 4.2. Campesino preparing to dress a chicken. Most life-cycle celebrations begin early in the morning, when chickens or turkeys are slaughtered and prepared for festive meals. *(Photo by Gabriela Zamorano)*

given by the guests, the hosts shoulder nearly all expenses, though customarily mezcal and cigarettes are provided by the matrimonial *padrinos*.

Over the next few days, the hosts serve three meals daily. Large tortillas (measuring approximately eighteen inches in diameter) are placed in the center of the table; strips are torn off by those seated at the table, who fold them and use them to ladle the food (Sp. *comida*) out of the bowl. If the dish has chunks of meat, they use the tortillas to lift the meat neatly from the bowl. Only small children and guests from the city are given silverware—if there is any at hand. Guests are served in groups or shifts; that is, ten to forty men or women are relieved of their tasks at once in order to eat. Each shift lasts approximately twenty to thirty minutes.

Weddings are especially symbolic events, and maize tortillas form an

Figure 4.3. Village band playing at a wedding party. In 1998 young women and girls were allowed to participate in the group for the first time. Membership in musical groups is as much a responsibility as it is a pastime. *(Photo by Rafael Colin)*

important part of the ritual known as the *derecho* (Sp.) or delivery of the bridewealth. On the day before the wedding, representatives of the groom arrive in the late morning at the house of the bride with turkeys,[12] tall stacks of tortillas, bread, cacao, and other items. The representatives, led by the groom's baptismal godparents, biological parents, older siblings, close neighbors, and *padrinos*, are received at the bride's house, and her parents offer mezcal, cigarettes, and frequently beer. The visit is often solemn for the family of the bride, who ideally will move to a site near the groom's family. The baptismal godfather delivers a short speech; after a brief period of respectful conviviality (perhaps thirty minutes), the representatives return to the groom's house, where the fiesta shifts into high gear.

After the Mass on the wedding day, the entire procession goes to the house of the groom (usually led by the village band or orchestra) for lunch.

Both tortillas and other food (*mole* is nearly universal at Talean weddings) follow a "human chain" from the kitchen to the serving areas: the items are passed from one person to the next until they reach the tables. In the late afternoon, after music, dancing, and drinking,[13] the bride's family and baptismal godfather formally take leave of the bride, the groom, and his family. Campesinos often deliver emotion-laden, eloquent speeches. Generally the bride's parents express a strong sense of loss even though their daughter may only be moving a short distance away.

It is tempting to romanticize *gozona* as a quaint cost-sharing arrangement, but for most people it is a survival strategy, a heavy obligation that simply cannot be avoided—and a way of guaranteeing *mantenimiento*. In a typical year a family may give away more than two *fanegas* (202 kg—nearly equal to the amount consumed in two months by a family of four) in *gozona* events. These gifts demonstrate how maize forms a critical part of exchanges linking families, fictive kin, and neighbors. Villagers frequently grow the crop using the mutual-aid arrangement for pooling labor; once within the village, it arrives as a food gift delivered by invitees. Maize is transformed into tortillas by a group of women working together, is consumed by the guests, and in the case of weddings is supplemented by more tortillas delivered by the groom's family. In the end it must be repaid as a debt. Thus maize is a good that flows back and forth between networks of social relations. It is part of a complex connecting immediate and extended families, neighbors and relatives, *padrinos*, village community, church, and deities.

Maize and Agricultural Revolutions

> Columbus did not realize that the gift of maize was far more valuable than the spices or gold he hoped to find. He had no way of knowing that the history of maize traced back some 8000 years or that it represented the most remarkable plant breeding accomplishment of all time. He might have been embarrassed if he had understood that then, as now, this plant developed by peoples he judged poor and uncivilized far outstripped in productivity any of the cereals bred by Old World farmers—wheat, rice, sorghum, barley, and rye.
>
> Walton C. Galinat,
> "Maize: Gift from America's First Peoples" (1992)

Early History: Mesoamerica's Agricultural Revolution

Maize was first cultivated more than 7,000 years ago, probably near the Tehuacán Valley of Puebla, south of Mexico City (Long et al. 1989). The earliest fossilized remains are no bigger than a human thumb. Not until about 2000 B.C. were higher-yielding varieties developed, and these were a fraction of the size of modern cobs.

By this time the ancestors of the Zapotec already had thousands of years of agricultural experience. Recent archaeological research at Guilá Naquitz, a cave site in the Valley of Oaxaca, has revealed that squash was cultivated long before maize; accelerator mass spectrometry (AMS) tests indicate that *Cucurbita pepo* was grown as early as 9000–10,000 years ago (Smith 1998; Flannery 1999). The earliest AMS date for the domesticated bean, *Phaseolus vulgaris*, in the Valley of Oaxaca is more recent (340 B.C.; Kaplan 1995), but there is evidence that by the time the first permanent settlements were established there (1500–500 B.C.) a host of food plants were already under cultivation: maize, squash, chile, avocado, cactus, the century plant, and others (Marcus and Flannery 1996).

Twentieth-century botanists have hotly contested the genesis of maize. The grain is related to a wild grass called *teosinte* (*Zea mexicana*), whose tough seeds grow in a single row and whose cob is about the thickness of a pencil. *Teosinte* does not have a husk, which means that, unlike domesticated maize, its seeds are easily able to propagate naturally. The debates center around the nature of the relationship between *teosinte* and maize. In the 1930s a popular theory was put forth by biologist George Beadle, who argued that maize had evolved from *teosinte* through genetic mutations. This was challenged by Paul Mangelsdorf (1974), who contended that *teosinte* resulted from a cross between an extinct form of wild maize and a distant relative, *Tripsacum dactyloides*. Only then did the *teosinte* backcross with its wild maize parent, leading to modern maize. By the 1970s this view dominated. Shortly afterward, however, botanist Hugh Itlis conducted studies that seemed to reaffirm the Beadle theory, which once again became the favored view among botanists—though many still hold to a version of Mangelsdorf's theory.[14]

Arturo Warman argues that no matter which hypothesis is true, the success of modern maize can only be attributed to human intervention in the collection, protection, and cultivation of a species that would otherwise have been doomed to extinction by an inability to disperse its own seeds. (A key feature of modern maize—its thick husk, which protects the fruit from harsh weather and predatory animals—makes its reproduction nearly impossible in the natural state.) Thus, whether maize emerged out of a catastrophic mutation or by selective breeding, humans helped it along, collecting its seed, replanting, and perfecting the crop over generations

(Warman 1988:47–48). In a thoughtful essay on maize and its origins, biologist Walton Galinat (1992:50–51) agrees:

> Though the first peoples of this hemisphere obviously lacked the knowledge of cytogenetics and biotechnology available today, there is every reason to believe that they possessed powers of observation and imagination equal to our own and sufficient to take this remarkable stride in plant breeding ... Armed with this intuitive understanding of inheritance and driven by experience with famine, aboriginal planters learned not to consume the best seed but to save and sow it.

Evidence indicates that maize cultivation was a transition that took place over a period of generations or even centuries (Ponting 1991:37–38). In central and southern Mexico, it appears that agriculture was carried out by hunting and gathering groups who settled down just long enough to plant and harvest their crops before moving on again.[15] This is different from the Old World sequence, in which sedentary life preceded agriculture.

Eventually the hunter-gatherers did settle (by about 1000 B.C.). Maize soon provided the alimentary base for large ceremonial centers where food surpluses were redistributed to priests, artisans, and administrators. Agriculturalists probably continued selecting the most productive varieties of maize, and the length of cobs more than doubled before the Classic period (300–900 A.D.) began. Ambitious architectural achievements such as the city of Teotihuacán, with an area of approximately ten square miles and a population of 100,000, illustrate how increases in food production made complex societies possible. There is no doubt that maize played a critical role in the creation of Mesoamerican civilizations. In the Valley of Mexico food crops were grown in productive "floating gardens" (Sp. *chinampas*) and collected in tribute from the peripheries of the empire. Teotihuacán collapsed about 700 A.D., only to be replaced by the military empires of the Toltec (at Tula) and the Aztec (at Tenochtitlán).

All of this goes to show that humans and maize were engaged in a dialectical relationship from the beginning: humans produced maize, but maize also produced human societies. If we think of the plant as a technology, this should come as no surprise; indeed, a recurring theme in the anthropological literature is that "man makes himself" (Childe 1951) through tools and other material artifacts. A more nuanced analysis might describe how maize, the environment, and humans influenced each other dialectically (Levins and Lewontin 1985:104):

> It is impossible to avoid the conclusion that organisms construct every aspect of the environment themselves. They are not the passive objects of external forces,

but the creators and modulators of these forces. The metaphor of adaptation must therefore be replaced by one of construction, a metaphor that has implications for the form of evolutionary theory.

Thus not only was maize reliant upon humans, but humans became reliant upon the material conditions and social structures that maize itself played a part in creating. Maize, in short, engaged human society dialectically.

If we think for a moment about the idea of a living maize, maize as a plant-person, we might hypothesize that the remarkable plant breeding done by the ancient Oaxacans and continued by their contemporary Zapotec descendants (and hundreds of thousands of other Mexican and Central American campesinos) might be directly related to maize's special status as a living being with a will of its own. It seems entirely plausible that the stunning achievements of maize's domestication and improvement and the urban societies that it made possible were stimulated by the very assumptions that would be considered incorrect by modern cosmopolitan scientists. The intense symbolic, religious, and emotional significance attached to maize—the respect and even love that these farmers had for maize the plant-person—may well have led them to spend more time caring for it, improving it, nurturing it, strengthening it, and adapting it to different ecological niches over the course of generations. The legacy of this expertise survives today: more than 20,000 varieties of maize exist in Mexico and Central America.

The "Second Agricultural Revolution"

The biological species that Columbus brought back to Europe from the New World probably had more long-term impact in global terms than all of the American gold and silver combined. Alfred Crosby (1972) notes that "the Columbian Exchange" of biota had significant demographic repercussions, specifically the Old World population explosion after 1492. In many parts of Europe, Asia, and Africa maize was generally successful because it did not compete with other grains—wheat, rice, and barley—but complemented them. Farmers often grew the new crop in the off-season or on plots that might otherwise have been lying fallow. Furthermore, it enabled them to grow food on marginal lands that had previously been considered too sandy, dry, or high for wheat or rice (Crosby 1972:171):

> Its advantage over equivalent Old World plants is that it will prosper in areas too dry for rice and too wet for wheat. Geographically it has fitted neatly between the two. Its supremely valuable characteristic is its high yield per unit of land which,

on world average, is roughly double that of wheat. For those to whom famine is a reality, maize has the additional benefit of producing food fast.

To drive home his argument, Crosby relates the story of how maize diffused across the Old World. Spain was the point of entry. People in many parts of the continent were slow to accept the crop, not only because the weather in most areas of Europe was too cold, but also because of a distinct cultural preference for wheat. Maize was used as animal feed in many places, and several observers considered the crop to be unhealthy, indigestible, and low in nutrition. Even so, millions of Europeans adopted it beginning in the 1500s, and many continue to do so today. Maize was particularly successful in Italy (as *polenta*), in Yugoslavia and Romania (as *mamaliga*), and, to a lesser degree, in France and Spain. The crop was popular among the poor, especially peasants, who would plant it for their own consumption—often mixing it with wheat flour to make bread—in order to sell more of their wheat crop to the rich (Crosby 1972:176-181). Maize was given a variety of names, nearly all of which disguised its New World origins: it was called "Rhodes corn" in Lorraine; "Spanish corn" in Tuscany; "Sicilian corn" in Provence; "Turkish grain" in Italy; "Indian corn" in England; and "Turkish corn" in Germany, Holland, and Russia (Braudel 1967:113). For many Europeans, the most incredible quality of maize was its astonishingly high yield. The crop produces many more kilograms of grain per hectare than other grains and a comparable amount of calories per hectare as rice and significantly more than wheat (see Table 4.2).

The fact that so many Europeans referred to maize as "Turkish" may indicate that the crop was rapidly accepted and made popular in Turkey, though this is by no means a certainty. In India terms for maize are related to the word "Mecca," which implies that it reached the subcontinent from an Islamic region. We cannot be sure of this since travelers to the Near East in the 1600s and 1700s make little mention of the crop. Today it is of secondary importance in the Near East, with the exception of Egypt, where it has been a staple crop since the eighteenth century. There is strong evidence that maize had much to do with Egypt's demographic growth (Crosby 1972:188-190).

Maize is a staple crop across many parts of Africa. As early as the 1600s observers reported that it grew "profusely" in the Gold Coast, where it was prepared by baking, and evidence indicates that it reached the Boshongo people of the south-central Congo basin in the same century. In the 1900s maize became even more critical in Africa; indeed, the crop "has become,

Table 4.2. Average Yields and Calories per Hectare of Maize, Rice, and Wheat

	KG/HA	MILLIONS OF CALORIES/HA
Maize	3,576	7.3
Rice	2,997	7.3
Wheat	2,064	4.2

Sources: Crosby 1972:175; Warman 1988:28.

for the first time, a mainstay of diet for most of east and central tropical Africa" (Crosby 1972:186–187).

But perhaps the most important Old World destination for maize and other New World crops was China, where it and other American crops were rapidly accepted: "while men who stormed Tenochtitlán with Cortés still lived, peanuts were swelling in the sandy loams near Shanghai, maize was turning fields green in south China and the sweet potato was on its way to becoming the poor man's staple in Fukien" (Crosby 1972:199). By the turn of the nineteenth century maize had become the primary food crop in many of the uplands of southwestern China. During the 1700s, the valleys of the Yangtse River and its tributaries had become heavily populated; the excess population, forced up into the hills and mountains, found that maize cultivation was an easy way to extract subsistence in the highlands. Thus, as rice cultivation in the lowlands approached its limit, "the various dry land food crops introduced from America have contributed most to the increase in national food production and have made possible a continual growth of population" (Ho 1959:191–192). Today maize is still spreading rapidly across China, often at the expense of more traditional crops such as sorghum and millet.

Warman (1988) adds a political-economic dimension to Crosby's analysis by arguing that maize was one of several elements related to the development of slavery. The wide geographic range of maize (and its high yields) meant that slaves could be fed more cheaply with maize than with wheat. The grain was also well suited for storage on slave ships. He concludes: "Slavery in Africa preceded the introduction of maize, but its expansion and exponential growth can only be explained by the American demand for it, and the role of maize in satisfying the growing market" (Warman 1988:65–80 [quotation on 80]; my translation). This might be extended to include other peoples subjected to domination in the age of empire: colonized peoples and internally colonialized peasants. In effect, Warman tells us, cheap maize subsidized the creation of empires.[16]

The diffusion of maize thus formed complex patterns over time and space. The crop spread across the Old World within a century, transforming diets, economies, and entire societies along the way. Maize and other food crops from the Americas became a part of European, Asian, and African culinary and agricultural traditions, just as coffee and sugar came to define many countries in the American tropics. Maize was used to feed slaves, peasants, and other colonized populations because it was geographically versatile and produced high yields. In the latter half of the 1800s maize cultivation was mechanized and subjected to other technological "improvements" in the United States; by the mid-twentieth century, ironically enough, it was exported to the so-called Third World by development agents.

Post-Conquest Agricultural Revolutions in the Americas

Crops flow in multiple directions. At the same time that maize spread across the Old World, local farmers in rural New Spain were incorporating Old World crops and techniques into their farming practices. In the Sierra crops such as sugarcane and techniques such as the use of metal tools, plows, and draft animals were adopted by farmers attempting to restructure their societies and economies in the wake of the Conquest. Other techniques and knowledges, particularly those related to the processing of new crops and animal husbandry, were incorporated by campesinos.

Much more recently, the Green Revolution swept Mexico, bringing with it hybrid seeds, chemical fertilizers, pesticides and herbicides, and mechanized farm equipment. In the Sierra most of these technologies either never arrived or else were rapidly rejected, with the exception of chemical fertilizers. A number of villages began experimenting with fertilizers in the late 1960s and in most cases continued using them for staple food cultivation (though in significantly smaller amounts than on factory farms, as we shall see) and even incorporated them into "hot/cold" classificatory schemes. There is some evidence that fertilizers are better at increasing yields in intercropped fields than in monocropped ones (Richards 1985:70-72). According to Talean campesinos, fertilizers effectively doubled maize yields and so played a crucial role in reducing land pressure at a time when the village's population was straining the land's capacity.

Maize and Mexican Land Reform and Food Policy

Mexican Agriculture and Food Policy

Maize, Mexico's most important food crop, has been at the center of federal agricultural and economic policies for many years. The Rincón has historically been relatively insulated from many of these policies, since the majority of villages have produced enough maize and beans for self-provisioning. Taleans turned to cash cropping early in the twentieth century, but even so most campesinos supplemented—and did not replace—their cultivation of subsistence crops with coffee. Still, to the extent that Taleans bought (and continue to buy) maize and beans from campesinos in surrounding villages or from merchants selling maize produced in the Valley of Oaxaca, they have been affected by Mexican food policies.

The topic of agriculture and food policy in Mexico is complex; indeed, entire books are devoted to the issues (Sanderson 1986; Fox 1992; Hewitt de Alcántara 1994; Barry 1995; Randall 1996). Here I have a more modest objective: to review the underlying themes in Mexican food policy that have surfaced over the course of the last century and to relate them to the Talean situation.

The balance between large and small farms has shifted over time. During the latter part of the 1800s, for example, in the era known as the Porfiriato (1884–1911), policies tended to favor haciendas at the expense of indigenous communities. On the eve of the 1910 Revolution, only 4 percent of Mexico's rural families owned land, and nationwide an estimated 92 percent of the total population was landless. Export-oriented agriculture was favored over staple food production: during the Porfiriato, export crops increased from 4 percent to 20 percent of total production, while land devoted to maize production decreased from 52 percent to 33 percent (Barry 1995:15–16). By the early 1900s Mexico was importing maize from the United States (Randall 1996:3–5). Between 1908 and 1910, for example, nearly one-third of Oaxaca's maize was imported from abroad (Ruiz Cervantes 1988:336–338). It is unclear to what extent the Sierra followed this pattern; because of the difficulties involved in transporting maize to the region, the quantity imported to the Sierra was probably lower.

The 1910 Revolution was largely a result of such inequalities. In the end, Zapatistas and Villistas insisted on land reform, formally guaranteed in Article 27 of the 1917 Constitution, which established the *ejido* system of inalienable (but inheritable) communal lands directly assigned to individuals by the federal government. Government officials reserved the right to expropriate and redistribute hacienda land for this purpose. Land redistribution was carried out halfheartedly by postrevolutionary presidents; but during the presidency of Lázaro Cárdenas (1934-1940) the process accelerated: some 20 million hectares were redistributed, more than twice as much as under his predecessors. His successors, however, slowed land reform, and new policies limited access to credit, infrastructure, and inputs for small *ejido* farmers. They helped create a two-tiered system of agriculture: small private farmers and *ejidatarios* produced staple foods (maize and beans) which provided cheap basic foodstuffs to urban consumers, while medium- and large-scale commercial farmers (including agribusiness concerns) produced export crops. Farmers on plots of less than five hectares generally continued operating at or below subsistence levels.

The Green Revolution had a number of effects in Mexico. It is clear that yields of wheat and potatoes increased dramatically, even on small farms. The problem, however, was that access to credit and technology (hybrid seeds, chemical fertilizers, pesticides, herbicides, machinery, etc.) was often much more difficult for small farmers.[17] Thus, in general, the Green Revolution favored Mexican agribusiness farms in the 1950s and 1960s.[18] In addition, maize yields did not increase significantly because in many parts of the country local farmers had already approached the biological limits of the crop. These problems were compounded by an urban bias in food policy which made consumer subsidies (such as the CONASUPO government-subsidized food stores initiated in the 1960s and 1970s) more forthcoming than producer subsidies.

Despite production increases linked to the Green Revolution, a number of factors made Mexico a net importer of grains by the 1970s. Specifically, the country's demographic explosion and land tenure and land use policies outpaced agricultural production gains. The Green Revolution's limits were reached in the late 1960s and 1970s; grain production began to stagnate, and Mexico once again began importing grains, including maize (Barry 1995).

Under the presidency of José López Portillo (1976-1982), efforts were made to reverse this trend through an ambitious program for national food self-sufficiency, the Mexican Food System (Fox 1992). It achieved moderate success for a few years but ended before the potential for increased production by small growers was realized (Barry 1995:99). Soon Mexico was again importing significant amounts of U.S. maize—approximately 20 to 25 percent of its total consumption—which served to demonstrate the "inefficiency" of Mexican agriculture vis-à-vis U.S. factory farming (Barry 1995:99). This would later be used to justify "free trade" policies.

By the 1990s president Carlos Salinas had embarked upon a series of dramatic neoliberal reforms. Opening up Mexico's grain market was a key priority for U.S. trade representatives in NAFTA negotiations in the early 1990s, and they succeeded. Even before NAFTA was signed, Article 27 of the Mexican Constitution was revised, halting land reform before it had even occurred in such places as Chiapas and opening the way for *ejido* privatization.[19] Significantly, many of the agricultural credits, fertilizers, and guaranteed prices for maize established in the 1980s came to an end, due in part to the terms of NAFTA.

Deregulation posed an immediate threat to farmers growing maize as a cash crop; indeed, the removal of trade barriers depressed prices as cheap U.S. corn poured into the Mexican market. To head off possible social unrest—especially in light of an imminent economic crisis—in late 1993 the federal government created the Program of Direct Support Payments for the Countryside (PROCAMPO), a system of direct cash payments to maize producers. Farmers are paid according to the amount of land they have planted for subsistence crops—about US$105 per hectare. In Talea campesinos wishing to participate in the program report the amount of land they have cultivated each year to government representatives and eventually receive checks in their names. PROCAMPO, which meets the terms of NAFTA because it is scheduled to be phased out over a fifteen-year period, has been described bluntly by some as a way of "fertilizing votes" in rural areas (Barry 1995; Bartra 1995). A number of Talea's campesinos see it this way as well. Others have been reluctant to participate, either because of the time commitment (it requires attending a number of meetings) or because they possess lands with no ownership title.

Who has benefited from "free trade"? In general, it seems that transnational corporations are poised to make big gains from economic liberalization. According to some critics, "free trade" policies are likely to increase foreign control over agribusiness in Mexico (Nader 1993), a process which began decades ago (Cockroft 1990). The Green Revolution promoted a growing dependence on inputs (notably hybrid seeds and farming machinery) supplied by foreign corporations. On another level, food processing in Mexico has been affected profoundly by the transnationals: by 1995 fully one-third of the country's food-processing industry was in the hands of U.S.-based companies, including Anderson Clayton, Ralston Purina, Pepsico, and Tyson Farms. Dozens of other European-based companies such as Unilever and Nestlé are also in the business (Barry 1995:62-64). Furthermore, new international intellectual property legislation eases the way for U.S. and European firms to claim proprietary rights over biotechnologies—a process likely to lead to "the intensification of international linkages, scientific disparities, and the furtherance of national market penetration" by transnationals (Buttel, Kenney, and Kloppenburg 1985).

Effects of Mexican Food Policies in the Rincón

In the Rincón land has not been subjected to *ejido* reforms, because for most of this century neither *ejidos* nor large haciendas have existed in the region. When the Rincón Zapotec have been exploited, it has been through labor arrangements in the region's mines or through commercial monopolies, particularly over cotton, cochineal, and coffee. Commerce, not land, has customarily been the source of economic power in the Sierra since the Conquest.

Most Taleans work plots of land smaller than five hectares, and most of those who farm grow maize—though not all households produce the maize needed for an entire year.[20] (There are notable exceptions; some growers in Talea produce more than enough maize for household consumption and sell a surplus at the local market.) Historically, extra maize has been purchased outside the community. For many years, it was bought from surrounding villages or from merchants bringing the product from the Valley of Oaxaca, but more recently it has been provided by CONASUPO government-subsidized stores. Locally produced maize is not sold to CONASUPO buyers; household surpluses are generally small and are sold in regional markets at a price 20–30 percent higher than at the CONASUPO store.

How have national food policies affected farming in Talea? Though profound changes have not shaken the village's economic core (as they have, for example, in Zinacantán, Chiapas; see Cancian 1992), there have been subtle shifts in the region that are probably related to Mexican agricultural and food policy. The delivery of cheap maize to the Rincón through CONASUPO stores probably lowered maize prices beginning in the late 1960s. Maize farming became a less profitable activity because surpluses destined for local markets were subjected to competition from cheap CONASUPO maize coming from outside the region. This probably encouraged campesinos in Talea and other villages to convert more milpas to *cafetales*, since coffee was more profitable throughout much of this period.

Another effect was the change in production that Green Revolution technologies—particularly chemical fertilizers—made possible. Though tractors, mechanical harvesters, and other such implements are scale-dependent technologies, fertilizers are not, and small farmers in Talea, Tanetze, and Yaeé began using them on a significant scale through a government program which provided credit for their purchase (Tyrtania 1992: 196). Informants report that yields nearly doubled as a result, freeing land for coffee cultivation.

Finally, informants describe a major decrease in the relative price of maize beginning in the late 1960s. In the early 1960s one *almud* (4.2 kg) of maize was worth two *mozos* (that is, two days' work at minimum wage), according to various informants old enough to have worked during that period. By the early 1990s an *almud* could be purchased with less than half a day's work.[21] In other words, relative to the daily minimum wage, the price of maize fell to one-fourth of its value within approximately a decade. This probably occurred as the combined result of increased outmigration which put a premium on labor, government subsidy programs (CONASUPO stores) which drove maize prices down, and higher yields owing to chemical fertilizers.

In light of these policies, the questions change considerably. Instead of asking "Why did campesinos in Talea (and across the Sierra) begin to shift away from maize farming?" another question seems more appropriate: "Why didn't the campesinos shift completely to coffee cultivation, since maize prices had dropped and coffee prices were at high levels in the mid-1970s?" Tyrtania (1992:194) gives us a partial answer:

> In this context [of low market prices for maize] producers . . . appear as irrational beings who are cultivating a ruinous crop, in economic terms. But the campesino who retains effective control over productive resources does not calculate his family's biological survival—derived from self-provisioning—in monetary terms.

Emiko Ohnuki-Tierney (1993) has made a similar point with regard to rice in Japan: the crop is not just a commodity consumed by a nation of economically rational actors; rather, rice carries with it a bundle of cultural, political, and symbolic meanings—including, for example, notions of food quality—that are elided by simple economic analyses. We shall consider these considerations with respect to Sierra maize later.

These things take on added meaning when put in the context of fluctuations in coffee prices. For many Taleans, the 1959 coffee price crash probably remained fresh in their minds, as did the saying "You can eat maize, but you can't eat coffee." As long as maize production was limited to household consumption, it was insulated from the effects of declining market prices, and there was little incentive to stop growing it. Generally, subsistence farming and cash cropping form a kind of assets diversification policy in which risks due to international price swings, weather patterns, illness, etc., are reduced by being spread out over a number of crops. If one crop fails, say the Taleans, we can always rely on the other. It seems that for most campesinos the answer to government maize subsidies was not to stop growing maize—it was simply to limit it to household consumption.

To summarize, in spite of the allure of cash cropping, maize farming has not disappeared in the Rincón largely because it holds special significance in the lives of the Zapotec. In the Rincón maize has a soul. Indeed, the customary Rincón Zapotec view of nature is remarkably different from the conventional view in industrialized societies, to the extent that the earth, the rivers, and individual crops such as maize are seen as animate objects and are personified. Many development initiatives might be said to "dehumanize" the earth, animals, and even human beings in the quest for raw materials (see Alvares 1992:64-69). In the Rincón humans are engaged in a reciprocal relationship with these entities, often using saints and mountain spirits as intermediaries or advocates. In historical terms, maize farming extends far into the past, and its centrality in many Native American societies is deeply rooted. Finally, activities associated with maize have as their objective the maintenance of the campesino household and lands worked by its members. Thus maize farming in the Rincón involves activities conducted by people enmeshed in complex reciprocal relationships with their families, kin, neighbors, fellow villagers, saints, spirits, and the earth itself. The very embeddedness of maize in these intricate networks has made it an essential part of campesino lives that seems unlikely to be given up soon.

CHAPTER 5

FROM MILPA TO TORTILLA

Growing, Eating, and Exchanging Maize

In this chapter the focus shifts from Zapotec conceptions of "nature" to the knowledges and techniques deployed by Talean campesinos—that is, to their science and technology. I begin with a brief discussion of land tenure, labor arrangements, and classificatory systems employed by the Zapotec, followed by a description of farming methods. I then look at the conversion of maize into food by women. The concluding sections focus on the relationship between Talean agriculture and factory farming, including several cases in which Talean methods have been subjected to inquiry on the part of cosmopolitan scientists trying to understand how U.S. farming might benefit from local farming traditions. In order to understand why Rincón farming is done in a specific way, we must understand what "growing one's own food" means from the perspective of Talean campesinos.

Growing Maize in Talea

Maize farming in Talea is generally carried out using a two- or three-field fallow system; that is, two or three fields are rotated so that each is farmed every other year or every third year, respectively. This differs from the slash-and-burn system employed in many other Rincón villages and probably is a result of Talea's limited land. As mentioned earlier, most of the village's land is considered private property, though land titles do not always exist. The forest above the village is considered communal property (Sp.

bienes comunales), and Taleans may cut firewood there for household use. Apart from private and communal lands, there are lands owned by mutual aid societies called *barrios*.

In the case of *barrios* land titles are held in common, taxes are paid to the municipal authorities in a single payment, and land tenure rights are passed down from father to child.[1] Nonmembers may rent land annually at the discretion of the association's elected officials, but coffee may not be planted on *barrio* land because coffee trees are long-term plants and groves would effectively become private property. The *barrio* land tenure system has broken down considerably; according to informants, as recently as a generation ago the *barrios* owned significantly more land than they do today. Much of it has been sold to private holders.[2]

When land was scarcer (that is, before chemical fertilizers and outmigration reduced land pressure), informants report that milpas were often rented (sharecropped). All work was done by the tenant farmer, who then turned over half of the harvest to the landlord. Today campesinos who rent land typically pay 1 *fanega* (24 *almudes* or approximately 101 kg) for each *almud* (¼ hectare) planted—a rate that ranges from approximately 16 to 33 percent of the harvest.

Another arrangement is *gozona*, the reciprocal work system in which labor is exchanged. For example, two campesinos might agree to pool their labor to sow each other's maize seed, with Monday through Wednesday devoted to the first field and Thursday through Saturday devoted to the second.

Talea's land ranges from a high altitude of approximately 1,800 meters above sea level to a low of approximately 1,000 meters. Land is considered "hot" (Sp. *tierra caliente*) if lower than 1,300 meters; "temperate" (Sp. *tierra templada*) if between approximately 1,300 and 1,550 meters; or "cold" (Sp. *tierra fría*) if above 1,550 meters. These classifications are supplemented by other criteria based on color, consistency, rockiness, temperature (due to shade from mountains or forest), and compactness. A nearly infinite number of permutations may be created by combining two or more of these classifiers or adding other descriptive terms.

In an interesting analysis of soil classification types in two Mexican municipalities, a geographer recently compared local classification schemes to those of modern soil surveyors. He found that in most cases "traditional and scientific appraisals correspond rather closely . . . What distinguishes local classification from regional [soil] reconnaissance is degree of detail;

Table 5.1. Soil Criteria and Types Identified by Campesinos

CRITERIA	SPANISH TERM	ZAPOTEC TERM
Color		
yellow	*tierra amarilla*	*yugách*
white	*tierra blanca*	*yuchiguích*
red	*tierra colorada*	*yushná*
black	*tierra negra*	*yugásj*
Consistency		
gummy, clay	*tierra gruesa*	*yugúna*
sandy	*tierra arenosa*	*yuyúsh*
Rockiness		
rocky	*tierra pedregosa*	*yuguiáj*
pebbly	*tierra cascajosa*	*yurój*
Compactness		
"tight"	*tierra apretada*	*yunálh*
loose, soft	*tierra blanda*	*yuyúla*
Altitude and Shadiness		
cold	*tierra fría*	*yuziág*
moderate	*tierra templada*	*yutempládu*
hot	*tierra caliente*	*yushlá*

farmers recognize subtle variations and combinations of soils overlooked by most surveys" (Wilken 1987:25). Talean campesinos also tend to describe soils in excruciating detail.

Talean farmers divide the year into three seasons determined by abrupt changes in precipitation. The timing of the three seasons varies from one year to the next—unlike seasons in industrialized societies, which are fixed by the calendar—and is a practical guide for farming. The "frost" season (Sp. *tiempo de helada*) begins once the annual rains end—typically in mid-October, but sometimes as late as November. The frost season ends when temperatures warm up and the days become longer, in late February or

March. This marks the beginning of the "dry" season (Sp. *tiempo de sequía*), the hottest season of the year. The "rainy" season (Sp. *tiempo de lluvias*) lasts from May, June, or July to October or early November. It is marked by storms from the Gulf of Mexico that bring downpours, often lasting for hours. Some weeks it may rain heavily for four or five days. Talea is located approximately halfway between the Gulf of Mexico and the Pacific Ocean, and its summer storms sometimes include heavy downpours and winds originating from hurricanes striking either coast.

The following data were recorded in a milpa lying at an altitude of approximately 1,600 meters above sea level. The agricultural cycle at higher altitudes tends to begin sooner and end later compared to lower altitudes.[3] Although I have attempted to include time estimates for the tasks, these are difficult to calculate because of differences among the farmers based on age and experience and also because many have animals which must be taken to water and pasture. This can easily take more than an hour daily (see Table 5.2).

Clearing Underbrush

The rainy season typically ends by the last week of October, and the campesino begins clearing the prospective milpa of underbrush. This work, called *rozando* (Sp.), is done entirely by hand with only a long machete (approximately 80 cm) and a custom-made hook fashioned from a tree branch. The campesino takes the hook in his left hand, uses it to grab a thick bundle of grasses and weeds to his right, and then neatly cuts close to the ground with a horizontal swing or two of the machete. He pulls the hook around in a semicircle without lifting it and continues swinging until the entire area before him—a semicircular area spanning nearly three meters—is cleared. The hook and machete are used as pincers to pull the freshly cut bundle of weeds to the right into a neat pile. The campesino moves forward to the edge of the cleared area and repeats the entire process. The series of intricately coordinated movements takes no longer than fifteen seconds and leaves the area beautifully cut less than an inch from the ground. An experienced campesino can complete a field of four *almudes* (approximately one hectare) in six or seven days.

The campesino begins at the lowest end of the milpa and works his way up, row by row. The job taxes the lower back, machete arm, and hands, yet is accomplished easily by farmers. "It's a game for us," I was told. "We've been swinging machetes since we were kids."

Table 5.2. Workday Schedule for a Campesino with Animals

March 29, Maize Planting

TIME	ACTIVITY
6:35 A.M.	place tinder in coals to start fire
6:35–7:05 A.M.	fill tank of water at watering hole
7:05–7:10 A.M.	place beans, coffee, and more firewood on coals
7:10–7:50 A.M.	plant maize while breakfast reheats
7:50–8:35 A.M.	eat breakfast
8:35–8:50 A.M.	take oxen and mule to fresh pasture areas
8:50–12:10 P.M.	plant maize
12:10–12:15 P.M.	place beans, coffee, and more firewood on coals
12:15–12:50 P.M.	plant maize while lunch reheats
12:50–1:40 P.M.	eat lunch
1:40–2:30 P.M.	take oxen and mule for water and then to fresh pasture areas
2:30–5:35 P.M.	plant maize
5:35–5:40 P.M.	place beans, coffee, and more firewood on coals
5:40–6:20 P.M.	plant maize
6:20–7:20 P.M.	eat dinner
7:20–7:40 P.M.	drink shot of mezcal, chat, smoke cigarette
7:40–8:00 P.M.	prepare pot of fresh beans for following day
8:00–8:35 P.M.	sharpen machete blade and chat
8:35–9:00 P.M.	smoke cigarette and chat
9:00 P.M.	sleep

First Plowing

Plowing (Sp. *arando*, Zap. *bládzu bedx*) ideally is done immediately after clearing underbrush. According to informants, a generation ago farmers plowed just after the fiesta of San Miguel (September 29), but in that era the rainy season began on time (May) and ended on time (August or early September), so delaying the task could be disastrous: it would mean plowing a hard-packed field. Now, I was told, the rains frequently begin late

(June) and end late (late September or even October), and farmers have adapted the timing of agricultural tasks to new weather conditions.

Farmers note that the objective of plowing is to protect the precious humidity lying below the surface of the ground by destroying weeds and grasses that "draw" or "pull" it (Sp. *jalar*) from the ground. Plowing therefore helps retain the soil's moisture by killing weeds that would otherwise rob it from below.

"Modern" scientific explanations for tilling soil are essentially similar to those given by campesinos: "Loosening the soil also reduces thermal conductivity, and consequently heat flows between surface and substrate. Tillage also breaks capillary connections in the soil and inhibits water transfers to the surface. The tilled layer dries out quickly by evaporation, but subsoil moisture is conserved" (Wilken 1987:235).

Campesinos note that plowing on time is a vital part of conserving soil humidity. Plowing too early (before the rains end) might make the task easy since the earth is soft, but is self-defeating because new weeds will simply crop up again due to imminent rainfall. Plowing too late is problematic because by this time weeds have "pulled" much of the water out of the ground and have left the earth packed tightly. Not only has the moisture been sucked out, but plowing itself becomes a difficult task when done on a dry, packed field. In short, plowing is a carefully timed activity that depends on extremely variable rain patterns in a wide range of soil types. Needless to say, it is very difficult to coordinate with the coffee harvest (beginning as early as November) and various obligatory fiestas from November through January. Fortunately, there are lunar indicators which, according to campesinos, are effective in helping predict rain patterns (see below).

Not all campesinos can plow. Even fewer learn to care for teams of oxen, which need daily attention.[4] Fewer still can construct plows, which require an extraordinary degree of calculation, coordination, and craft. The skills needed to build good plows, to use them properly, and to maintain draft animals are characteristic of "red-blooded" campesinos in Talea, and these tasks are well remunerated.[5] Campesino families take pride in their animals, which are the subject of endless conversations between farmers, their wives, and their children. (Their importance was made strikingly clear to me when an informant sold his oxen to a butcher after nearly a decade of service. The entire family was visibly moved. "Poor animals," said his wife. "They put food on our table for many years." He solemnly agreed: "They were my right hand. I hope the new ones will work as well as they did.")

Young boys are especially enthusiastic about oxen, learn about them at an early age, and may be given responsibilities for their care (watering and pasturing) before they are ten years old.

Young campesinos are sometimes asked to raise oxen "in company" (Sp. *en compañía*), as a joint venture with a wealthier patron. In this arrangement the patron (often a merchant) purchases two young bulls and turns them over to the campesino, who is charged with their upkeep. The team provides several years of service and may be contracted out for extra cash by an enterprising campesino. If he cares for them well, within five or six years they will have matured into large, healthy animals and may be sold to a butcher for perhaps twice their original cost. The two partners then evenly divide the money paid by the butcher. It is a relatively low-risk investment for both parties, assuming that the animals do not die of disease or mistreatment, and both stand to gain from the arrangement. This is one way in which campesinos and merchants are linked together economically.

During the plowing season, oxen must be carefully attended to. Before breakfast, the plowman takes them to drink water to cool them down after the energy-packed nighttime meal of dried corn husks, a "hot" substance. He then gives them more husks to eat before sitting down for breakfast. (He may sprinkle the husks with salt water to make them more palatable.) This takes the better part of an hour. Next the plowman lays out two long leather connecting straps (Sp. *coyuntas*), the assembled plow and beam (Sp. *arado* and *timón*), and the yoke (Sp. *yugo*), slips the braided leather loop that connects plow to yoke (Sp. *barzón*) into place, and brings out the oxen one by one. He lashes each tightly to the yoke with the strap, takes a switch in one hand and the plow handle in the other, and begins shouting orders at the animals. The entire team crawls forward steadily: the plowman coaxing with catcalls, hisses, tongue clicks, trills, and occasionally oaths; the plow and yoke creaking and groaning under the stress; and the beasts pulling mightily, lifting fresh earth and stone, huffing in the morning mist, intently listening for commands.

A skilled plowman seldom needs to strike his animals with the switch—as an informant said, "These *pendejos* [fools] know that *I'm* the one who gives the orders." It is critical that the animals be disciplined in the first few days after being purchased:

My ox Negro gave me trouble once, and once only. I was taking him to the watering hole off to one side of La Cumbre [the Crest] and I had noticed that recently he had been getting a little brave, he was getting a little too close to me. That's why

Figure 5.1. A Talean campesino plowing his field for maize planting. The milpa below will lie fallow the following year. Plows and draft animals offer a number of advantages over mechanized farm equipment. *(Photo by Gabriela Zamorano)*

you should always have the rope in your hand formed into a loop, doubled up, so you can hit him if he gets too close. Well, this time I thought I'd carry a wooden stake along with me to educate him in case he got too close. And sure enough, on this day he got too close and then tried hooking me. I nailed him right on the snout, but really nailed him . . . it hurt me to do that, because I was striking my own animal, and blood even started dripping out of his snout. But Negro hasn't given me any trouble since that day.

Young bulls are castrated once they begin displaying aggressive tendencies. Sometimes a specialist from Zoogocho, a Cajonos Zapotec village, carries out this dangerous task; he is known throughout the region as *the* expert in castrating work animals. In other instances a campesino may do the task himself (following a different procedure in which the testicles are crushed rather than severed), with the assistance of a number of companions.

The most difficult part of plowing is "throwing down the first line" (Sp. *tirando la primera raya*), which involves breaking open earth for the first time. The first few rows can be plowed more efficiently if a second campesino guides the team along a straight path. The plowman begins at the bottom of the milpa and continues upward, row by row. The spacing between rows should be, in the words of one campesino, "not so far that you are unable to fold all of the newly plowed earth into the previous row." When pressed for a specific distance, he recommended that new rows be cut a *cuarto* (the distance between the thumb and little finger of a splayed hand, approximately nine inches) above the previous row.

Experienced plowmen know that the only way to endure plowing is to use the tip of the plow as a pivot point and the handle as a lever. It is critical to keep the plow angle at about 75 degrees with respect to the ground and to take special care when large rocks are unearthed. One must work the handle from side to side so that the rock is unearthed without getting caught in the plow. Good plowmen use the switch sparingly and only to strike an animal that falls far behind; it is preferable (and less tiring) to shout commands.

If a second campesino is helping, he may gather weeds into rows using a forked stick. During the course of the day, these are moved out of the team's path so that they do not impede plowing.

All day long the plowman constantly moves over rough, rocky earth. Perhaps more than any other task, plowing requires concentration (on the earth, the animals, the contour of the land) and coordination (of commands to the animals, of the switch, of the plow, of one's body). Well-trained animals make the task infinitely easier. Campesinos who are able to plow often describe it as the most relished of farming tasks, and it is perhaps the one in which they take the most pride. Others describe it as exhilarating, probably because of the incredible physical power that the plowman directs.

At the end of the day (usually an hour before twilight), the plowman unhitches the animals, takes them to fresh pasture, and pours out fresh husks before each one. He leaves the plow and beam in the milpa, covers them with weeds so that the sun does not dry them out, rolls the long leather straps into a bundle, and drops them into a hole near the ranch house, where the ground's humidity keeps them flexible.

An experienced campesino can plow one hectare in approximately ten days. From Talea children can look eastward and see their father's progress, for the earth-colored stripe formed by the plow grows wider with each passing day, transforming a pale green field into a rich band of mahogany-colored earth.

The initial round of intercropping is also carried out during the first plowing; a squash known as the *yútu gu* (Zap.) or *támala* (Sp.) is either planted or broadcast in the recently plowed milpa. (This squash can be planted any time a field has been recently plowed.) It grows quickly, and campesinos often boil its tender leaves or orange flowers, called *cuan setugú* (Zap.) and *yaj setugú* (Zap.), respectively, in a pot of boiling beans to add flavor to meals. They may also boil them in salted water with chile. The flower is especially delicious and is considered a delicacy throughout Mexico.

Campesinos often broadcast black beans as well. In late December or January they collect several dozen bundles of the growing bean plants (Zap. *cuan za*) for household consumption or for sale in the weekly market, where off-season winter greens can be sold for a high price. The *quelites* (Sp., greens) are boiled in water and eaten as vegetables.

Data collected by geographers, agroecologists, and other scholars emphasize the advantages of intercropped milpas: maize-bean-squash intercropping significantly increases the total amount of food that can be produced in a given area, greatly inhibits the growth of weeds, reduces the risk of certain pests, and helps retain soil humidity (Richards 1985; Altieri 1987; Wilken 1987).

Second and Third Plowings

Under ideal conditions, the second plowing begins any time from mid-January to late February. It proceeds more quickly than the initial plowing because the ground has already been broken and is still soft. Within seven days an experienced campesino can cover one hectare. At this time remaining weeds are pulled out by hand. The campesino intercrops a squash variety called *yutuchína* (Zap.) or *chompa* (Sp.) in the richest part of the milpa. We planted seeds approximately eight paces apart, using a short-handled hoe to dig into the earth until the humidity line was reached, then deposited four to five seeds in each hole and covered them with a thin layer (2–3 cm) of soil.[6]

The third plowing takes place approximately two weeks before cultivation begins, in mid-March for a field in *tierra fría* that will be planted with large white maize. The campesino pulls out remaining weeds by hand; if there are any large dirt clods (larger than a cabbage), he breaks them open with a wooden mallet so that they do not roll over tender maize shoots. The clods, called *chekoyú* (Zap.), have "natural fertilizer" in them, according to some campesinos, because they are filled with rich, humid earth. At this time campesinos sometimes intercrop a squash that produces pear-shaped gourds; in December or January they are picked and used to form ladle-like bowls called *zhríga* (Zap.) or *jícaras* (Sp.). They serve as ladles or as containers for seed.

With both second and third plowings the campesino must take care to keep oxen from trampling or munching growing squash vines. If this becomes a problem, he smears a bit of manure over the plant's broad leaves, which nauseates the animals.

One difference between the first, second, and third plowings has to do with how oxen are fed. By the time the second plowing takes place, the previous year's maize crop has been harvested, so it is common for the bare stubble to be used as feed, assuming that the rotated fields are close together. The campesino tethers the beasts with stakes so that they can reach the edges of the milpa and gradually lets them further in as the days pass so that the stubble gets nibbled away. It is critical to do this so that the oxen do not leave any stalks uneaten.[7] The animals clear away the following year's milpa while simultaneously fertilizing it.

The last step before planting is the burning of the underbrush, now thoroughly dried after five months and piled into rows. Leaving the underbrush and the previous year's stubble is a form of mulching that offers advantages: it helps to retain the soil's humidity and when done before the rainy season can improve the reception of seed (Wilken 1987:235–236). After the last plowing, campesinos carefully burn the dried vegetation.[8] Ash is said to "fertilize" or "feed" the milpa.

Planting Milpa

In Talea campesinos leave maize seed (Zap. *yhznini*) on the cob throughout the year, suspended from a net bag in a cool, dry place so that maize weevils (*Sitophilus zeamais*) do not infest the kernels. They plant different varieties of maize which they classify according to color, size, ecological criteria (altitude), and maturation period (see Table 5.3). Several days before planting, campesinos shell the kernels from the cobs by hand, row by row. The kernels at the tip of the cob are not removed; they are said to be

Table 5.3. Varieties of Maize Grown by Talean Campesinos

---

1. Large maize (Sp. *maíz grande, maíz de monte*; Zap. *zhua zhn*)
   a. White (Zap. *zhua chiguích, zhua guí'a*)
   b. Yellow (Zap. *zhua gach*)
2. Small maize (Sp. *maíz chico*; Zap. *zhua cuídi*)
   a. White (Zap. *zhua chiguích cuídi*)
   b. Yellow (Zap. *zhua gach cuídi*)
3. Black maize (Sp. *maíz negro*; Zap. *zhua gasj*)
4. "Zoochila" maize (Sp. *maíz zoochila*; Zap. *zhua sochíl*)
5. Red maize (Sp. *maíz colorado*; Zap. *zhua shná, zhua zhíb*)*
6. Spotted maize (Sp. *maíz pinto*; Zap. *zhua pínt*)

---

*The Zapotec word *zhíb* means "injured"; red maize has the color of blood or, in the words of an informant, "the color of the heart of Jesus."

too small to bear fruit. This work is frequently done before and after dinner in the village, by all able family members, including some children less than two years old.

In *tierra fría* large white corn (Zap. *zhua chiguích*) is planted during the last few days of March and into April. Customarily, planting begins after the new moon or *luna cresciente* (Sp., literally "crescent" or "growing" moon) phase. Campesinos should fell trees which cast shadows over the milpa, for they "chill" the field and impede the development of maize.

Once these preliminaries are out of the way, planting may begin. The farmer mixes maize seeds together with those of a squash known as *huíche yútuyak* (Zap.) and frequently adds a type of polebean seed (Zap. *za láya*) as well. Then he fills a carrying container made of either an armadillo shell or a gourd with the seed mix and ties it about his waist. Armadillo shells are preferred; indeed, they are said to have a "secret" which gives them an advantage over the gourds, though I was unable to determine the nature of the secret.

Planting, unlike plowing, begins at the top of the milpa so that recently planted rows of maize do not get trampled. Campesinos use only a single tool, a straight metal-tipped digging stick called a *sembrador* (Sp.) or *yegúz* (Zap.). The technique is simple: the farmer sweeps away rocks and large dirt clods with the stick, then uses the pointed blade as a tiny shovel to perforate the ground and rapidly scoop out earth until the humidity line is reached, usually 10 cm or so below the surface. (The humidity line, distin-

Figure 5.2. A campesino planting maize. This work is done with little more than a metal-tipped digging stick. As the campesino plants, dry weeds smolder. *(Photo by Gabriela Zamorano)*

guishable by its dark color, marks the point at which the dry topsoil ends and the moist earth begins.) Next he drops the seeds into the center of the hole, four maize seeds and whatever additional bean and squash seeds happen to get pulled out. (I was told that many years ago five maize seeds were planted, but this is no longer the case.) He uses the tip of the digging stick to knock a very thin layer (approximately 1–2 cm) of earth over the seeds. A striking characteristic of maize selection and cultivation is the careful attention given to each individual seed, as opposed to the bulk seed selection practices associated with Old World crops (Lipp 1991:15).

Sometimes parts of a milpa are covered with rocks. Maize thrives in such areas, because the rocks trap a greater degree of subsoil humidity than bare earth does. In these cases larger rocks may have to be moved aside by hand before planting.

According to campesinos, plants should be spaced at a distance of one *vara* (approximately 80 cm). This falls within the typical range for southern Mexico and Central America, particularly in those areas where space is at a premium (Wilken 1987:245–246).[9] Time and again, farmers told me that spacing plants more closely than the prescribed distance would result in their "robbing force" from each other. In some other parts of Latin America farmers explain spacing as a wind reduction strategy (Wilken 1987:248). Another advantage of spacing plants far apart has to do with intercropping, for low-lying bean and squash plants have a better chance of receiving sufficient sunlight—otherwise, say campesinos, they will "freeze."

It is common for campesinos to pool their labor during the planting season in *gozona* arrangements.[10] In this way the planting gets done twice as fast (or three times as fast, if there are three campesinos involved) and the otherwise monotonous work—carried out during one of the hottest times of the year—takes place in a convivial setting with plenty of joking and gossiping. Normally campesinos work at the same speed, so that a line of two, three, or four people proceeds uniformly across the milpa. Four campesinos can plant one hectare in approximately three to four days. The maize sprouts or "is born" (Sp. *se nace*) within a week, unless heavy rains soak the seed. In this case the maize must be replanted.

Replanting

Approximately two weeks after planting, campesinos take note of areas where seeds have not sprouted and plant four new seeds in each of these holes. They replant only the parts of the milpa which have failed, using the same kind of seed as in the original planting.

Several weeks later—and the timing varies greatly depending on the weather, the lunar phase, and the work schedule of the campesino—the second replanting takes place. Farmers use small "three-month" maize varieties (Zap. *zhua huén*) when they replant more than three weeks after original planting. These varieties appeared in Talea approximately ten years ago and have an important advantage: they mature extremely quickly. Small maize can be planted successfully until early July in *tierra fría*. Some campesinos prefer a medium-sized, violet-colored "black" maize (Zap. *zhua gasj*) for replanting, which takes slightly longer than the "three-month" maize to mature.[11]

Weeding

Within six weeks after planting, assuming that the rains arrive "on time" (in May), the maize plants should be a meter tall or more. Rain also promotes weed growth, however, and the milpa soon requires "cleaning" or weeding with a short-handled hoe in order to keep its "force" from being "robbed." The only tool campesinos use for this task is a curved hoe handle (Sp. *garabato*, Zap. *bládzu*) fitted with a metal blade. Though this work is among the most physically taxing, it is not uncommon for women to perform the task alongside men, particularly young wives or daughters. It essentially involves "pulling" the hoe around the left side of the plant two or three times so that all vegetation surrounding that side is uprooted and drawn toward the base of the growing stalk; repeating the process for the right side; and moving back to repeat the first step in the area just below the plant. No space in the milpa goes untouched. Campesinos must keep from destroying growing squash vines, though the campesinos I worked with had an almost instinctive sense of their location.[12] The campesino not only removes growing weeds but also forms a mound of earth (perhaps 10–15 cm high) which binds the growing stalks together at the base and, they say, prevents winds from blowing them over. As in plowing, the campesino begins at the bottom of the milpa and gradually works uphill.

There is a troublesome pest known locally as the *belyá* (Zap.) or *gusano cogollero* (Sp.), probably the armyworm larva (*Spodoptera frugiperda*) or perhaps the corn borer (*Diatraea grandiosella*). It lives in the growing stalk and eats the plant's fresh leaf sprouts. A strong rain sometimes fills the funnel-like stalk with water and drowns the pest, but most campesinos do not wait for the weather to do this task. During weeding, they remove and destroy as many as possible. Moles can be a problem as well. If they cause excessive damage, campesinos sometimes douse the openings to their tunnels with gasoline or Fulvidol (Bayer's Mexican brand of pesticide). Most, however, avoid using pesticides, partly because of the restrictive cost, but also because of their preference for "clean," high-quality foods.

Though weeding is not as complicated as other tasks, it can be nerve-wracking for novices. The work takes place during the hottest, driest part of the year, and the sun beats down relentlessly during the final days of the dry season. Fleas are also particularly bad at this time; one may wake up in the morning with dozens of bites. While working in rocky areas, it is easy to scrape one's knuckles raw on the jagged stones while hoeing, and it is not unusual to find black widow spiders (*Latrodectus mactans*)—some with swollen bodies as large as olives—hiding beneath the rocks. In a single day a companion and I encountered twelve black widows. One is also more likely to encounter snakes (sometimes boldly stretched out in the middle of a milpa) during May and June than at any other time of the year.

Weeding progresses slowly. The May 15 fiesta of San Isidro is a convenient marker for starting work in *tierra fría*. A lone campesino who begins weeding after the fiesta is likely to finish before the end of the month—a total of nearly two weeks for one hectare. Typically, a second weeding is carried out in late July or August, depending on the amount of rain that has fallen. Campesinos make an even taller mound of earth (approximately 20 cm high) around the bases of the plants since they are much taller and subject to wind damage.

Applying Fertilizer

When I commented on the intense physical demands of his work, an informant told me that conditions were much worse a generation ago:

My father, may he rest in peace, had to plant sixteen *almudes* [approximately four hectares] to keep his wife and four children from going hungry. Land was scarce then; he didn't have any land of his own, so he worked *en compañía*—he had to give half the crop to the landowner. Plus there was no fertilizer so it took eight *almudes*, not four, to keep us fed. There were some years when we just didn't have enough to eat.

Many attribute the improvement in yields to chemical fertilizers, introduced in the late 1960s. In 1966 or 1967 guano arrived packed in large paper bags. Within several years this was replaced by commercial factory-produced bulk fertilizers similar to those used today. According to Tyrtania (1992:96), this process accelerated under government programs promoted in Talea, Yaeé, and Tanetze. For the time being, it appears that fertilizer—the only Green Revolution artifact that has made a significant impact in the Sierra—has become indispensable.

The most critical question in fertilizing a milpa is timing. Many describe fertilizer applications as a "charge" of "force" or "energy" that must be injected at critical moments determined by weather patterns. The charge is good for about three months, according to a campesino with a keen eye for observation:

This year it rained on time [mid-May], so the ground has plenty of humidity. The maize will take off by itself, without any "charge," and the roots will extend slowly and broadly. Fertilizer can be added later, once the roots have had time to grow, so that the plant "carries" [that is, develops full, long cobs with large kernels]. Putting fertilizer in before the rains will cause problems: it gives a plant that looks healthy and tall from the outside, but with no roots. It has no foundation. The first heavy rain or strong wind that comes along will blow the stalk over because it is too tall and has tiny roots. One must be very careful.

Most use fertilizers sparingly: the rule of thumb is one heaping teaspoon per planting. The work is best done by two people, one who perforates the ground with a stick at a distance of about a *cuarto* (nine inches) from the stalks (on the uphill side) and the other who follows with a bucket and teaspoon, depositing the fertilizer. Within three or four days a team of two can fertilize one hectare.

According to campesinos, a single 60-kg bag of fertilizer is good for an *almud* of planted maize. This is equivalent to 240 kg per hectare, which compares favorably to North American and West European farms, where dosages of more than 400 or even 500 kg per hectare are common (Goodman and Redclift 1991:211). When asked why they apply it so sparingly, campesinos note that fertilizer is too expensive to waste. During the 1996 planting season, a 60 kg bag cost N$110—nearly four days' wages. (For many, it is the biggest single expense of the year.) But they also note that too much fertilizer "burns" the earth and can ruin a milpa. "I've seen a few milpas that were so heavily fertilized that white spots were left in the earth, like salt," reported a farmer.

It is common to hear Taleans describe the technology in ambivalent terms. "The soil here has become accustomed to the fertilizer," some say, "and now earth doesn't give [food] without it." This assessment is echoed in other Rincón villages. In Yagavila people claim that once used "it poisons the soil," and from that moment forward harvests fail without it (Tyrtania 1992:193). As of the early 1980s, the Yagavilans had not adopted fertilizer; informants from the village of Yovego (approximately fifteen hours' walk from Talea, in the Rincón Bajo) reported in 1997 that it had never been used in their village. Some Taleans claim that certain weeds were unknown to them before the introduction of chemical fertilizer: a prolific weed called *yajkúza* (Zap.; species unidentified) that oxen refuse to eat and a long, stringy grass (unidentified) that grows in sugarcane fields are among the most troublesome.

Fertilizers have had other effects as well. "Counterseasonal" planting (Sp. *contratiempo*) was a rare phenomenon until the advent of fertilizers, according to informants.[13] This resulted in a rapid reduction in land pressure and later in the virtual abandonment of some farms lying far from the village (near the border with Villa Alta).

When speaking in Rincón Zapotec, Taleans use the term *beb* (garbage) to describe fertilizer, which implies that organic refuse was once used for that purpose. A number of older informants verified this, stating that quantities of coffee pulp, ash, and other organic materials were previously used to "feed" the milpa. Campesinos have incorporated fertilizers into local classificatory schemes based on hot/cold concepts. They describe fertilizers as hot and explain that this is why they are effective. This is also one reason why fertilizer is used sparingly—to keep from "burning" the plant.[14] It is kept away from work animals, for they are attracted to its salty taste but can die of "heat" if they ingest it.

In the mid-1990s a government program, PROCAMPO, provided participating campesinos with direct cash subsidies, as mentioned earlier, and a 1997 program called Alliance of the Countryside (Alianza del Campo) provided fertilizer at reduced cost. Though some Taleans joined the program, others refused:

Why am I going to get into programs? My father always avoided them, and I understand why. They just take away time. One must attend meetings before they give any money, and they say that this year the fertilizer isn't going to arrive until June 9—what good is the fertilizer then? I need it much earlier! Better to be free and not depend on somebody else to deliver you things.

Recently some regional organizations have attempted to promote the use of organic composts for coffee and maize cultivation. With technical assistance from agronomists and biologists, they are trying to convince growers to adopt the methods in order to produce coffee that can be organically certified (see Chapter 7). Organic compost bins are labor-intensive, however, and many claim that they do not have the time to construct them. It may well be that more farmers will eventually incorporate organic composting, though, particularly if organic coffee prices continue rising.

There is evidence that the effects of chemical fertilizers are maximized when they are applied to intercropped fields. Specifically, returns tend to increase significantly when chemical fertilizers are used in conjunction with intercropping techniques (Richards 1985:71-72). We shall explore this in more detail shortly.

Removing Maize Spikes

Maturing maize plants in *tierra caliente* form spikes during the months of July and August, which are cut off with a machete shortly afterward. Most campesinos use these to feed their oxen and beasts of burden for several days, though care must be taken because spikes are "cold" and may upset some animals, particularly younger ones. Removing spikes apparently promotes the growth of large, fat cobs (Lipp 1991:15).

Collecting Fallen Maize

Severe weather often knocks down a significant portion of a milpa. In order to minimize losses, campesinos may make one, two, or three trips to collect fallen stalks beginning in September. Even if no maize is damaged, it is customary to pick baskets of sweet corn for household consumption, as it is considered a treat. Campesinos may make more trips, depending upon the quantity of maize that has been damaged. They carry out this work with urgency, for fallen maize is quickly consumed by deer (*Odocoileus* sp.), harvest mice (*Reithrodontomys fulvescens*), rabbits (*Osyetolofagus cuniculus*), javelinas (*Tayassu tajacu*), coatis (*Nasua narica*), weasels (*Mustela frenata*), and other animals. Even if they do not reach the maize, morning dew and rain soon saturate the cobs, which begin to either rot or sprout.

Harvesting

By the end of the rainy season the milpas turn a deep golden color, and the drying stalks and leaves rustle loudly in the wind. The precise date of the maize harvest varies from year to year, but these two signals are the best indicators of when to begin *la pisca* (Sp.). Depending upon the altitude and the year's rainfall pattern, maize may be sufficiently toasted as early as November or as late as January. The most precise way of determining when to begin harvesting is to select a few samples from various parts of the milpa, shuck them, and test them. If individual maize kernels are easily indented with a thumbnail or the maize is difficult to shell, the milpa needs more time.

Once maize passes the test, harvesting can begin. Large straw baskets with tumplines tied across the handles are placed across the forehead and hung over the shoulders. Campesinos walk carefully between the rows of maize, taking two rows at a time, moving horizontally across the field starting at the top. Each spot has from one to four stalks, though typically two or three of the four planted seeds successfully bear fruit and some stalks have two ears. If we assume that rows and plants are evenly spaced at one *vara*, a hectare of land in Talea produces approximately 37,500 plants, well within averages calculated for southern Mexico and Central America (Wilken 1987:246-247).

The technique is simple: the campesinos take the stalk in one hand (bending it down and breaking it if it is too tall) and the ear in the other. They twist it until it breaks from the stalk and toss it over their shoulder into the basket.[15] (For this reason the task is called *quebrando mazorca* in Spanish: "breaking the ear.") Once the basket is full, they take it to the ranch house, dump its contents into a pile, and repeat the entire process.

The task is more difficult than it sounds. The basket grows heavy after a few ears have been harvested, and by the time it is half full the work becomes a delicate balancing act. Transporting a full basket to the ranch house is a skill in itself (especially if it lies uphill), for the weeds—which grow up to two meters tall—are covered with dew in the mornings. Campesinos routinely fall, and by the time a novice reaches the ranch house his or her neck muscles are throbbing and legs are rubbery. At times, no sooner is an ear detached from the stalk than a squadron of angry black ants scrambles up the campesino's arm. Wasps make their nests in other ears, particularly in large dry ones. Needless to say, after a few encounters young campesinos learn to harvest deftly and with great care.

Occasionally one comes across a caramel-colored grasshopper (unidentified species; perhaps *Sphenarium histrio*), known as *yajzách* in Rincón Zapotec. This is considered a delicacy, though children are sometimes squeamish about eating them. Farmers wrap the inch-long insect inside a bit of husk, thus keeping it fresh until roasting time. Then they place the bundle at the edge of the coals and turn it over a few times to ensure even cooking. They remove the wrapper and detach the long hind legs of the grasshopper before eating it. The insect tastes much like a roasted peanut. Grasshoppers are consumed across much of southern Mexico, though most of the *chapulines* (Sp.) sold in Oaxaca City are seasoned with chili powder and lime juice and toasted on a griddle rather than roasted.

Sometimes hailstorms strike during the harvest months. Campesinos consider this to be a good omen—a sign that the following year's harvest will be abundant.

Harvesting is often done in the company of a neighboring campesino using *gozona*. It proceeds relatively slowly; a few campesinos might take four days or more to cover one hectare.

Husking and Separating Ears

The inconveniences of the harvest are made up for by the massive pile of freshly picked ears. Each night before dinner campesinos shuck maize. They continue after dinner for perhaps two or three more hours, talking about oxen or "the old days" or experiences in Los Angeles or Mexico City or simply exchanging jokes. Seated on wooden stools, they take the ear at its widest part (near the base), pierce it near the top (where it is all husk and no maize) with a sharpened metal pin, and pull the tool away from the ear, splitting the husk in two, rather like a banana that has just been opened. One half is taken in each hand; by forcefully pulling apart one's fists the ear is freed of its wrapping. The campesinos throw the husk into a pile (far from the campfire, for if a coal lights the husks it could quickly set fire to a thatched-roof ranch house), allowing the maize's fine reddish "beard" (tassel) to drop to the ground along with the fruitless tip of the ear. They then toss the ear into the storage bin.

The pin is an ingenious tool, elegantly simple and perfectly suited for husking. A generation ago, they were fashioned from the sharpened thigh bones of turkeys. Today these are rare: it is more common to use a thick grade of copper wire. The pin, which is about one *huíni* (Zap.) long (approximately six inches), is perforated or twisted into a loop near the top like a sewing needle, so that it may be attached to the wrist. Campesinos also use them for "cleaning" the ears (picking out rotten kernels).

Sometimes unusual ears appear. *Zhua shunásh* (Zap., "Maize of the Virgin") are ears with an uncommon mutation characterized by a row of neatly aligned kernels which disappears halfway up the cob. They bear a remarkable resemblance to the image of the Virgen de Guadalupe, Mexico's patron saint, whose apparition was seen by a campesino named Juan Diego in 1531. *Zhua shunásh* are said to bring good luck to campesino families until the next harvest and are placed at the family's altar for the duration of the year.

Another ear has a startling appearance, for it is covered with a dark splotchy growth that gives it a bizarre texture—almost furry. (It looked more like an animal than a plant to me the first time I saw it.) This is especially prized because the growth—a fungus—is considered a delicacy throughout much of the country. Many Taleans prepare it in a kind of pie (Sp. *empanada*). Later I learned that the fungus is called corn smut (*Ustilago maydis*) in the United States and is considered a blight by North American farmers. Lamentably, these ears are typically destroyed in the United States,[16] even though they are low in fat and cholesterol and high in carbohydrates, protein, and fiber (Wiegner 1995).

Campesinos either toss the rest of the ears into a large bin assembled especially for the harvest (if they are in good condition) or put them in bags (if they are substandard: that is, if they are mildewed or have begun sprouting as a result of waterlogging or are too thin and wrinkled for human consumption). The bin is always located inside the ranch house (sometimes covering most of the floor space) and is made out of bamboo or, less frequently, planks of wood. Farmers line its floor with several layers of plastic sheets, torn fertilizer bags, or old straw mats so that the earth's humidity does not seep into the maize. Some sprinkle the floor of the bin with powdered lime to keep field mice and insects away. Unfortunately, others use Fulvidol (Bayer's Mexican brand of pesticide) for the same purpose.

At bedtime campesinos often spread the husks out over the sleeping area, cover them with *petates* (straw mats), and use them as a mattress. The mattress is relished for its "heat" (since *totomozle* [corn husk] is "hot"). In the morning the campesinos select the best husks to take back to the village, where they are used as tamale wrappers throughout the year. They take the rest outside and pack them into wooden molds which they use to create tightly bound bales. Campesinos feed this to both beasts of burden and oxen during the dry season, when fresh pasture is scarce.

Cleaning, Separating, and Threshing

The next step involves cleaning and separating the maize for threshing. After campesinos pick the rotten or infested kernels out of each ear, they separate them into groups: (1) *Yhz níni* (Zap.), seed maize, consists of the biggest, fullest ears with the largest kernels—none of which should be blemished or infested in any way. (2) *Yhz dxí'a* (Zap.), good ears that are eaten rather than kept for seed, are further separated according to variety; in our case this included *zhua chiguích* (Zap.), large white maize, and *zhua cuídi* (Zap.), small (three-month) maize. (3) Thin "second-class" ears, *yhz las* (Zap.), which have wrinkled kernels, are either fed to beasts of burden or else mixed with "first-class" maize in small quantities for sale in the market. Many campesinos proudly note that they never eat "second-class" maize because "we campesinos only consume the best things: the best maize, beans, and coffee, clean food, none of it rotten or dirty." Finally, there is (4) a pile of maize ears unfit for human consumption, consisting of mildewed ears or *yhz núdzu* (Zap.), fed to pigs and chickens, and sprouting ears, fed exclusively to pigs. If there is good black maize or red maize in the mix, these may also be separated to avoid mixing with white and yellow varieties.

Campesinos clean the ears while sitting inside the bin (perched atop hundreds of cobs), so that the rotten kernels remain within the bin or else outside of it on *petates*. At the end of each day farmers collect the rotten kernels, store them in bags, and sun them on old *petates* the following day to keep them from mildewing further and to reduce their weight. The campesinos note that nothing goes to waste—all of the maize is meticulously taken care of.

After the campesinos have cleaned and separated a sufficient number of ears, they are ready to thresh them. They prepare a special place for the task by draping plastic sheets around an area approximately two meters square (this prevents stray kernels from flying across the ranch house) and line the floor with corn stalks and then *petates*. They stuff a tightly woven net bag made of *ixtle* fibers (produced from the century plant) with "good ears" and tighten the bag by tying it in place with a drawstring. The campesinos then place the heavy net atop a *burro* (short four-legged wooden table with widely spaced slats) or some other object (for example, a fertilizer bag) and begin beating it with a short pole.[17]

The ingenuity of the net bag—produced in Yohueche, a Cajonos Zapotec village where many century plants are grown and *ixtle* items are still crafted—becomes apparent as soon as the beating begins. The kernels instantly pop from the ears and begin pouring out of the net onto the *petates* below. The bag's weave prevents cobs from slipping out but allows the tiny kernels to pass freely. Every few minutes the campesinos must rotate the bag so that cobs lying near the center get beaten. Within twenty or perhaps twenty-five minutes (depending on the size of the net bag) they are done threshing. A small bag holds as little as four *almudes* of grain, a large one perhaps twice as much or even more. The campesinos take the bag outside and remove the few remaining kernels by hand. They then pile the empty cobs outside the ranch house; over the course of the year they may serve as fuel, as corks for bottles, or as toilet paper.

Soon the campesinos are ankle-deep in maize, which must now be "lifted" (Sp. *levantado*) by scooping the kernels into plastic fertilizer bags with a gourd. Bits of cob broken in the threshing process are carefully removed.

Campesinos never mix maize varieties when threshing. Large white maize, small yellow three-month maize, black maize, and red maize are all done separately. One reason is aesthetic: the Taleans consider splotched tortillas (for example, white tortillas with blue swirls or spots) to be ugly, though these will occasionally appear at social functions where gift maize is mixed in the same basket. There is also a less obvious reason. The small three-month maize has a tendency to get "pierced" or "stung" (Sp. *picado*) by weevils more quickly than the large varieties. "It grows faster, but it gets pierced faster, so we set that maize apart and it is the first to get consumed or sold," explained an informant.

Cleaning, separating, and threshing the maize is laborious, and it might take a campesino with a helper up to two weeks to process the harvest of a one-hectare field. When the last cobs are done, the campesinos carefully sweep the ranch house; each kernel of maize—including rotten maize— is meticulously collected and put into a bag for transport to the village. Campesinos who work far from Talea may hire a muleteer to do most of the work. The rate in January 1996 was N$50 to N$60 per *fanega* (approximately 20 percent of the maize's market value).

Theoretically, the harvest could be delayed for weeks without much damage to the milpa. Once processed and stored in the ranch house, it could be left even longer (though mice and other rodents would certainly take some). However, many campesinos impose a deadline corresponding to the beginning of the Dulce Nombre celebration (third Sunday in January) to get the maize back to the village. In part this is because maize is needed during those days to feed visiting relatives and pilgrims requesting food and lodging. In addition, the Taleans have strained relations with the neighboring village of Tabaa, and many fear that Tabaa's villagers—who, as mentioned earlier, are said to have burned the granaries of villagers from nearby Yojobi in the mid-1950s—may do the same to Talea's maize stocks.

Sun-Drying

Once the maize is in the home, women typically assume responsibility for it. A critical task is "sunning" (Sp. *soleando*) the kernels, that is, laying them out over *petates* placed on a patio so that they dry completely. Weevils are likely to "pierce" humid maize more quickly than dry maize; therefore, sunning is often done in the days immediately following its arrival. It is repeated throughout the year as needed; women must constantly monitor the maize by touching it to determine whether it is too "cold" (humid) and in need of sunning. As soon as sunlight reaches the patio in the mornings, the maize is laid out; it remains there all day, until just before the evening shade reaches the kernels. Women lift the maize from the *petates* if rain threatens. They usually sun-dry a batch of maize for two or three successive days, until it "heats." This technique is effective; within minutes hundreds of weevils can be seen crawling away.

It is common knowledge that maize keeps longer in traditional homes with dirt floors, adobe walls, and tile roofs than in modern homes with cement floors and walls and corrugated sheet metal roofs. This was explained

Figure 5.3. A woman and her daughter spreading maize for drying. This is the most important step in maize preservation. *(Photo by Gabriela Zamorano)*

to me in the following way: "Concrete and *lamina* [sheet metal] homes heat up too much, and this is why the maize gets 'pierced' [Sp. *se pica*] so quickly there, especially the three-month maize. Adobe stays cool, and the maize is more secure where it is cool."

Converting Maize to Food
Making Tortillas

In Talea tortilla making is a time-consuming activity that is almost always done by women. It is divided into two stages: "grinding" or "milling" (Sp. *moliendo*) and "throwing" or "tossing" tortillas (Sp. *echando tortillas*). The first stage begins when a campesina measures maize into a metal bucket containing a lime solution. She boils the maize, allows it to cool, and then rinses it in a tub perforated with drainage holes. This process loosens the tough hull of the kernels.[18] She then takes the product, called *nixtamal* (Sp. from Nahuatl), to one of the village's four electric mills, where it is rapidly ground into dough. Before the first mill was introduced approximately forty years ago, women "broke" the maize kernels with hand grinders, then ground them into a fine meal on a grinding stone (Sp. *metate* [from Nahuatl] or Zap. *guich*).

The campesina forms tortillas on a metate, overlaying the surface with a circular piece of banana leaf or, more commonly, a sheet of plastic. She places a ball of dough slightly larger than her fist in the center and, with her fingers and palms, "stretches" the dough evenly outward until it reaches the edges of the circle. It is important that the tortilla be of uniform thickness and that the dough not get too gummy. A bowl of water is kept to one side of the grinding stone to keep the dough from drying out. Then she carefully lifts the tortilla (which measures nearly eighteen inches across), along with the plastic, with both hands and flips it onto a clay griddle (Sp. *comal* or Zap. *dxil*). She peels the plastic from the top surface of the tortilla, which is turned over once the first side is slightly toasted. An *almud* of maize yields twenty-four tortillas, each weighing approximately 0.17 kg.

On average, men consume five to six tortillas daily while women consume slightly less, approximately four to five. A ten-year-old child might consume half as much and a child younger than five one-fourth. Approximately 75 percent of the daily caloric intake of Talea's campesinos is derived from maize, which matches data recorded for other parts of Mexico, Central America, and the Northern Sierra (Ghindelli 1971; Williams 1973; Lipp 1991). Though total consumption may sound high (for adults, approximately 3,500 to 4,000 calories a day and even more during fiestas), it should be borne in mind that the villagers burn much of this energy during daily work routines. Simply getting to the farm may require a two-hour hike with steep uphill grades. Daily chores and farming tasks are also physically taxing.

Many Taleans are particular about their tortillas and have individual preferences. In general, "soft tortillas" (Sp. *tortillas blandas*)—which have just come off the griddle for the first time—are preferred. Others prefer "recooked tortillas" (Sp. *tortillas recocidas*). All the people I spoke with were disgusted by the taste of tortillas that had "sucked up" excessive amounts of lime. Although some children do not like the blue tortillas made from "black" maize, adults often claim that they are slightly "greasier" or

Figure 5.4. Electric *molinos* (mills) are used for grinding boiled maize into meal. The first *molinos* were introduced in Talea more than forty years ago, replacing hand grinders and grinding stones. Many say that *molinos* have significantly reduced the labor time and effort needed to make tortillas. *(Photo by Gabriela Zamorano)*

Figure 5.5. Making tortillas in a Talean kitchen. The tortillas are formed on a *metate* (grinding stone) and prepared on a clay griddle. Most households continue to use firewood for cooking tortillas. In the background bundles of *panela* are stored on a shelf. *(Photo by Laura Nader)*

"heavier" than tortillas made from the lighter maizes—and tastier. I never saw black maize or red maize used in tamales, *memelas* (toasted maize cakes), or *atoles* (Sp. from Nahuatl, maize-based drinks).

There is a clear preference for "creole" or locally produced maize (Sp. *maíz criollo*), rather than maize from the Valley of Oaxaca or from CONASUPO outlets. Maize from the valley is considered too "light" and, according to many informants, includes kernels that are still green. Furthermore, it is generally agreed that maize from outside the region is "dirtier" and contains thin, underdeveloped maize and rotten kernels:

We like to know where our maize comes from! We plant it, weed it, harvest it, we know that dogs don't urinate on it and that it hasn't been irrigated with sewage [*aguas negras*]. Who knows where that "outside" maize comes from, where it's been? We can eat happily, knowing that we're eating the best maize.

Not surprisingly, locally grown maize costs 20–25 percent more than CONASUPO maize, and it is almost always in short supply. A relatively well-to-do Talean, an electrician, told me that he was gladly willing to pay the extra money because "the campesinos really suffer to grow the crop—how can one deny them?" Another noted that his wife much preferred buying local maize because "it is embarrassing for someone who has money to go into the CONASUPO store . . . people talk, and will say that you're *codo* [Sp., miserly]." A young Talean man who worked for a time at a CONASUPO warehouse in Oaxaca City vowed never to purchase subsidized maize after his work experience: "I know what happens to it . . . They have the whole floor of the warehouse covered with maize, and the men who work there just piss in the corners, right on the maize, and spit on it, and blow their snot on it. Why should they care? They don't have to eat it!" Others claim that they have seen decomposing mice in recently arrived maize shipments from Oaxaca. Once again, the Taleans act not as "economic men" and women, but as consumers who are acutely conscious of the quality of their food—not unlike Japanese who consider imported rice to be substandard (Ohnuki-Tierney 1993).

In Talea maize is almost never wasted. When tortillas become stale or dry, they are broken into chunks and soaked in black bean sauce to make *chilaquiles* (Sp.), used in a soup, or put in a bowl of coffee. If they begin to mold, they are placed directly on hot coals, which turn spores into specks of dust that are then easily brushed away. Extremely old tortillas or those that have been nibbled by mice or other animals are fed to pigs or dogs or else broken up, soaked in water, and fed to chickens or turkeys.

Caring for chickens, turkeys, and pigs is an important responsibility undertaken almost exclusively by women in Talea. Most avoid giving their chickens and turkeys store-bought feed because it is greasy and, they say, gives the meat an offensive taste and texture. According to several informants, such chickens are so fatty that "they leave a layer of grease floating in the soup . . . the meat is even slippery." (Many who have been to the city note that the meat served there is repulsive because of its greasiness.) Instead they mix a small amount of feed with *quebrajada* (Sp.), cracked maize made from slightly mildewed or substandard kernels that have been hand ground. Chickens that come from the city are called *engordas* (Sp.) or "fattening" chickens. The local creole varieties, called *criollos* (Sp.) or *bhráj ba* (Zap.), by contrast, command a significantly higher price because their meat is lean and of superior quality.

Older Taleans remember a number of instances in years past in which the maize crop failed due to bad weather or plagues. To get the most out of their limited grain stocks, they mixed maize dough with dough prepared from a special variety of banana, *yhla gasj* (Zap.) or *plátano guineo* (Sp.). To prepare these tortillas, the bananas are ground into a fine mush and leached

Table 5.4. Foods and Beverages Prepared from Maize in Talea

yht za'a (Zap.)	tortillas made from sweet corn
lan za'a (Zap.)	black beans with sweet corn, flavored with *poleo* (pennyroyal)
yht yhla (Zap.)	tortillas made from a maize dough and banana dough mix
yht guza'a (Zap.)	tamales made from sweet corn
chilaquiles (Sp.)	tortilla chunks soaked in bean or tomato sauce
atole (Sp.)	maize gruel drink flavored with chocolate, sweet corn, or *panela*
memelas (Sp.)	thick patties of toasted maize laced with large bean (*zatópe*)
yht gu (Zap.)	tamales filled with *chipil* herb, beans, or *mole*

by hand, so that the water drains away and the dough thickens. Then it is mixed with maize dough and prepared in the same fashion as maize tortillas. Though the caloric content of *yht yhla* (Zap., "banana tortillas") is somewhat lower, in past years the food has served as a valuable stopgap during the months immediately preceding harvest, not to mention during the disastrous rains of the late 1960s and the insect plagues of the mid-1940s reported by older informants.

Many studies of Mexican and Central American farming and food systems stress the role of the maize-bean-squash complex or triumvirate,[19] but in many regions the nutritional role of wild greens or *quelites* is absolutely vital (Messer 1972). In addition, they often form an essential part of ecologically sound farming systems (Bye 1981). The Rincón is certainly one place where such plants are collected and consumed, whether in cultivated areas, in forests, or along the edge of streams and rivers. They provide riboflavin (vitamin B2), niacin (vitamin B3), and ascorbic acid (vitamin C)—nutrients which are deficient in a maize diet—and may be used as a source of vegetable protein in a diet which may be low in proteins. Members of campesino families often relish boiled greens seasoned with chile and salt, especially when the weather is hot; such meals are considered light and refreshing compared to the heavier dishes prepared from beans. A list of *quelites* collected in Talea is included in Appendix B.

Talean women are experts in converting maize into many nutritious, tasty foods (see Table 5.4). A Talean cookbook would undoubtedly make a valuable addition to dietary anthropology and is worth a complete study in itself. Appendix E contains two recipes given to me by Talean women to illustrate typical Rincón cooking.

Figure 5.6. Campesina cutting banana leaves. *Tamales de mole* (maize rolls filled with a rich sauce of chiles, chocolate, and spices) are wrapped in banana leaves before they are steamed. *(Photo by Gabriela Zamorano)*

The Kitchen Garden

In spite of the nutritional complementarity of maize, beans, and squash, Talean food would hardly be palatable without the addition of many spices and flavorings (Sp. *sabores*) grown by campesinas in kitchen gardens (see Appendix B).

Kitchen gardens include not only flavorings, but also medicinal plants and ornamental flowers. Among the most popular domesticated remedial herbs grown in Talea are *rosa de Castilla* (*Rosa chinensis*), *manzanilla* (*Matricaria chamomilla*), and *ruda* (*Ruta graveolens*). There are literally hundreds of kinds of flowers grown in village gardens, which are important for ritual purposes; they are offered to Catholic saints and the deceased. The most popular flowers include orchids, chrysanthemums, gardenias, roses, and lilies. Kitchen gardens thus serve three vital functions: they provide much of the flavor and nutritional value in the traditional diet of maize, beans, and squash; they are a source of medicine; and they provide gifts for supernatural beings. A full analysis of Oaxacan kitchen gardens would make a fascinating and important research project in plant biology, cultural anthropology, or nutritional science.

Kitchen gardens often look untidy because many different plants are grown in close proximity, but they are functional and tend to follow a logic that reveals a sophisticated knowledge of plant biology. As in intercropped fields, plants in kitchen gardens grow at various heights, have root systems of different depths, and mature at different times of the year. But what makes them perhaps more challenging than intercropped milpas is the extremely limited space available for most household gardens. As a result, it is not at all unusual to find a kitchen garden with perhaps forty or fifty different species of useful food, medicinal, and ornamental plants growing within a tiny plot of two square meters.

Finally, kitchen gardens serve an educational function not often recognized in the village. Most children are taught their first lessons in cultivation by mothers tending their gardens. When they are infants, still strapped to their mothers' backs, they observe how seeds are deposited in the earth. As toddlers, they take note of sprouting seeds and watch their mothers water and compost them. By the time they are three or four years old, many children have successfully cultivated their first seeds in kitchen gardens.

"Pushing the Tortilla": Sauces and Relishes

Utensils are rarely used in the Rincón; instead, the tortilla serves as both fork and spoon. Indeed, a meal is not considered a meal unless tortillas are part of it. Talea, like many places worldwide, is a "starch-centered culture" in which maize forms the core and a wide variety of "relishes"—known as *comidas* (Sp.) or *zhwh* (Zap.)—form a culinary periphery (Mintz 1985). This is strikingly apparent in the way meals are served in the village: a stack of tortillas is piled in the center of the table, and strips are torn off and eaten by individuals, who have their plates of food near the edges of the table. A clear symbolic and spatial division is thus made between tortillas, *yht* (Zap.), and relishes, *zhwh* (Zap.).

Relishes serve to "push the tortilla" (Sp., *empujar la tortilla*) into the body by making it more palatable. Beans, soups, vegetable dishes, chile peppers, stews, and sauces with such names as *amarillo, colorado,* and *mole* (all Sp.) add zest to meals in Talea. But maize is the critical element that provides satiation (Mintz 1985). In the words of an informant:

Here we are accustomed to eat more tortilla than *comida* — not like in the city, where they give you mostly *comida* and a tiny tortilla or two on the side. But too much *comida*, especially with lots of meat, can cloy [Sp. *hostigar*] the body. The tortilla is what fills, what leaves one satisfied. That's why if one is hungry and doesn't have *comida*, a tortilla with a little salt and chile will be enough to give strength.

The idea of satiation is further illustrated by the Rincón Zapotec invitation to eat, *gáguru yht* (let's eat tortillas). Some, particularly campesinos, note that bread, though tasty, puffs up or expands in the stomach; therefore it is unwise to eat too much of it before eating a meal. Others say similar things about soft drinks: that the gas fills the stomach and takes away one's appetite. Snacking or eating between meals is considered a "vice" by many campesinos. Snack foods, whether in the form of factory-produced *sabritas* (Sp., taken from the brand name of Mexico's potato- and corn-chip conglomerate) or locally made *pepitas* (Sp., roasted squash seeds), are seen as indulgent.

Though processed factory-made foods and drinks are probably consumed more now than at any other time in Talean history (particularly soft drinks, beer, milk, sardines, pickled jalapeños, and snack foods), most families continue to rely upon maize and many other locally grown plants and animals for the bulk of their diets (see Appendices B and C).

Maize, Local Markets, and Cash Income

A few Talean campesinos produce enough of a surplus to sell their maize in the weekly market. It commands a premium price, often 20 percent more than the price of CONASUPO maize. Campesinas are typically responsible for selling maize and other crops grown and collected by their husbands and fathers in the fields, including beans, squash, *panela*, coffee, avocados, chayotes, green beans, bananas, and many other kinds of fruit. Women also sell items that they produce, collect, or craft. These products include kitchen garden crops, wild greens and medicinal plants collected from the forest, bread, *posonque* (a frothy cacao-based beverage), poultry and poultry products, and pork. Often a woman specializes in the sale of only a few different products and in many cases eventually establishes a regular clientele.

Figure 5.7. Talea's weekly market is held every Monday. Women play a vital economic role as the principal buyers and sellers of agricultural produce. Although the market has gradually declined over the last forty years, it is still an important site for economic and social interaction between people from different Rincón villages.
(Photo by Gabriela Zamorano)

By selling these items, women make an essential monetary contribution —and sometimes the only monetary contribution—to household maintenance. With money earned at the Monday market, they purchase food that their families might need during the course of the week and buy school supplies, clothing, medicine, and other items for their children. They might also invest their money in livestock, especially chickens or turkeys. When coffee prices are low, the economic support provided by campesinas is especially strong; their relative contribution to the family's cash income may easily surpass that of the male household members.

A Word on Yields

How much maize can a campesino family expect to harvest? According to one source, maize production in Mexico averaged approximately 1,100 kg/ha in the early 1980s (Tyrtania 1992:156), but more recent estimates indicate a range of 1,270–2,180 kg/ha (Barry 1995:103).

In Talea campesinos typically harvest 3–5 *fanegas* per *almud* (1,210–2,016 kg/ha).[20] This often varies radically from year to year: in an exceptionally good season harvests as high as 6 *fanegas* per *almud* (2,419 kg/ha) may be realized, while in a bad year harvests may be as low as 2½ *fanegas* per *almud* (1,008 kg/ha). Although weather patterns create a great deal of uncertainty, a conscientious campesino can minimize damage by planting "on time" (so that root systems have sufficient time to develop), by collecting fallen maize regularly, and by weeding at least twice during the season. It is said that a good farmer can consistently produce at least 4 *fanegas* per *almud* (1,613 kg/ha), and my observations confirmed this during two consecutive harvests. This is approximately 20–25 percent of what a U.S. factory farmer—with the entire arsenal of chemical, mechanical, and biogenetic inputs available—can expect to harvest in an average year.

Upon closer inspection, "yield" and related categories such as "productivity" and "efficiency" may be ideologically tainted, like the category "Gross Domestic Product." All these terms provide some information, but they obscure a great deal more. My argument is essentially that campesino maize farming does not lend itself well to such a reductionist analysis because it is not a reductionist activity. We have seen that maize farming in Talea is a subsistence activity, not a commercial one; it is a strategy for maximizing the use of family labor, not for reducing labor costs through mechanization; it is carried out by neighbors who exchange work, not by wage workers; it is a way of caring for a willful living being, a genuine plant-person; it is a direct link to the deities; and nearly all of its fruits are eaten directly by farmers' families.

Talean campesinos have notions of U.S. corn farming, its gargantuan scale, and its high productivity because many of them have gone to the United States to work—some in the mid-1940s during the Bracero Program (a wartime emergency guest-worker arrangement), others much more recently. Some worked in the Midwest, shucking corn. Though they seemed to admire certain aspects of U.S. farming, many came to realize that the farmers were few in number and that most of the workers were simply hired hands. When I informed other campesinos about the high quantity of chemical fertilizers and pesticides commonly used on U.S. factory farms, they were in virtual disbelief: "How could the land keep producing food after being injected with so many 'hot' things?" they asked. Still others asked what kinds of crops the North Americans intercropped in their big fields. They were surprised to learn that intercropping hardly existed in the United States and certainly not on large corn farms. On several occasions I was asked how the *gringos* can grow such large ears of corn. I explained that they planted hybrid seeds that are hardy but sterile—and so had to be purchased every year. This seemed particularly outrageous to them, for seeds are, without exception, public domain in the Sierra: a conscientious campesino should never have to purchase them. (Other than hiring a plowman, the only other input that a Talean campesino purchases is chemical fertilizer.) The price of farm machinery is far beyond the economic reach of the farmers; once they thought it through, they realized that their land is much too steep or rocky and their milpas too irregularly shaped for tractors anyway. Finally, regarding the use of corn as animal feed in the United States, I am certain that most of the campesinos thought I was misinformed or perhaps mad when I told them that more than two-thirds of the U.S. harvest is used to feed livestock.

In short, "yield"—the volume of grain that a campesino family is able to coax from a given unit of soil—is only one consideration among many from the farmers' perspective. In this sense, Talean campesinos are not unlike others in Latin America:

> The returns from traditional agriculture cannot be measured only by productivity, the preferred standard in market-dominated farming systems. In addition to quantity, traditional farmers may feel that variety, quality (also locally defined), risk reduction, and even prestige justify labor expenditures although such investments may not have tangible returns. (Wilken 1987:8)

Indeed, the process of using such exclusive standards as "yield" or "productivity" seems to evoke an evolutionary scheme in which "inefficient producers" are doomed to what Michel-Rolph Trouillot (1991) calls "the savage slot" while ("civilized") "First World" farmers will invariably head up the list. The ideological use of "machines as the measure of men" (Adas 1989:404–405), combined with flawed notions about "Indians," has generally tended to describe a civilizational scale ex post facto:

> The myth that the Indians were hunters rather than agriculturalists, which persisted despite considerable evidence of their skill as farmers, and the vision of

America as a land of abundant resources that the Indians had not begun to tap because their technology was too primitive, buttressed the arguments of those who sought to justify the Indians' subjugation and displacement. The settlers' association of civility with human domination over nature and their view of the new continent as a sparsely populated wilderness led thinkers on both sides of the Atlantic to the conclusion that its Indian inhabitants were savages, much in need of assistance from the "industrious men" and "engines" that only Europe could provide . . . Technological development was increasingly equated with the rise from barbarism to civilization, and machines were viewed as key agents for the spread of this civilization in the New World wilderness. Prominent politicians, writers, and artists of the day caricatured the Indians as slothful, technologically poor, and unprogressive vestiges of savage societies that must either adopt the white man's ways or perish. The shortcomings of the Indians were set against the virtues of the expansive European pioneers whose hard work, discipline, thrift, foresight, and technological ingenuity were transforming the undeveloped wilderness into a land of unprecedented prosperity.

It is ironic that in the early years of the westward expansion in the United States farming by Europeans was not very different from that of the Native Americans. It was nonmechanized, done on small plots, and often used New World techniques. Indeed, as Alfred Crosby (1986) has shown, it was the relative immunological advantage of the Europeans—not their technology—which led to their demographic success in North America. It is probably more accurate to interpret Adas's "politicians, writers, and artists of the day" as development apologists.

Another assumption, still alive and well in the modern era, involves the *tabula rasa:* the assumed technological, cultural, and intellectual void or blank slate often thought to characterize "traditional" societies. We have already seen how inaccurate this assumption is with respect to maize, because this American invention promoted European imperial expansion (just as it had for the prehispanic Mesoamerican civilizations) and a rapid demographic explosion. The following section addresses the notion of the *tabula rasa* by analyzing some of the key elements in Rincón Zapotec maize production and consumption that cosmopolitan scientists have only recently begun to understand.

Recovering Knowledge: The Case of Talean Maize Farming

Anthropologists have historically played at least two important roles with respect to local knowledges. On the one hand, "salvage anthropologists" have diligently recorded volumes of data about quickly disappearing lifeways for the anthropological record; on the other, they have attempted to understand why "peasant" peoples and other "primitives" cling "conservatively" to quaint but unproductive practices and customs. Rarely has it occurred to them that local methods might be highly effective or translated into the terms of modern science, much less that "Other" societies might have solutions to technical problems that have vexed "us" for centuries. A quick review of techniques, knowledges, and issues related to maize farming can serve to illustrate some of the things that cosmopolitan scientists have begun to learn from local farmers.

Here my focus is on how certain techniques and knowledges present in one agricultural system (that of the Rincón Zapotec) have become validated by cosmopolitan scientists. They have generally described the local knowledges and techniques in cosmopolitan terms; that is, they have explained techniques developed in one society using the terminology and concepts of another. Instead of employing explanations of "hot/cold" or "food quality," for example, they use terms such as "nitrogen fixing," "land-equivalency ratio," and "dietary amino acid requirements." This section therefore might be seen as a comparative analysis of epistemologies, in which the concepts used by one set of experts (local farming specialists) are translated into those used by another (agronomists, geographers, etc.) through specific techniques or practices.

Intercropping

Taleans plant several kinds of beans and squash between their maize plants before, during, and after planting. They most frequently explain this in terms of food quality—specifically, they talk about the advantages associated with having a variety of foods available counterseasonally.

For some cosmopolitan scientists, intercropping is valuable for different reasons. They note that intercropped bean plants fix nitrogen in the soil and reduce damage caused by a pest known as the corn earworm (*Helicoverpa zea*), according to recent studies (e.g., Downs 1984). Squash has its own distinct advantages. Within weeks squash plants form a dense network of thick broad leaves which, if extended properly, can inhibit weed growth and help retain humidity by covering large patches of ground (Altieri 1987:214–216, 289–290). In addition, "chemicals washed from its leaves enter the soil to act as herbicides" (Ortiz de Montellano 1990:95). The naturally occurring chemicals, *cucurbitacins*, poison both insects and weeds.

In general, intercropping helps minimize soil erosion by binding together topsoil with extensive root systems. Furthermore, a maize, squash, and bean mix makes efficient use of sunlight, water, and soil nutrients because of the multilayered structure of the intercropped system—maize forms an upper canopy, beans an intermediate layer, and squash a broad layer of vegetation near the bottom. Each root system operates at a different depth, drawing nutrients from specific subsoil levels. Cosmopolitan scientists are just beginning to understand the complexity of such systems. And the variety of foods produced by a mixed field of maize, beans, and squash mix can provide all the essential amino acids for a healthy diet (Wilken 1987:66-69).

A particularly surprising phenomenon described by entomologist Miguel Altieri exposes the shortcomings of standard agronomic analyses that attempt to assess intercropped fields in the inappropriate terms of yield for monocropped farms. An alternative measure known as the LER (Land Equivalency Ratio) has been used to gauge the productivity of intercropped fields more accurately, since a monocrop yield analysis renders one or two or more crops irrelevant (Altieri 1987:113):

Total yields per hectare [in intercropped plots] are often higher than sole-crop yields, even when yields of individual components [i.e., maize and beans] are reduced . . . [Land Equivalency Ratio] expresses the monoculture land area required to produce the same amount as one hectare of polyculture, using the same plant population. If LER is greater than one, the polyculture overyields. Most corn/bean dicultures and corn/bean/squash tricultures studied are examples of overyielding in polycultures.

It should be added that individual components in intercropped fields do not always suffer decreased yields; indeed, other studies conclude that in some maize mixes yields of maize alone may *increase* as much as 40-50 percent (Downs 1984; Richards 1985).

Maize Mounds

The construction of maize mounds is practiced across much of Mexico, Central America, and parts of the southwest United States (though with notable variations in mound size, spacing, etc.) and is ancient enough to have been mentioned in literature from the Conquest period and in the codices. Apparently it has not been conclusively shown by cosmopolitan scientists that maize mounds are beneficial, though Talean campesinos claim that the primary advantage they offer is protection from wind.

Mound-building, the campesinos say, also gives them time to search individual plants for pests.

According to Wilken (1987:143), maize mounds may promote "tillering," improve soil drainage and aeration, decrease soil moisture losses from evaporation, and perhaps most obviously inhibit weed growth. Since uprooted weeds form a part of the base, where they eventually rot, the mounds also add organic fertilizer to the earth surrounding the plant.

Wilken is puzzled by the inability of many campesinos to explain why they build mounds. In particular, he points to a confusing phenomenon he witnessed in the region: many "traditional" farmers agreed with a number of "modern" critics who disparage the technique and claimed that it was probably not useful. Still, he notes, even the doubting campesinos built mounds, perhaps driven by "traditional wisdom, supported by centuries of practice and uncounted man-hours of effort" (Wilken 1987:144). Wilken (1987:144) cites Paul Weatherwax's "appealing but unsatisfying explanation":

Why does the Indian persistently cling to this way of doing things? And does it [mounding] have any merit? The Indian's answers to such questions are vague. It prevents the water from standing around the plants; or it supports the plants and prevents wind damage. But his clinching answer usually is that he gets more corn when he does the thing this way, and that this is the proper way to grow corn anyhow.

Two rounds of mound-building were practiced by most Talean maize growers, and I was never told that it was a useless activity. It is clear that this is one of the most strenuous tasks in the maize cycle (taking more than twenty days of labor per hectare). The campesinos' explanation seems relevant in Talea, for winds are certainly a problem in some years; indeed, in August and September 1996, even with the mounds, many Taleans lost a considerable amount of maize to wind. High winds seem to accelerate in the mountains' many folds, creating localized whirlwinds that can quickly knock down many plants.

Local Uses of "Modern" Technologies: The Case of Fertilizer

It would be naive and inaccurate to claim that "modern" technologies have not made an impact in Talea. Chemical fertilizers are by far the most important. Fertilization seems to be a "modern" version of an old technique in which *beb* ("trash" in Rincón Zapotec) was used to fertilize individual plants.

Even so, fertilizer application seems to have undergone changes in translation. Most significant is the quantity applied to a milpa; as we have seen, a cautious Talean farmer uses less than half of what factory farmers in northern Europe and the United States use per unit of land. Cost has a great deal to do with this, but so do ecological concerns which are explicitly stated by campesinos. When I asked why they did not add more fertilizer, many replied, "We don't want to 'burn' our lands." This would seem to represent a desire to maintain the viability of the milpa for future generations.

For agroecologists, limited chemical fertilizer applications on intercropped plots have an important advantage not mentioned by campesinos. In studies carried out by Nigerian researchers, it was found that fertilizers, when applied to fields cultivated using local techniques (i.e., intercropping), showed significantly higher returns than when applied to fields cultivated using "improved" methods (monocropping). The results offer an unexpected situation in which a "modern" input, chemical fertilizers, when blended with local techniques such as intercropping, offers superior results (Richards 1985:71–72):

[This case] shows a thought-provoking set of results. The best returns were for a "modern" input, applied on intercropped farms, but using local management procedures. It is precisely on evidence of this sort: the ability of West African smallholders to get best results from a combination of so-called "modern" and "traditional" techniques, that the case for the "populist" strategy rests... Science should be the servant not the master of this kind of inventiveness.

The Effects of Lunar Phases on Agriculture and Weather Patterns

Another set of knowledge used by Talean campesinos revolves around the effects of the moon on various agricultural activities. Similar knowledges exist in Asia and other parts of the New World, and it is difficult to ascertain to what extent they evolved independently. Lunar effects, in the terms of Talean campesinos, can be divided into three categories: accelerated growth effects related to the "growing phases" of the moon (first crescent to last quarter), wood-hardening effects related to the full moon, and effects on rain patterns.

Older campesinos in Talea think that maize planting should begin during the phases of the *luna cresciente* (Sp., "crescent" or "growing moon") whenever possible. When I asked why, I usually received some variant of the following: "That is when the moon grows, so plants will grow too." In fact, many growing things are said to grow faster under the influence of the waxing moon (including plants and even human hair); thus crops, particularly maize, are planted during this lunar phase.

Trees felled for constructing homes or plows and branches for fashioning hand tools are cut during the full moon phase. According to the campesinos, this hardens the wood by preventing the implements or boards from becoming infested with termites. Saplings (for example, coffee saplings) are said to be more successfully transplanted during this phase as well. Many people were not able to give me a reason why it worked; they simply claimed that it was effective. Others noted that it was only logical; wood cut during this phase will be tough or firm (Sp. *maciza*) because the moon is tough, solid, and firm, which is why it is called the *luna maciza* (literally "solid moon" or "firm moon" in Spanish).

A number of rules of thumb exist concerning lunar phases and rain. The first is that, in general, it is likely to rain during the new moon phase. If rains do occur during this phase, they are likely to continue into the "crescent" or "growing" phases of the moon and possibly (but not as likely) into the full moon phase. It is always very unlikely to rain during the waning lunar phases (Sp. *luna menguante*).

Like other anthropologists (e.g., Foster 1960:115), I initially thought these were superstitions, but now it is evident that the question is still open to debate. As early as 1923, a study published in *Nature* concluded that maize and other seeds germinated under polarized moonlight exhibited intensified hydrolysis and an increased yield as opposed to seeds exposed to ordinary light or darkness (Semmens 1923). Although this study has since been discredited, recent work in the field of photobiology has demonstrated a relationship between moonlight patterns and the germination of certain seeds (Hartmann and Nezadal 1990; Ensminger 1996). The new reports imply that planting crops during certain lunar phases may effectively inhibit weed growth.

The possibility of lunar effects on meteorological patterns is also an open question. More than a century ago the journal *Science* reported that scientists found that, although the influence of the moon on weather is slight, there is a tendency for rain to decrease during the full moon phase and to increase following the new moon (Merriam and Hazen 1892). These old reports have been discredited, but researchers have recently discovered a correlation between lunar cycles and temperature patterns on earth (Shaffer, Cerveny, and Balling 1997). In the 1960s scientists discovered correlations "between lunar phases and the frequency of thunderstorms, daily changes in atmospheric pressure, hurricanes (which are more common during a full moon), cloudiness and the size of ice nuclei in the atmosphere" (quoted in Szpir 1996:59). The moon's effects on the earth's climate also appear to include precipitation and atmospheric circulation patterns (Balling and Cerveny 1995). Although the mechanisms are not yet entirely clear, future research may help us understand the importance that many Latin American campesinos (and other cultivators) place on lunar phases.[21]

Maize and Nutrition: Exploding the Myth of the "Inefficient Producer"

Dietary anthropologists and nutritional scientists have analyzed the diets of ancient Mesoamerican peoples. In a fascinating reconstruction of the Aztec diet on the eve of the Spanish Conquest, Bernard Ortiz de Montellano (1990) notes that the Aztec exploited a wide variety of foods which, even in small amounts, would have remedied the shortcomings of a maize-based diet. Though they derived more than three-quarters of their calories from maize, they supplemented it with beans, squash, tomatoes, chiles, amaranth (*Amaranthus* sp.), *tecuilatl* (protein- and vitamin-rich water algae processed into cakes and eaten like cheese; scientific name *Spirulina geitlerii*), fermented *pulque* from the century plant (*Agave* sp.), mesquite pods (*Prosopsis* sp.), and other foods. Animal proteins were an important part of the diet as well: the Aztec "ate practically every living thing that walked, swam, flew, or crawled," including armadillos, pocket gophers, weasels, rattlesnakes, mice, iguanas, domesticated turkeys and dogs, fish, frogs, tadpoles, fish eggs, water beetles, grasshoppers, ants, worms, and an aquatic salamander known as the *axolotl* (*Ambystoma* sp.) (Ortiz de Montellano 1990:115). Significantly, small animal species can contribute substantially to human nutrition, and their meat does not differ significantly from beef or pork in dietary terms except that the fat content is much lower. This is because smaller animals eat less, grow faster, and more efficiently convert plants into protein. Smaller game animals like armadillos, iguanas, and grasshoppers use the ecosystem more efficiently than do large grazing animals. Ortiz de Montellano (1990:119) concludes that

the Aztecs were neither malnourished nor suffering from protein or vitamin deficiencies. In fact, they were probably much better fed than the modern Mexican population, which in 1973 had an average diet of 1787 calories per capita per day [compared to an Aztec average of approximately 2,200 calories] and some 45 percent of which never eats meat.

Though it is not identical to the pre-Conquest diet of the Aztec, we have seen that the maize-based diet of the Rincón Zapotec includes a wide variety of vegetable and animal foods that are consumed efficiently. To put it differently, the Talean campesino families—who by the standards of many economists are inefficient producers of maize—are remarkably efficient food consumers.

This point deserves some reflection. The separation of agricultural production from food preservation, processing, and consumption serves to tilt the analytical scales in favor of large-scale factory farming by limiting the range of scientific inquiry. Looking at food in narrow terms of yield or productivity glosses over the fact that food is, essentially, consumed— or, in other words, is "both a means and an end of production" (Wilken 1987:17). If we looked at the entire food cycle as a process, we might come to the conclusion that Zapotec campesinos use maize more efficiently than, say, North Americans, because the former do not waste maize, they consume 95–98 percent directly rather than feeding most of it to livestock, and so forth. In short, some methods of evaluating agricultural systems might, by their very design, valorize factory farming over local farming.

But the Rincón Zapotec have a different set of criteria for explaining why they eat what they eat. One is based on the notion of food quality.

The Question of Food Quality and Taste

Many campesinos frequently stated: *Somos pobres pero delicados* (We're poor but picky [literally "delicate"]). What they mean by this is that the quality of maize they produce is just as important as quantity (if not more so). Locally produced maize is held as the standard by which "outside" maize (for example, that sold at the CONASUPO stores) is judged, and the latter is deemed inferior; indeed, they argue, why else would the maize get sold if it were not "second-class" maize laced with green corn, bits of empty cobs, and perhaps even rotten kernels? The issue of "knowing where it comes from" is important in this context; long food chains, to many campesinos, mean more intermediaries, more places where "trash" can be added to the maize to increase profits, and a greater likelihood that maize will be contaminated with unclean substances (human waste in particular).

Indeed, one group of campesinos extended the debate to critique urban consumers in general by noting that "they eat like pigs" because they are not picky about what they eat, nor about when they eat it. Those who had been to the city for brief periods often told of falling ill quickly with intestinal problems and blamed the city food. They noted that people in the city eat food prepared by strangers, food that comes from "God knows where," and "cold" foods like lemonade, beer, turkey, and pork at night, a practice that is considered to be physically risky because it can upset the body's humoral balance. They were also intrigued by people who apparently had no qualms about eating at all hours (including late at night, like the *gringos* in the central plaza of Oaxaca) or skipping meals. One concisely summarized this perspective:

Here we have clean water, clean air, clean food . . . No need to boil our water—even though we may have to collect it from a watering hole, we know it's clean. Some city folks don't like the idea of drinking water from the same watering hole that an ox or a mule drinks from, but the animals are cleaner than we are. All they eat is greens. Apart from that, we at least know that we grow our humble food ourselves, we know it's "first class."

It would be hard to imagine these campesinos accepting notions of "improvements" in "yields" or "standards of living" if that meant compromising the quality of the most fundamental human necessities: air, water, and food. Their view of the savagery of urban life became a powerful, if unexpected, critique of "modernization."

I have reviewed some of the details concerning maize production and consumption in Talea and what they mean with respect to factory farming and the curiosity of some cosmopolitan scientists. A number of interesting points are worth noting: certain Talean farming techniques appear to be environmentally benign compared with factory farming; Talean farmers who practice supposedly "inefficient" production techniques are highly efficient consumers of the crop, owing to a pattern that directs nearly all of the maize toward human consumption; and, for Taleans, "knowing where one's food comes from" is generally more important than maximizing its quantity. From the point of view of Talean farmers, maize farming is a logical activity that is well suited to their subsistence needs. In other words, it can help to ensure the maintenance of their households and future generations.

CHAPTER 6

SWEETNESS AND RECIPROCITY

Sugarcane Work

The history of sugarcane in Talea is remarkable because over the course of nearly 500 years the crop has been integrated almost seamlessly into the diets and farming routines of villagers. It is a staple food taken with literally every meal, since it is used to sweeten coffee.

Most farmers agree that sugarcane work is significantly more difficult and dangerous than that associated with maize, beans, or coffee. The cane leaves' serrated edges have reportedly cut the faces and even eyeballs of careless farmers; poisonous snakes easily hide in the fine, long grass which grows between the plants; and, on rare occasions, workers' hands and arms may get mangled in the steel rollers of the sugar mill. The end product, *panela* (Sp., unrefined brown sugar), is the result of many days of physical labor, but well worth it, according to many campesinos. When wrapped and stored properly, it can keep for years and in emergencies can be sold at the weekly market. Thus, a stockpile of *panela* serves as a nest egg for families who might otherwise have difficulty procuring cash, especially during periods of low coffee prices.

Talean sugarcane is characterized by two ironies. The first is notable because it represents an exception to a general rule outlined by Sidney Mintz (1985:193–194):

Even in the case of societies that have been sucrose consumers for centuries, one of the corollaries of "development" is that older, traditional kinds of sugar are being gradually replaced with the white, refined product, which the manufacturers like to

175

call "pure." In countries like Mexico, Jamaica, and Colombia, for example . . . the use of white sugar has spread downward from the Europeanized elites to the urban working classes, then outward to the countryside, serving as a convenient marker of social position or, at least, aspiration; the older sugars are meanwhile eliminated because they are "old-fashioned" or "unsanitary" or "less convenient" . . . Notions of modernity enter in strongly as more processed sugars diffuse to wider circles of consumers.

Yet in Talea (and in the Rincón more generally) there is a distinct preference for *panela*, in spite of the fact that white sugar has been much cheaper since it was made available in state-subsidized CONASUPO stores. In fact, Leonardo Tyrtania (1992:228) reports that in the early 1980s a single day's wage was enough to supply a family with a week's worth of white sugar, while an equivalent amount of *panela* cost more than twice as much. Even in 1997, similar price differences characterized the products; but, whether rich or poor, Taleans continued to rely heavily upon *panela*, using white sugar only when there was no *panela* available.[1]

Sugarcane in Talea also violates another norm: unlike the situation in most parts of the world, *panela* is grown on small plots, not large ones, and is produced primarily for household consumption and regional markets, not national or international ones. This is unexpected because the complex process of producing, processing, and packaging sugarcane often requires large amounts of capital and labor; indeed, it is a technology that generally needs an authoritarian control structure to finance, coordinate, and manage the sundry phases of growing and processing. In the words of Eric Wolf (1959:179), "the strategic factor in this process is the high cost of a large-scale mill needed to grind the cane," which has historically disqualified smaller growers in the Americas. Mintz (1985) has noted that sugarcane requires a tightly controlled, disciplined labor force and that because of this African slaves were imported to carry out sugarcane work. By contrast, in Talea sugarcane operations are carried out by improvised work teams, often kin, neighbors, or close friends, who meet capital and labor requirements by pooling labor, equipment, and animals in arrangements that combine discounted wage labor with reciprocal exchanges and food gifts.

In this chapter I attempt to answer the question of why sugarcane cultivation is different in Talea and how it is that the ends of production are related to the means. We shall see that sugarcane work seems to have little to do with economic rationality; indeed, Talean notions of quality and taste on the one hand and reciprocity on the other much more adequately explain the persistence of today's thriving regional *panela* industry.

A Brief Social History of Sugarcane

There is evidence that sugarcane is native to New Guinea; it was probably domesticated there between 8000–10,000 B.C. before diffusing to mainland Asia. Processing cane juice into a solid form did not occur until approximately 350 B.C. to 350 A.D., probably in Indo-Persian Khuzestan, when people succeeded in using controlled heat to create a taffy-like, partly crystalline solid. The crop spread rapidly across Asia and beyond: it reached Egypt before the end of the 600s, Cyprus and Morocco around 700, Crete around 818, and Sicily around 827 (Mintz 1991:117).

Sugar production in the West can be divided into three overlapping stages: the Mediterranean phase, from 700 to 1600; the Atlantic phase, from 1450 to 1680; and the American phase, from 1500 to the present. Sugarcane spread rapidly after the Moors brought it to Europe in the period of Islamic expansion through Spain, in the late ninth or early tenth century. In the 1400s Spain and Portugal produced sugar in the Atlantic islands, specifically Portuguese São Tomé and Madeira and the Spanish Canary Islands. Within the first few years after Christopher Columbus's arrival in the Americas, sugarcane was cultivated in the New World (Mintz 1991).

Columbus brought sugarcane to the Americas on his second voyage. The first sugar produced in the New World was cultivated and processed in Santo Domingo and was shipped to Spain by 1516. Mintz (1991:117) notes that "everything connected with the production of sugar was imported to the New World from the Old, as the cane itself had been; the Old World center from which plants, technicians, and technology all came was the Spanish Canary Islands." Even the labor force came largely from the Old World: African slaves were imported for plantation work. According to Mintz (1991:122), no other industry relied so heavily on slave labor as did the sugar industry: "Sugar and slavery traveled together for nearly four centuries in the New World. The demography of peoples of African origin in the New World was, until the twentieth century, largely isomorphic with the geography of the plantation system." Native American populations were used as well, though catastrophic epidemics were probably a major reason why Spanish labor codes prohibited the practice in the early

1600s for many tasks associated with sugar processing (Harris 1964:14–16). After emancipation, workers from China, South Asia, and Java were imported to replace African slaves.

Production expanded rapidly; by the mid-seventeenth century American sugar increasingly went to European markets, where demand was booming. This made the Atlantic island colonies less important as suppliers of the sweetener. By around 1700 Dutch, British, French, Danish, and Swedish colonists were growing sugarcane in the Americas. They shared much in common (Mintz 1991:119):

> Cane production would be undertaken primarily in subtropical areas; the enterprises themselves would be predominantly large-scale; the cane would be grown mainly on coastal alluvial flatlands and interior valleys; the production schedule would be seasonal. Because of the milling and processing phases, the plantation would be industrial, even though based on agriculture. Cane must be cut when it is ripe, ground when it is cut. Hence field and factory phases are tightly linked, and timing is vital; a tightly controlled labor force is needed.

As we shall see, it is not impossible for smallholders to grow, harvest, and process sugarcane; this is exactly what has been done in Talea and other Rincón villages for many years. Even so, it is generally rare for the crop to be grown by smallholders; the machinery and labor requirements tend to make sugarcane an "authoritarian technology" (Mumford 1964) in most cases.

In colonial New Spain sugar refining became the biggest agrarian industry. Theoretically, sugarcane could have been grown and processed by small-scale operators, but sugarcane cultivation became a large-scale capitalist form of enterprise—simply put, it was more profitable that way. Hernán Cortés appears to have been the first to plant the crop on the American mainland, in the Tuxtla region of Veracruz in 1528. Many others followed. From their beginnings in New Spain, sugar refineries required considerable capital. Even a small mill needed at least fifteen workers to keep it in continuous operation, as well as animals and field hands (Chevalier 1963:75). Throughout the early decades of the colonial period, Spanish colonists were lured away from producing wheat on their plantations because its price was capped by colonial officials. Sugar, however, was considered a luxury item and was therefore subject to the laws of the free market. Prices rose sharply in the late 1500s as European demand increased; consequently, production increased throughout the colonial period.

The archival record indicates that sugarcane operations existed in the colonial period in the northern Sierra. A group of Antequera-based Augustinian clerics owned a mill near Zoochila as early as the 1590s, though not much is known about it (Chance 1989:156). A sugar hacienda apparently existed near the Cajonos Zapotec town of Yalálag for more than 150 years. It was owned by a resident of Villa Alta and is first mentioned in archival documents in 1643. By 1700 ownership had passed into the hands of absentee owners in Antequera, and it is last mentioned in 1791 (Chance 1989:92). In the Zapotec village of Lachirioag, approximately six hours' walk from Talea, *panela* and *aguardiente* (Sp., distilled sugarcane liquor) were being processed by 1716. A number of individuals—probably well-to-do Zapotec—ground the cane in their own mills (Chance 1989:112).

Sugarcane does not grow well above an altitude of 1,600 m or in areas with scarce rainfall, so it has not been popular in all Sierra villages. The Rincón has offered the most suitable climate for its cultivation; indeed, Richard Berg (1968) claims that in the 1960s a great deal of *panela* was sold in the Zoogocho plaza by Rincón vendors. In the 1940s only a few communities in the Sierra produced sugarcane, including Talea, Villa Alta, Cacalotepec, Betaza, and San Mateo Cajonos (Alba and Cristerna 1949a:480). During the same period, de la Fuente (1949:84) noted that Yalálag farmers also cultivated the crop. Today a number of Rincón villages are important producers of sugarcane, including Yaeé, Lalopa, Yatoni, Otatitlán, and others. Yatoni is an especially important source of *panela* for Taleans who do not grow sugarcane.

Growing Sugarcane and Making Panela

Planting Sugarcane Cuttings

Sugarcane is planted during the first few months of the dry season (after the danger of frost has passed), in late February or March, and new fields are plowed one or two weeks before in preparation for planting. The crop does not grow well in *tierra fría* (above 1,600 m).

In Talea campesinos use cuttings from larger sugarcane plants to seed new fields. They cut rods of approximately 20 cm and deposit them in rectangular beds reaching just below the soil's humidity line. The campesinos use a hoe to dig these beds, which are spaced at a distance of approximately one *metro*. They lay the cuttings in the ground and cover them with approximately 6–8 cm of soil. The cuttings are positioned so that the "buttons" (Sp. *botones*) from which the plants will sprout are aligned in a plane

Table 6.1. Some Varieties of Sugarcane Cultivated in Talea

VARIETY	PROPERTIES
jaba	exceptionally tough stalk, leaves well suited for thatch roofs, very juicy
blanca	relatively tough stalk, less juice than the *jaba*, well suited for some soils
negra	purple-colored stalk, excellent for chewing
castilla	relatively rare
brasil	relatively rare

parallel to the surface of the cane field. Some campesinos frown upon the practice of fertilizing sugarcane, noting that it "overheats" the cane and results in an inferior dark-colored *panela*. Once a field is planted, it can keep producing sugarcane for approximately twenty years; as the cane is cut every other year, new shoots sprout. Thus planting cane is a relatively sporadic activity.

There are a wide range of sugarcane varieties, including *jaba, blanca, negra, morada, campeche, quijote, castilla,* and *brasil* (all Spanish names; see Table 6.1). Each has distinct characteristics; in general, *jaba* is considered the best for *panela* production because its stalks contain more juice than those of the other varieties.

"Cleaning" or Weeding Sugarcane

Many campesinos describe weeding a sugarcane field as the most difficult task in the agricultural repertoire. Though technically it is carried out no differently than the weeding of milpas, it is complicated by the fact that sugarcane leaves have serrated edges, the stalks have hairlike thorns, and the "weeds" consist of a single species of grass that can grow up to two meters long. Informants note that this grass, which is difficult to clear away because of its deep roots, did not appear until a generation ago, when farmers began using chemical fertilizer. Snakes and bees make sugarcane weeding even more inconvenient and dangerous.

Even so, many campesinos claim that it is much better to weed the fields by hand than with fire, as is done in Veracruz. Some Talean informants had spent several seasons working on sugar plantations along the coast and noted that the methods there were much more dangerous. Setting fire to the fields was like stepping into hell, they said: the heat was unbearable, wild animals of every sort (including snakes) rushed out of the burning fields, often in the direction of workers, and there was always the danger of getting completely surrounded by fire. In Talea, where great precautions are taken to control fires, forest land is too scarce to risk destruction.

Campesinos weed their sugarcane fields once or twice a year, usually after the rainy season has ended in late October or early November and again in the early spring. If a field has just been planted or harvested, it may require more frequent weeding in the following weeks and months.

In *tierra caliente* (below an altitude of approximately 1,300 m) sugarcane plants produce a spike which flowers in the spring. Campesinos typically remove this with a machete; otherwise the sugarcane fails to produce the maximum quantity of juice (Tyrtania 1992:224–225).

Coordinating Sugarcane Work: Reserving the Mill and Cattle

One and a half to two years after planting, the sugarcane has developed enough so that it can be cut. Once it is cut, campesinos must mill and process it into *panela* without delay; otherwise, the juice will begin to ferment within the stalk. Ideally, campesinos carry out this work in February, March, or April.

Converting sugarcane into *panela* requires a high degree of planning and coordination to guarantee that all the necessary equipment arrives when it is needed. There are fewer than twenty mills in Talea, so the first step taken by the grower is to schedule a date to ensure that one is available. This does not occur randomly; usually one asks kin, a neighboring farmer, or a close friend to provide a mill. This must be done several weeks in advance so that all parties can schedule other tasks around the dates. The campesino also reserves a team of oxen or cows from another campesino if necessary and might contract a different campesino to cut down a dry tree with a chainsaw to supply firewood for the furnace. Sometimes still another campesino is drawn into the task indirectly by providing a ranch house for those who do not have a milling area close at hand or wooden molds for pouring the processed molasses. Transporting the *panela* back to the village often requires hiring yet another campesino. Consequently, sugarcane binds together four, five, or more campesinos through the pooling and rental of equipment, animals, and ranch houses—weeks before the milling ever begins. If *mozos* (Sp., hired laborers) are employed, this may draw even more people into the task, sometimes even people from neighboring villages. Thus the campesino effectively plays the role of contractor, coordinating the activities of several different subcontractors who provide equipment, space, and services.

The subcontractors typically price their service on a two-tiered scale, according to whether or not the contractor is a relative. Kin may be given a discount of approximately 20 percent, though a portion of this is often paid back in the form of a gift of *panela* logs. Non-kin are charged at the regular rate.

This peculiar system of labor contracting is mixed with other modes of exchange. Quite often, cane grinding is done with a version of *gozona* (reciprocal work)—for example, a farmer owning a mill may exchange five days of equipment and services for five days of service from a companion and his oxen. Or he may provide five days of equipment and services but only charge for two days' work, reserving the rest for *gozona* repayment. Similar arrangements may be made with the labor. Furthermore, special gifts are often a vital part of the sugarcane operation: the owner of the sugarcane (that is, the work coordinator) customarily gives those who help during the work period several logs of *panela*, bottles of fresh sugarcane juice or *tepache* (Sp., fermented sugarcane juice) or *melada* (Sp., syrup or molasses), and other "goodies" and expects to receive similar gifts when paying back the *gozona*. All of these features—which essentially are forms of reciprocal exchange between kin, neighbors, and close friends—make *panela* production a markedly different activity than the strict cash contract arrangements characteristic of large sugar refineries.

Relatively egalitarian production processes have been able to develop in part because mills can be acquired at significant discounts—therefore, there is less of an imperative for mill owners to recoup their investment. Mills are most frequently purchased secondhand from plantations in Veracruz, where they can be procured for approximately N$15,000 (US$2,000 in April 1996)—about four times the price of a young team of bulls. This is yet another striking example of the resourcefulness of Talean campesinos. The mills are, without exception, cast-iron implements consisting of three rotating drums encased in a thick shell, with openings used to insert and extract the stalks of cane. An axle protrudes from the largest of the drums; this is coupled to a wooden beam that is connected to a yoke with a length of rope, which the animals pull.

Several generations ago, most mills were produced locally by Talean carpenters out of regional hardwoods. Older informants note that with the wooden mills sugarcane has to be run through the mill twice to extract the juice. For this reason, the task of milling cane is referred to as *torciendo caña* (Sp., literally "twisting cane"), even though the stalks are no longer twisted back into the mill a second time. According to campesinos, the cast-iron mills are more durable than the old wooden models and much safer. They say that in past generations several campesinos had hands and arms mangled in the process.

Preparing the Milling Area

The area in which the sugarcane is milled must be a level patio-like surface long and wide enough for the animals and farmers to work comfortably. A three-*brazada* square (approximately 8 m by 8 m) is a reasonable minimum. Additional space is needed for the furnace, the firewood, and the wooden molds. A number of other things must be kept in mind when selecting the site. Most importantly, the site should lie as near the sugarcane field as possible, preferably at the same level or even slightly downhill, since the stalks must be carried in bundles by workers. Campesinos do the milling during the dry season (frequently as late as April or early May), so they select work areas in places that are shady during the afternoons, since the animals' work is grueling and continuous and there is always a risk that they will "overheat." Finally, it is advantageous to locate the furnace on a second terrace, just below the mill, so that the freshly extracted cane juice can easily be pumped to a lower level using gravity.

According to informants, most milling areas were constructed long ago, when *panela* (not coffee) was the major cash crop. Because they are used annually, established sites require little preparation.

Gathering Firewood and Assembling the Mill

Campesinos must do this task weeks in advance so that there is a sufficient stock of dry wood available for fueling the furnace. A dry oak tree (*Quercus* sp.) is ideal, and the task may take several days if no chainsaw is available. Campesinos must fell live trees at least three weeks in advance so that the wood is dry enough to burn and light enough to be transported. The tree is cut, chopped into firewood, and carried to the milling site.

Before the animals arrive, the campesinos assemble the mill on a wooden stand planted in the center of the milling area. They fasten it into place with bolts or thick nails and do the same with the beam used to turn the axle. Both the stand and the beam are typically retained for use year after year.

Once the mill has been connected to the beam, the campesinos drive meter-high stakes into the ground near the center of the milling area. These serve as racks for stalks of sugarcane.

Cutting the Cane

When these preliminary steps have been taken, the work may begin. The campesinos cut stalks of sugarcane, taking them by the hand, cutting them as close as possible to the ground with a machete, and then cutting off the top (the portion with green leaves). They shave away dry leaves, which cling to the stalks, with a brush or two of the machete's unsharpened back edge. Then they toss the stalks into a pile. This work is often done during the morning so that milling can begin immediately after lunch.

Technically speaking, cutting cane is not difficult, but it is uncomfortable work, since the stalks are covered with a fuzzy surface of fine, hairlike thorns that stick to workers' fingers, hands, and arms. The inconvenience is minimized if they work as a team, for the task is light enough that joking and gossip can occupy their thoughts.

Workers leave small shoots (less than the diameter of a thumb) or *retoños* (Sp.) in place so that they may develop fully. They do not cut the shoots until the following season (two years later).

"Twisting" or Milling the Cane

When the workers have transported ten or twelve loads to the milling area, milling may begin. The campesino lashes the animals to a special yoke (shorter than the one used for plowing) and connects it to the beam using a loop made of leather or thick rope. Then one person—often a woman or a child—takes a switch and begins goading the animals to pull. Once they start circling, the mill begins grinding away, and a second person begins feeding stalks of cane into the mill, one at a time. The steel drums pull the cane through the mill, crushing the stalk, extracting its juice, and forcing it through the other side, much like a clothes wringer. The sweet juice streams out of the mill and into a large metal tub below. The stalk's pulp drops out of the other end. In the meantime, the workers continue cutting cane so that the pile of stalks is replenished during the course of the day.

The work is not physically tiring, but it is repetitive to an extreme not found in any other kind of farming task. For one worker it involves pacing in a circle literally thousands of times a day, while for the other it requires continuous feeding of cane stalks at the rate of perhaps five per minute—

Figure 6.1. Talean campesinos grinding sugarcane. It appears that this work has changed little in the Rincón in the last 400 years. *(Photo by Laura Nader)*

a short cycle time for agricultural work. In its rhythm this work is much more like industrial factory work than like any other farming tasks, in the sense that the campesinos make a limited number of movements centering around a cast-iron apparatus that dictates work speed. Because the animals are moving continuously, both workers must remain focused on their respective tasks, particularly the one feeding the stalks into the grinder. A careless moment might result in a bump to the head caused by the heavy wooden beam rotating inches above. Indeed, even the milling area itself starts to take on a geometric, factory-like appearance after several hours. The space begins taking the form of a set of concentric circles: the mill in the center, then the path trod by the animals, and at the outer perimeter a ring-shaped mound of white pulp.

All workers take periodic breaks, and they may gather together to drink the cool sweet juice or a gourdful of *tepache* (which is slightly alcoholic) from a fermenting tub.

After 1½ to 2½ hours (depending on its size), the tub fills with the greenish juice. With a length of flexible plastic hose, a worker gravity-pumps the liquid to a vat placed on a furnace used to process the juice into molasses.

Converting Cane Juice to Molasses

Campesinos use a specialized furnace for converting sugarcane juice to molasses. It is a subterranean furnace fitted with two or three ventilation shafts which allow the intense heat to escape. The shafts allow oxygen to pass through the furnace; consequently, extremely high temperatures may be reached and maintained.

The workers shove dried pulp into the furnace to ignite the coals; then they place firewood inside so that the vat can begin heating. As it warms, one worker fills a second tub with juice and drains it into the vat. In our case two tubs filled a vat, and by 4 P.M. both had been emptied into it. More wood is added to make the liquid reach boiling temperature quickly. As the fire begins heating up and late afternoon wears on, workers may go to the cane field to cut and transport stalks so that work can begin early the following day.

In the late afternoon, as other campesinos return to Talea, they are inevitably invited over for a gourdful or two of *tepache*, candied bananas, and papayas. They almost always accept the invitation. It allows farmers to catch up on the latest news and gossip and allows those doing sugarcane work to send messages back to their wives in Talea. Specifically, they may ask for more tortillas to be sent with a neighboring campesino so that the work routine will not be interrupted. Once they have begun grinding sugarcane, farmers often do not return to the village until the work is completely done—which may be ten days later or even longer.

When evening approaches, the pace slows and workers often begin to unwind. They may have several gourdfuls of *tepache*, sometimes with a shot or two of mezcal "for the cold" (since *tepache* is considered a "cold" substance that can make one ill), and the mood tends to lift as dinner is prepared. The functions of this conviviality should not be underestimated. The grueling mechanical work done during the day and often well into the night becomes a kind of shared suffering and requires a great deal of self-discipline on the part of the workers—the machinery dictates bodily movements and rhythms to an extent not found in any other kind of agricultural work. The release in the evenings is part of what makes the daily grind (so to speak) bearable.

For this reason, perhaps, sugarcane work is often described in an upbeat way by farmers. There is plenty of food and drink on hand, some of it alcoholic, nearly all of it sweet, and lots of companions to chat and joke with in the evenings. Unlike other phases of the agricultural cycle, *panela* processing is always done in the company of others. Children especially look forward to accompanying their brothers and fathers to the fields, for they are allowed (and indeed encouraged) to eat as much taffy, candied fruit, and sugarcane juice as their stomachs can hold.

After dinner, once the juice begins boiling (after two to three hours), the campesinos skim off bits of "trash" (shredded cane stalk fibers, thick suds, and other matter) from the surface of the vat with a ladle-shaped strainer fashioned out of a bamboo pole and a perforated metal bowl. The trash is deposited in a bucket and used as pig feed.

The campesinos then drop green bananas and papayas into the boiling juice to make candied fruit. As they stew in the vat, they absorb some of the molasses. The workers eat candied fruit as a between-meal snack or else take it to their children in Talea. Bananas are peeled before being placed in the vat; papayas are not.

Next the campesinos add two cupped handfuls of ash to the vat; this serves to congeal the molasses once it is poured into molds the following day. They sift it little by little through a strainer, mixing it slowly into the vat so that it does not clump together. Throughout the period following dinner, some of the men gather about the furnace and often take turns adding more dried pulp and firewood to the furnace to keep it at maximum temperature.

In the meantime, the rest of the campesinos begin wrapping the cakes of *panela* processed during the day. They retire when they have finished this task, usually at about 10 P.M. Someone—typically the owner of the cane field—takes responsibility for adding fuel to the furnace every few hours so that the vat boils continuously all night long. It must boil for approximately twelve hours. Assuming that boiling temperatures are reached by 6 or 7 P.M., the vat should be ready after breakfast the next day.

There is a test for determining whether or not the vat is ready to be pulled off the furnace. The campesino gathers up some of the molasses on the end of a short stick, allows it to drip into a gourd of cold water, then fishes out the product. If it is congealed, the molasses has "reached its point." If it is still sticky and pliable, it must be boiled longer.

Older informants note that before coffee became an important crop the first few months of the year were devoted nearly exclusively to *panela* production. Some families worked night and day, producing two vats of molasses daily, one during the day and another at night, with only short breaks for sleep in between. An informant reported that some years a full month of work might be devoted to sugarcane and as many as sixty vats might be processed during that time. This was possible because coffee harvesting was still a relatively unimportant activity.

Converting Molasses into *Panela*

Four men lift the heavy, hot vat. They take it into the ranch house and place it upon a layer of dried pulp which has been set upon the floor. Two men working at opposite ends of the tub stir the liquid vigorously in order to cool the molasses. The first round of stirring goes on for twenty to twenty-five minutes; then the workers wait for several minutes. The edges of the cooling molasses begin creeping up the side of the vat as it coagulates. After several minutes they begin stirring once again, for five to ten minutes. At this point the molasses is ready to be poured into the molds.

The molds typically come in two different sizes: a standard size that yields cakes approximately 4 inches in diameter and a "gift" size that yields cakes approximately 2½ inches in diameter. Campesinas sell the former in the village market; the latter are reserved as gifts for visitors from the city.

The campesinos use a bowl-shaped gourd to ladle the rapidly drying molasses into plastic buckets, which in turn are used to pour the stuff into molds—nothing more than thick planks of wood with cake-shaped holes chiseled into them. As they fill the holes, the molds are stacked on top of one another, separated by wooden blocks. They conduct the work quickly so that all of the molasses is scooped out of the vat before coagulating. With small wooden spatulas, campesinos diligently scrape clean the trace amounts of molasses left along the sides of the vat and promptly consume them on the spot or else empty them into a bowl, where they dry to form a kind of taffy. They eat an extraordinarily high amount of sweets during the sugarcane season.

The campesinos allow the molasses to cool for several hours; molds filled at 10 A.M. are ready by 3 or 4 P.M. They then knock the cakes of *panela* out of the molds with a wooden mallet and stack them for wrapping in the evening.

Wrapping *Panela* for Storage

Campesinos refer to one cake as a "face" (Sp. *cara*), two cakes as a "head" (Sp. *cabeza*), and four heads, stacked end to end, as a "pulp" (Sp. *bagazo*). They intricately wrap the logs in a sheath fashioned out of dried cane leaves. This work is done at night after dinner, as the campesinos exchange jokes, stories, and gossip. They completely cover the *panela* with the wrappers and thus protect it from rats, mice, ants, and humidity. Then they transport the bundles back to the village for storage above the hearth, which is the driest part of the house. A pair of *bagazos*—referred to as a *pancle* (Sp.)—is the standard unit of sale; when I left the field, it could be purchased for approximately N$45, just under US$6.

The Cultural Preference for *Panela:* The Predominance of Quality

A remarkable phenomenon in Talea is the villagers' distinct preference for *panela* over white sugar. This seems counterintuitive, especially in light of the fact that over much of the colonial period, in many parts of the world, "the whiter the product, the more expensive it was" and the more preferable it was for lower classes trying to emulate people higher on the class scale (Mintz 1985:83). It seems even more counterintuitive when we consider the fact that today store-bought white sugar is available at a fraction of the cost of *panela*.

In spite of this, Taleans of all socioeconomic categories consume *panela* rather than sugar, which may have to do with its complementarity with coffee—which is served as a beverage at practically all meals. Sweetening coffee with sugar is not the same, according to many Taleans, because the consistency of the drink is too watery. *Panela* is what "fills" the body, they say, while sugar leaves one unsatisfied. *Panela* is used in other foods, including *atoles*, bread, and a number of special candied dishes prepared for the Todos Santos fiesta (All Saints' Day) at the beginning of November. Others note that sugar is "cold" and therefore represents a health risk, while *panela* is "hot" and safe to consume.

Some campesinos say that in spite of the fact that coffee is generally a more lucrative crop than sugarcane, they are reluctant to give up the latter because purchasing *panela* often involves sacrificing the guarantee of quality represented by household production. Making *panela* is a complex process, and it is not easy to tell whether or not it has been produced cleanly. Growers from other villages who sell their product in Talea, for example, are sometimes said to allow their dogs to lick the molds, "since it's going to the Taleans anyway, and in Talea they eat anything." Similarly, another informant told me the story of a group of Juquilans who inadvertently let a live chicken fall into a boiling vat of molasses. He claimed that the campesinos simply fished out the bird and continued boiling the molasses, which was destined for sale in Talea.[2]

In general, a light-colored *panela* is said to be of high quality, while a dark-colored *panela* is considered to be of lower quality. Both, however, are valued more highly than white sugar, perhaps because the source of the latter (and therefore the quality and cleanliness) is unknown.

Another probable reason for the persistence of sugarcane cultivation in Talea is that it represents an alternative cash crop for supporting households in times of severe coffee price crashes or for buying maize and beans in case of crop failures. In this sense, it might be seen as an important element in a four-way diversification strategy (maize, beans, sugarcane, coffee) for helping to maintain households in the wake of periodic crises. Indeed, like bags of green coffee, logs of *panela* can be seen as savings accounts or nest eggs to be sold in times of need.

The Meanings of Panela: *Todos Santos and Reciprocity*

The manufacture of *panela* is a social process as well as an economic one, linking campesinos in relationships that are based on a combination of wage labor, reciprocal work arrangements, and food gifts. Furthermore, for Taleans the food represents a high-quality product when grown for one's own family and satisfies an acquired taste. It also serves as a kind of life insurance policy or savings account that can be traded in during household crises. But sugarcane has another special meaning in Talea, as it does across many parts of Mexico: it is a ceremonial food that is ritually offered to the deceased each year on November 1, during the celebration of Todos Santos, a Catholic holiday honoring the dead. It is likely that the celebration has prehispanic roots: according to Robert Ricard (1966 [1933]:31), the Aztec offered pastries to the deceased.

In preparation for this fiesta, all Catholic households in the village, no matter how humble, clean and prepare the space around the altar and arrange shelves to hold special foods. The most important is *panes de muertos* (Sp., "breads of the deceased"), sweet breads that are molded in the shape of human bodies. These "stand" before the altar and are surrounded by other (mainly sweet) offerings on all sides: plates of squash, papayas, and apricots, candied in sugarcane molasses; heavily sweetened bowls of chocolate *atole;* stalks of sugarcane arrayed at the edges of the assemblage; bowls of *mole* (Sp., a chocolate-chile sauce); peanuts; fresh fruits (especially loquats); and occasionally turkeys or chickens. These foods are arranged several days before the holiday; afterward they are eaten by household members, once the deceased have had the opportunity to partake of the special meal.

Todos Santos in Talea culminates in the *recorrido de los bhni hue'* (Sp. and

Figure 6.2. Altar for Todos Santos. Sugar is an important ritual food associated with death. *Panes de muertos* (breads of the dead) are sweet breads formed in human shapes (bottom row). Altars are adorned with many other foods and drinks, including *mole*, tamales, tortillas, candied fruits, peanuts, *atoles*, chocolate, mezcal, and fresh fruits. Tall stalks of fresh sugarcane are used to adorn the edges of the altar. (Photo by Gabriela Zamorano)

Zap., "the journey of the wounded/wailing men"), in which groups of men, boys, and (beginning in 1996) some women visit each Catholic home after nightfall on the evening of November 1 to sing solemn prayers in Spanish and Latin. The prayers, which are addressed to the Catholic deities, are petitions for the release of the family's deceased souls from purgatory into heaven.[3] Each group includes a leader, a treasurer responsible for guarding the cash offerings, a person who rings a ceremonial bell announcing the group's arrival, and a person who carries a bucket of holy water which is sprinkled in each of the homes visited (the latter two are often young boys). One or two of the village's policemen escort each group, "to maintain order." (Occasionally one of the *bhni hue'* has too much to drink.) After the group leader says a blessing, holy water is sprinkled before the family's altar and the group begins singing. Afterward the head of the household offers gifts of cash and food to the visitors. The treasurer delivers the cash to the *jefe del templo* (Sp., elected director of church affairs) at the cemetery the following morning, where the groups of *bhni hue'* meet when all the homes have been visited. During the night, the group leader distributes the food gifts (typically sweet bread, peanuts, store-bought candy, or fruit) among the participants, who vary in number from four or five to as many as twenty-five or thirty per group. Occasionally, the *bhni hue'*—who may take as long as ten to twelve hours to cover their section of the village, often traversing muddy unlit footpaths—enjoy a meal prepared especially for them by the members of an individual household; in other homes mezcal may be offered to members of the group, for the long, cold night ahead. As they traverse the town, one or two men toll the church bells throughout the night; sometimes the group leaders send them food and mezcal. Except for the *sacristanes* (Sp., elected church stewards), participating as a *bhni hue'* is voluntary.

On November 1 nearly all Catholic families go to the cemetery to visit the deceased, clean their grave sites, and leave them fresh flowers and candles. Many families spend much of the day there and may take tamales or other foods to eat. In the afternoon the musical groups arrive and play funeral marches in honor of the deceased.

The use of sugary products in the Mexican Todos Santos celebration seems to follow a pattern described by Mintz (1985:172-173), in which "sugar [turns] into something extraordinary, ceremonial, and especially meaningful . . . [while it simultaneously becomes] something ordinary, everyday, and essential." The *panes de los muertos*, candied fruits, and other goodies are made exclusively during this time of the year and so carry more symbolic weight, but the Todos Santos customs have reached even the humblest villagers.

A discussion of the various aspects of Todos Santos in Mexico, regional variations, and similarities and differences in comparison to Spain and other parts of Europe is beyond the scope of this study.[4] Suffice it to say that, as in many other parts of Mexico, foods derived from *panela* and sugarcane are important vehicles for communion with the deceased, a gift to the dead in exchange for the petitions made by the living (Parsons

Figure 6.3. Music is an essential part of the Todos Santos celebration in Talea. At the cemetery the village orchestra plays a funeral march in honor of the deceased.
(Photo by Gabriela Zamorano)

1936:531). It is striking that nearly all Taleans, when asked how many children or siblings they have, will include the deceased. A trip to the cemetery may be described as "a visit to see my mother." Along with offerings of flowers, votive candles, and prayers, food offerings appear to form the basis of a reciprocal mode of exchange with the deceased. In this respect it is useful to recall Mauss's definition of *The Gift* (1954 [1924]:1):

prestations which are in theory voluntary, disinterested, and spontaneous, but are in fact obligatory and interested. The form usually taken is that of the gift generously offered; but the accompanying behaviour is formal pretence and social deception, while the transaction itself is based on obligation and economic self-interest.

To summarize, cultivating and consuming sugar in Talea appears to be done quite differently than in many other parts of the world. In spite of a tendency for sugarcane work to favor large-scale (industrial or plantation) farming and processing, the Rincón Zapotec have kept it simple: they use ox-driven mills and improvised work teams. These teams are sometimes bound together by wage agreements, but are more frequently linked by reciprocal exchanges. Campesinos fully expect to collect generous gifts of *gozona* labor, *tepache, panela* logs, sugarcane juice, and molasses in exchange for their giveaways of the same goods and services. Although *panela* is more expensive than white sugar, Taleans are willing to pay more for it (either in cash or in the production process), partly owing to notions of food quality and a clear preference for its taste and satiation value. Furthermore, sugarcane products, as we have seen, represent a medium for gift exchanges with the deceased. *Panela* is also a classic gift for visitors and, like beans and maize, is an acceptable gift for those hosting life-cycle functions. In this regard sugarcane much more resembles maize and beans—which enter reciprocal modes of exchange—than coffee, which is more strictly treated as a cash commodity. Rincón Zapotec notions of reciprocity and quality, in short, have overridden the tendency for sugarcane to take the form of a crop subject to the profit motive and authoritarian labor controls over the course of nearly five centuries, even with price subsidies and economies of scale taken into account.

CHAPTER 7

THE INVENTION OF "TRADITIONAL" AGRICULTURE

The History and Meanings of Coffee

This chapter is about the adoption and appropriation of coffee by Zapotec campesinos over the last century. History is part of the story. Nearly 1,000 years ago the crop began traveling across the globe, preceding the "Columbian exchange" (Crosby 1972) of New and Old World organisms by centuries. At each site where coffee was adopted (or coercively imposed), it combined with distinct land tenure systems, labor arrangements, and agricultural practices; consequently, knowledge about the crop was transformed as it crossed geographical boundaries. Coffee was introduced in Latin America against the background of older systems of farming, and agricultural techniques were developed which made sense in terms of existing theoretical and classificatory schemes. Talea is a case in point, for it is clear that coffee, in spite of its recent arrival and its Old World origins, is grown in a way that might be referred to as "traditional." An exploration of new coffee cooperatives in Oaxaca's Northern Sierra is also part of the story. The cooperatives, institutions structurally rooted in Sierra Zapotec, Mixe, and Chinantec communities, promote organic coffee cultivation based on local techniques that are ecologically sound.

Coffee through Time: The Social History of a Global Commodity

Coffee in Motion: From the Horn of Africa to Southern Mexico

Coffee has traced an intricate path over the centuries, which has extended from East Africa and the Near East to Southeast Asia and Latin America in the eighteenth century by way of Europe. The complex movement of the crop across continents and oceans makes it difficult to characterize as "Western" or "Eastern," "modern" or "traditional."

According to one popular account, coffee was discovered in the ninth century by Islamic shepherds caring for goats near the Red Sea in present-day Ethiopia. (Coffee still grows wild in and around Ethiopia's Kahve province.) They were concerned about the fact that their goats suffered insomnia and soon discovered that the animals' restlessness was caused by their feeding upon the sweet red fruit of a mysterious shrub. The stimulating coffee bean was a rapid success. In spite of prohibitions against its consumption, it became popular throughout the Ottoman Empire during the 1500s and covered nearly the entire Islamic world by the end of that century. Coffee was cultivated on a limited scale at sites in and around the Arabian city of Mocha, and the set of knowledges and technologies associated with coffee cultivation, processing, and consumption was an exclusively Muslim possession (Ukers 1935; Paredes, Cobo, and Bartra 1997).

Through contacts with the Near East, coffee reached Europe by the late 1600s. Its European success was secured in Paris, where the coffee house was imported along with the beverage. Similar patterns followed in Germany, Italy, and Portugal (Ukers 1935; Wolf 1982:336–337). Fernand Braudel (1967:186) notes: "If there was such an increase in consumption — and not only in Paris and France [but across Europe] — from the middle of the eighteenth century it was because Europe had organized production itself." For many years the world market depended exclusively on supplies from the Islamic world, particularly from Mocha. The Dutch appropriated the crop, however, and by 1712 had begun growing coffee on large Indonesian plantations. Production started booming and prices began dropping, and within years coffee became the Dutch East Indies Corporation's major export crop.

The significance of this event should not be understated. The Dutch had effectively transformed coffee cultivation because they combined a biological artifact (coffee seeds) from one place with a technological system (plantation-based capitalism) from another. Coffee became a pawn in a global "botanical chess game" (Kloppenburg 1988:154):

Plant germplasm was appropriated and shifted across the continents and archipelagos of what is now the Third World as the European powers sought commercial hegemony . . . Because most of these plantation crops were of tropical or subtropical origin, the movement of germplasm tended to be lateral, among colonial possessions, rather than between the colonies and the metropolitan center.

Agricultural crops, in other words, were collected by the imperial powers and then redistributed horizontally, across different tropical regions.

If the initial move in this "chess game" was coffee's transplantation from the Near East to Indonesia, the following move was across the Pacific, to the Americas. The plant arrived in the New World in the early 1700s, reaching Guayana in 1714 and Haiti, Santo Domingo, Brazil, and other colonies shortly thereafter. When the Indonesian plantations suffered a blight in the 1880s and 1890s, Brazil was poised to dominate the world market. By 1900 it supplied three-quarters of the world's coffee (Wolf 1982:337; Paredes, Cobo, and Bartra 1997).

In Mexico coffee cultivation was concentrated in the gulf coastal state of Veracruz. From there it spread to Chiapas, Oaxaca, and other states in the 1800s, though Veracruz remained the most important coffee-producing state through the 1880s. To this day, much of the nation's coffee cultivation is concentrated in these states. Between 1890 and 1920, however, the Soconusco region of southwestern Chiapas emerged as the most important Mexican coffee zone, in part because legislation passed in 1856 prevented indigenous communities from retaining rights over territories granted to them in the Spanish colonial period. Mexican and foreign investors established large coffee plantations there during the dictatorship of Porfirio Díaz (1884-1911), and German, English, and U.S. firms were given concessions to convert millions of hectares of virgin forest into *cafetales*, sometimes for as little as a few cents per hectare. According to one estimate, one-fourth of the land that passed into the hands of the foreign investors pertained to indigenous communities. Mayan workers were coerced into working on the plantations through a system of debt peonage, and labor conditions were hellish. (The situation has changed little in some parts of Mesoamerica.) Men, women, and children were taken to plantations, where they were forced to work long hours, seven days a week. Similar processes occurred in various parts of the country, and things did not change significantly until president Lázaro Cárdenas (1934-1940) instituted land and property reforms (Wolf 1982:337; Paredes, Cobo, and Bartra 1997: 33-35).[1]

Thus the systems of coffee cultivation that spread to plantations in the New World in the eighteenth century and reached Mexico in the latter

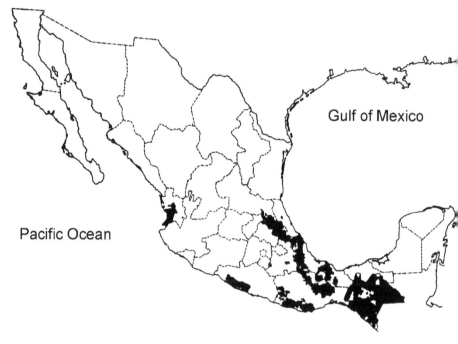

Figure 7.1. Major coffee-producing regions of Mexico.

half of the nineteenth century were organized according to a "colonizer's model of the world" (Blaut 1993) that relied on the exercise of raw imperial power to convert entire regions into zones of capitalist production.

The Introduction of Coffee in the Northern Sierra

From 1890 to 1900 Mexico quadrupled its production of coffee and emerged as an important international exporter of the crop (Nolasco 1985: 169-170). The pattern in Oaxaca paralleled that of the nation. Driven by a soaring demand in the United States in the late 1800s and by free trade policies during the Porfiriato, Mexicans in various parts of the country began cultivating coffee. There are different versions of how coffee arrived in Oaxaca; given the nineteenth-century importance of Veracruz as a coffee-producing center, it seems likely that the crop came from the north, through the districts of Tuxtepec or Santiago Choapan. According to one source, coffee entered the Northern Sierra via Choapan in 1883 before diffusing southward to Villa Alta, Yalálag, and the Mixe zone (Berg 1968: 70-71).

A more colorful account relates how coffee was introduced in the Rincón in the late 1860s in nine villages situated along the border of the Ixtlán and Villa Alta districts: Tepanzacoalco, Cacalotepec, Yaneri, Zoogochi, Yotao, Yagavila, Teotlaxco, Yagila, and Josaa. It is worth translating in full:

Before the middle of the nineteenth century, the inhabitants of these villages were backward in all senses of the term: they retained primitive customs, were malicious, untrustworthy, disobedient, and idolatrous. Because of the nature of their lands, they were only able to cultivate limited quantities of maize, enough for only four or five months annually . . . The rest of the year they relied on a diet of fruits, which they really only half-cultivated due to their indolence . . . Conscious of this situation, the activists of that era known as the Liberals, among them the General Fidencio Hernández, who in that period made frequent trips between Oaxaca and Veracruz, observed that the weather in these villages was identical to that of Córdoba and Orizaba, and they began collecting coffee seeds from the Veracruz plantations for cultivation in the Rincón region.

Because the seeds were planted out of season, they tried to bring saplings for the Rincón plantations, but since the locals were unintelligent people [*gente inculta*], the trees died. In the meantime, General Hernández acquired more preeminence throughout the region, and in his capacity as political chief and military commander of the district he decided to treat the affair as a legal issue so that coffee cultivation might be realized in spite of the stubbornness of the Rinconeros: General Hernández decreed that whoever did not present 25 coffee trees within a year was to be fined 100 pesos per family.

The argument was a convincing one and left the villagers with no option but to obey . . . Within a few years the Rinconeros began to collect the first harvests . . . The rapid economic change surprised them, above all because the coffee buyer came up to the villages looking for the crop and did not pay them in cash but rather with merchandise. During the harvest season, the coffee buyer exchanged maize, bread, salt, cotton cloth, thread, and *aguardiente* [sugarcane liquor] for the Rinconeros' coffee.

They say that in that era the people of the villages realized that they were indebted to General Hernández for their newfound economic success and that upon waking up every morning and after each of their three meals they would religiously pronounce the following Zapotec words: *Schalenu diuci schalenu Fidenciu Hernández, va gutagutu!* [Thanks be to God and thanks be to Fidencio Hernández that now we eat!]

In the corridor of the municipal palace of San Pedro Yaneri the following inscription was preserved until 1908: "To the unconquerable General Fidencio Hernández, who was a benefactor of these villages and introduced them to coffee." (Pérez García 1956:273–274; my translation)

This passage is remarkable for a number of reasons.[2] It appears in a two-volume study conducted by Rosendo Pérez García, who was a schoolteacher for many years in the Sierra Juárez. He claims to have collected his material between 1917 and 1949 and to have gained fluency in Sierra Zapotec and Cajonos Zapotec (Pérez García 1956:9). His own background and status as a mestizo in a region of indios—*gente inculta*, in his own words—undoubtedly had much to do with his pejorative description of the Rinconeros.

Although the account describes the introduction of coffee into the region as the result of a "white man's burden" borne by General Hernández, there may well have been other motives. It is possible that Hernández and other caciques were in cahoots with the Rincón's sole coffee buyer during this period. They all stood to benefit from the large profits that could be made by trading underpriced coffee beans for overpriced goods—as well as from the state subsidies on coffee production, trade, and export that existed at the time. If Hernández was indeed playing the role of cacique—and it seems that he was dishonest in his dealings with at least some Serranos who filed a lengthy legal complaint against him (Ruiz Cervantes 1988:370)—then initially, at least, coffee functioned as a hegemonic mechanism of control in the Rincón.

In spite of its ethnocentrism, Pérez García's account matches the information collected from informants in Talea more closely than any other. According to villagers, coffee saplings were first brought to Talea from the villages of Yagavila and Teotlaxco, to the northwest (in the direction of Veracruz). Still another source reports that the Yagavilans claim that their grandfathers "planted coffee as if it were maize" (Tyrtania 1992:215).

The Introduction of Coffee in Talea (1905-1945)

After its introduction, coffee was probably accepted quickly in Talea because of its small land base and large population. Talea, one of the few towns in the Northern Sierra founded after the Conquest, slowly acquired what little land it could from neighboring villages, and most of it became private property. As mentioned earlier, when an important silver mine near the village was abandoned in the early 1900s, hundreds of mine workers and their families were left unemployed, and many of them turned to agriculture. Coffee was an attractive option for families attempting to survive on plots of less than five hectares. Throughout the first decade of the century villagers could use coffee to purchase maize, beans, and merchandise;

thus, from the beginning, coffee helped Taleans survive difficult economic circumstances.

Much of the growth in coffee production was stimulated by a rise in global consumption led by a rapid increase in U.S. coffee drinking at the turn of the century, when it was transformed from an expensive, exotic beverage enjoyed only by the elite to an inexpensive drink consumed by the middle and working classes. In this period U.S. per capita annual consumption increased from three pounds to more than twelve (Jiménez 1995).[3]

In the decades following its introduction coffee was traded directly for products from outside the Sierra. As noted by Pérez García (1956), maize, bread, salt, cotton cloth, and thread were among the most popular items in the late 1800s. In the 1920s and 1930s muleteers from Talea, Betaza, Solaga, Yalálag, Zoogocho, and other villages transported the region's coffee to Tlacolula—the economic link between the Valley of Oaxaca and the Northern Sierra—via Zoogocho. Others, particularly the Mixe, used tumplines to carry as much as five *arrobas* (nearly 60 kg) of the precious cargo from Zoogocho to Tlacolula. They returned with merchandise from the Valley of Oaxaca, and in this way factory-produced goods began to circulate through the region.

According to informants, Taleans were reluctant to dedicate too much land to coffee cultivation in the early years. (Other Rincón villages held out even longer; farmers in Juquila Vijanos, for example, did not cultivate coffee until the late 1950s [Laura Nader, personal communication, 1997].) Coffee was planted in the most marginal areas—on slopes too steep for planting maize or within the village itself, where plowing was not possible. Choice land was too precious for coffee, and it seems likely that a decade-long price crash during the Great Depression convinced growers to be cautious about placing faith in the new crop. Throughout this period relatively little currency changed hands; instead coffee was bartered directly for merchandise or maize. As demand for coffee (and its price) skyrocketed in the post-World War II period, however, the Rincón's coffee began to exceed the value of the merchandise shipped into the region, and money started to circulate through the Rincón.

The Postwar Coffee Boom (1946-1959)

After World War II record levels of coffee consumption were reached in the United States and several European nations. The reasons for this, according to one survey, were "a pent-up demand resulting from wartime scarcity, an increased number of newly founded homes, [and] a more active social life" (U.S. Department of Commerce 1961:7). Perhaps the most significant factor, however, was the institution of the coffee break, which allowed millions of U.S. workers to drink several cups daily in the workplace, at little or no cost. The coffee break permitted blue- and white-collar workers to consume a physiologically stimulating "nerve food" to help deal with "the pressures of the machinery of civilized life" (quoted in Jiménez 1995:47). Like sugar (and frequently in the same cup with it), coffee served as a "proletarian hunger killer" (Mintz 1979:70). There was also something intensely pleasurable about the coffee break. For a few minutes each day it gave workers a release, an opportunity to step away from the assembly line (or the shop floor, the news room, etc.), engage others in conversation, and enjoy the simple pleasures of a steaming cup of coffee and a cigarette. Coffee breaks, which had originated before World War II, became even more popular during the war (both in the armed services and in industry) and in the postwar decade. By 1959 more than three-quarters of all U.S. workers had coffee available at their place of work (U.S. Department of Commerce 1961:23).

The effects of the boom in the Northern Sierra were extensive. Rapidly rising coffee prices led to an influx of more money in the region from 1946 to 1950 in particular. Consequently, more consumer goods were in demand: tools, kerosene lamps, hand grinders, cloth, soap, beer, and biscuits (Berg 1968:79-80, 289). In Talea (as in coffee-producing regions worldwide) many people began planting more coffee trees, hoping that prices would remain high for at least a few years. This later would have profound effects.

Roads for shipping coffee out of the Northern Sierra became a high priority in the wake of the price boom. In 1953 a road to Zoogocho was completed, which made it easier to transport large quantities of the crop into Oaxaca City. Until the end of the 1950s Zoogocho was the most important coffee trading center in the Northern Sierra; but in 1959 the situation changed radically, following a sustained lobbying effort on the part of several Taleans with coffee interests to complete a road to Talea. Almost overnight Talea surpassed Zoogocho as the most important coffee outpost in the Sierra Juárez. Three or four Taleans promptly took the place of an aging Zoogocho buyer but apparently wound up in structurally similar positions: they effectively became agents of Oaxaca City-based trading houses who collected coffee from local growers on market days in exchange for maize, merchandise, or cash supplied by their affiliates in Oaxaca (Berg

Table 7.1. *Coffee Production in Oaxaca, 1945–1963*

YEAR	TONS OF COFFEE	YEAR	TONS OF COFFEE
1945	6,400	1955	15,100
1946	6,500	1956	14,800
1947	6,400	1957	18,900
1948	6,300	1958	21,000
1949	9,300	1959	19,000
1950	10,100	1960	19,200
1951	10,200	1961	19,800
1952	11,100	1962	15,300
1953	12,800	1963	23,100
1954	12,900		

Source: Ruiz Cervantes 1988.

1968:75). In spite of this dependent relationship, many Taleans remember the period as one of cultural fluorescence, a glorious time in which their village emerged as the jewel of the Northern Sierra.

Mexican Government Initiatives: The 1960s and 1970s

Although more people in the United States were drinking coffee than ever before (consuming more than three cups a day per capita in 1962; see Mathews 1994), the early 1960s was a time of lower coffee prices due to the fact that trees planted in the late 1940s worldwide were now bearing fruit. World production of coffee reached record levels, and the cumulative effect was a flooding of the market with coffee beans and a sharp decline in prices (Table 7.1 illustrates the rapid rise in production in the state of Oaxaca). In response to the crisis, the most important coffee-exporting and -importing nations crafted the International Coffee Agreement (ICA) in 1959, which established export quotas in order to limit world supplies. The Mexican government played an active role in the ICA, which successfully helped raise prices again. Versions of the ICA would be renewed for the next thirty years.

Individual nations took a number of additional measures. For example, the Mexican Coffee Institute (INMECAFE) was established to improve the cultivation, processing, and marketing of Mexican coffee for export and to stimulate the internal market (Nolasco 1985). INMECAFE would later assume an even broader role.

As prices rebounded in the late 1960s and 1970s, more land was converted from subsistence cropping to coffee cultivation across the Northern Sierra. Informants note that in the 1970s the pace of conversion accelerated especially rapidly—though as a rule maize farming was not abandoned. There were a number of reasons: chemical fertilizers, first used in 1967 or 1968, quickly doubled maize yields, easing land pressure and freeing up land for coffee. (The cost of fertilizer also required that campesinos plant more coffee, which led to a need to convert more milpas to *cafetales*, which in turn required more fertilizer; thus, a vicious circle began.) A boom in world coffee prices in 1974 was also a factor.

Government initiatives helped expand production as well. As early as the 1950s engineers from the Papaloapan Commission, a federal development project, promoted new coffee varieties and techniques (Nader 1964). Furthermore, government-subsidized maize and beans were sold at CONASUPO stores beginning in the 1970s, and some campesinos converted more land to coffee production, confident that income derived from coffee production could be used to purchase cheap maize. All of these efforts can be seen as attempts on the part of the Mexican government to promote economic development in rural areas through the cultivation and improvement of export crops.

Making Money Grow on Trees: Cultivating Coffee in Talea

Over the course of the twentieth century Talean campesinos absorbed coffee into their agricultural repertoire and developed successful techniques for growing it on their lands. Knowledge about coffee cultivation has evolved over a period of decades; while the Taleans have appropriated the crop, they have adapted it through their own frameworks and institutions.

In the Northern Sierra of Oaxaca coffee is cultivated on small plots, unlike the situation in much of Chiapas and Guatemala or the Southern Sierra of Oaxaca, where plantations were (and still are) the rule. Indeed, it appears that the crop was grown on small privately owned holdings (as is the custom in the Rincón) from the time of its introduction. In Talea there are only a handful of campesinos who own more than five hectares of coffee land, and none own more than eight. In Yagavila, five hours' walking distance northwest of Talea, a similar situation exists: fully two-thirds of the households have *cafetales*, but none have more than three hectares (Tyrtania 1992). By no stretch of the imagination could these plots be referred to as plantations. In historical terms, coffee cultivation represents a much less brutal activity than silver mining, cotton weaving, and cochineal production—all harshly imposed by the Spaniards in Oaxaca during the colonial period. As far as we know, coffee was never coercively imposed in Talea, as it was in the Soconusco region or in Yagavila by Fidencio Hernández.

Planting

Coffee is planted in the early summer, when torrential rains fall several days each week. Some Talean coffee growers do this by uprooting small saplings (approximately 50–60 cm tall) which have sprouted below larger coffee trees and then transplanting them to new sites. According to informants, those saplings with the longest, most vertical central roots are preferable, for they develop the most extensive root systems. The sapling is placed in a hole one half-*brazada* deep (the *brazada* is equivalent to approximately 170 cm in Talea); then the tip of the central root is pressed into the deepest part of the hole and dirt is packed tightly around it, so that the root grows straight down. The rest of the hole is filled with earth, which should be kept loose so that the roots can extend outward.

Other campesinos prefer to grow coffee from the seeds of their best trees.[4] In this case thick plastic bags (about 10 cm wide and 30 cm long) are filled with humus-rich earth, and a coffee seed is planted a few centimeters below the surface. The seeds should be planted in late spring (at approximately the same time that the coffee tree drops its seeds); in this way rains are most beneficial. The bags should not be exposed to direct sunlight, but rather be placed in a shaded area, either underneath trees or below a canopy of palm leaves or branches. Campesinos transplant the young saplings when the bags can no longer contain the growing roots.

Once the saplings have been transplanted, they will only survive and prosper if they are placed in an area in which shade trees have been planted.[5] The most popular kind of shade tree is the *guajinicuil* (*Inga guinicuil*), a rapidly growing tree native to the New World that produces pods and sheds leaves every six months.[6] Within four years after planting it can reach a height of three to four meters. With time the trees grow even taller, and their branches spread horizontally, creating a canopy of green shade several meters above the tops of the coffee trees. The *cafetal* thus has several layers of vegetation: *guajinicuiles* form an upper layer of foliage, the coffee trees form a secondary layer, and smaller food plants and vines (chayotes, chiles, chives, etc.) are grown at the lowest levels. The *guajinicuiles* offer protection from unfavorable weather such as high winds and freezes. They give the *cafetal* its characteristically olive color when seen from afar. The tree's fallen leaves cover the floor of the *cafetal*, providing a natural compost, inhibiting the growth of weeds and grasses, and serving to retain soil humidity.

Another kind of shade tree is the *bocayo* (unidentified species). It is particularly useful because it provides a litmus test for prospective *cafetales*: if campesinos plant a *bocayo* sapling in a field and it develops well, they can be sure that coffee trees will also prosper in the location. Informants explained that both *bocayos* and *cafetales* thrive in humid soils.

Other trees are commonly grown in the *cafetal*, especially avocado, orange, lime, banana, and *aguacatillo* (*Persea* sp., which produces tiny edible avocados and fragrant leaves that are used to season tamales). Cactuses and maize may be grown at the edges of the *cafetal* or near the patio, where sunlight is strong. Peas, beans, and other crops may also be grown near sunny patios. In the case of one informant, the trees and cactuses provided food for his family and the avocado trees produced so many avocados that his wife was able to sell several dozen at the weekly market during the season.

Another important detail involves the site chosen for a *cafetal*. According to an informant, the land best suited for coffee is *tierra fría* (above an altitude of approximately 1,550 m), because at lower altitudes the plant does not produce as much fruit owing to a greater spacing between clusters of fruit. The preferred soil consistency is *tierra gruesa* (Sp., literally "dense earth"). ("Denseness" is characteristic of *tierra colorada*, a reddish, clay-like earth found in many parts of Talea.) This is because *tierra gruesa*—unlike *tierra arenuda* (Sp., "sandy earth," typically yellow or white)—retains much humidity and thus "feeds" the *cafetal*. Rocky soil should be avoided because it impedes the growth of the coffee trees' roots. In the "old days," when maize cultivation was still the primary occupation of Talean campesinos, coffee was planted only in marginal areas—near riverbanks or cliffs, within the village limits, or on ground too steep for planting subsistence crops. Now coffee is often planted on the choicest land; during the 1970s milpas were converted to *cafetales*, in part because of easing land pressure (due to outmigration and chemical fertilizers) and high coffee prices.

Water is another consideration when selecting a site. Cultivating coffee too far from a spring or a waterhole is a bad idea, for relatively large quantities of water are necessary for "washing" the coffee after its skin and pulp have been removed. The alternative is to ship the entire fruit back to the village or to another site for depulping.

It may be helpful to plow the prospective site before planting *guajinicuil* or coffee trees in order to loosen the soil. I was told that previously some campesinos applied chemical fertilizers to coffee saplings to stimulate growth, but this practice has disappeared since coffee prices dropped in 1989.[7] One informant commented on a case control experiment carried out by his father-in-law:

> He tried adding a spoonful of chemical fertilizer to each tree in a row of young saplings he had recently planted and left the others alone. What he found was that the fertilized saplings grew no faster than the unfertilized ones. So he decided not to do it again. Why spend the money on fertilizer if it's not going to have an effect?

Many campesinos use an "organic" fertilizer: after the harvest they take a bucketful of the coffee bean pulp and deposit it 50 cm or so from the base of the coffee tree (on the uphill side, so that it will wash down and around the trunk in the rainy season). Irrigation is also avoided by campesinos, who claim that adding water may help a sapling grow quickly, but tends to shorten the productive life of the tree because it "rots" the roots.

The recommended spacing between coffee trees is a point of contention and reveals different technical perspectives. One informant told me that two *brazadas* is the preferred distance between trees; this allows those harvesting ample room to pull down and extend the highest branches of the coffee tree during the harvest season. It also allows the trees room to expand; that is, they do not "choke" each other when spaced out in this fashion, and the root systems have adequate space to develop.[8]

One of the "richest" (land-rich) Taleans—who probably owns between six and seven hectares of coffee land—told me that a better method is to plant trees a distance of one *brazada* apart. After further questioning, I found out that this campesino is among the few who plants the newer varieties (the so-called *caturra roja* or *caturra amarilla* varieties—introduced by either the Papaloapan Commission or INMECAFE engineers), which grow to a short height (2-3 m), tend to produce fruit three to four years after planting, but have a life span of only fifteen years.[9] The creole or arabica variety, by contrast, grows much taller and can produce fruit for more than fifty years if cared for.

When I brought these points up with the first informant, he countered by arguing that the *caturra* varieties, though easy to harvest because they grow vertically and to a modest height, produce an inferior bean that weighs less than that of the arabica. He also noted that the branches of the *caturra* are brittle and prone to splitting, unlike the pliant branches of the arabica. His criticisms appear to correspond to the conclusions reached in a comprehensive study of Mexican coffee cultivation:

> It is naive to think that by simply substituting a "high-yielding" variety like the *bourbón*, for example, in place of the *criollo* (arabica), yields will increase automatically. On the contrary, replacing a highly adapted variety like the arabica—which has adapted to specific climatic, ecological, and agricultural conditions over the course of nearly 100 years—with new varieties can be dangerous and might actually result in lower yields. In the best-case scenario, such a strategy will produce similar yields to those produced before the change, and this is without considering the losses incurred in the process of replacing one variety with another. (Nolasco 1985:148; my translation)

The author continues: "In reality, we do not know if the 'improved varieties' that currently are used for the production of seeds are superior to the *criollo* (arabica) varieties" (Nolasco 1985:108).

Apart from the biological shortcomings of the *caturras*, there is also the labor question. Specifically, one wonders whether the short-lived *caturra* might represent a heavy burden for a relatively poor family (dependent almost entirely on household labor), since every fifteen years a major effort is necessary to transplant new trees. The "rich" Talean, who relies almost exclusively on hired help and has relatively large cash resources, might see an advantage in planting a tree with a shorter lifespan and higher yields. The *caturra* thus appears to produce a quicker return but requires a great amount of labor time as well—a requirement that many maize-producing families are unable to meet.

This brings up an important point about *ricos* and *pobres*, rich and poor Taleans.[10] The *ricos* (typically those who own five or more hectares or harvest 300 *arrobas* [approximately 75 *quintales*] or more annually) often have a different outlook on the economics of coffee cultivation. Specifically, they do not seem to place the same importance on *mantenimiento;* indeed, as I mentioned earlier, bookkeeping, not householding, appears to be a fundamental concept for these farmers. For them, coffee cultivation is a business enterprise, not a survival strategy, and they are more likely to experiment with new techniques and to buy machinery (like electric motors for their depulpers—see below). *Humildes* or *pobres*, in contrast, are more likely to work *como los abuelitos* (Sp., "as the grandfathers did"), as one young *rico* said, using no electric- or gasoline-powered machinery.

"Cleaning" (Weeding) the Young *Cafetal*

After approximately four or five years arabica trees begin bearing fruit. Before this time (while the shade trees are still too small to cover the entire *cafetal*) it may be necessary to weed or "clean" the floor of the *cafetal* with a hoe once a year to keep weeds and other grasses from robbing nutrients and humidity from the soil. The work is best done after the rainy season once the weeds have reached their maximum thickness. This task involves "rasping" the top surface of the soil to uproot weeds with short roots and then piling them into small mounds where they can quickly dry. Though laborious, "cleaning" may be discontinued once the *guajinicuiles* begin shedding sufficient foliage to cover the floor of the *cafetal*.

Diseases and Plagues

Coffee trees may be damaged by a number of plagues and diseases. One of the most common in Talea is a fungus called *ojo de águila* (Sp., literally "eagle's eye"; scientific name *Phyllosticta coffeicola*) because it is characterized by circular brown splotches on the leaves of the tree, which may grow larger than an inch in diameter. Though this is not fatal, the yield of an affected tree may be reduced significantly. Informants said that it was the result of too much sunlight—the disease is explained in terms of the "hot/cold" concept as a "burn." The remedy is obvious to Talean farmers: to ensure that the *cafetal* receives sufficient shade (this is confirmed in an important study of Mexican coffee [Nolasco 1985:143]). Another disease also affects the leaves of the coffee tree, leaving their surfaces smudged with a charcoal-like substance that appears to be a fungus (unidentified species). It is rare, however, and according to informants the disease only affects a handful of *cafetales* in an area known as Yazhdúlu and a neighboring area called Sudo'. Two of the most damaging blights, known as *roya* (coffee rust, scientific name *Hemileia vastatrix*) and *broca* (coffee bean borer, scientific name *Hypothenemus hampei*), have not affected Talean coffee.

"Cutting" Coffee: Harvesting the Beans

The coffee plant flowers in May, first in *tierra caliente* and later at higher altitudes. The fragrant white flowers bloom for several days and then fall away, revealing a tiny green coffee bean about the size of a BB pellet. The button grows larger during the rainy season, until it reaches the size of a grape, and begins changing color in October or November, from green to brown to yellow to bright red. By late November and December the crimson fruits—known as *café shna* (red coffee) in Rincón Zapotec—are ready to be picked in the lower-lying *cafetales*. At higher altitudes the harvest may not begin until January. It is preferable to pick fruits when they are dark red or maroon in color, once they are fully mature.

The fruits of trees exposed to the most sunlight—those nearest footpaths or the edges of the *cafetal*—ripen more quickly than others and are picked first. The picker wears a palm basket or some other container (gourd, armadillo shell, etc.) tied around the waist, pulls down the pliant branches of the coffee tree with one hand, and picks ripe beans with the other, dropping them into the basket handful by handful. Once the basket is nearly full, its contents are dropped into a 60-kg nylon fertilizer bag.

At the outset it is important to recognize that women play critical roles in the coffee harvest; in fact, they are responsible for picking at least half of the harvest and probably more. Campesinas often begin harvesting coffee earlier in the season than men (who are often occupied with the maize harvest in December) and frequently work in the *cafetales* of land-rich Taleans to help their families earn extra income. Although mothers often take their small children with them to the fields, most campesinas pick coffee as rapidly as men if not more so.

In some instances men are unable to supervise the coffee harvest in their groves. A merchant may be occupied in business affairs; others may be required to serve a year-long term in a municipal office; still others may have agricultural tasks to perform or may fall ill. Under these circumstances, campesinas of all ages may be called upon to coordinate and execute the coffee harvest in a given year.

There are major differences in the way coffee is picked. To begin with, there are a number of distinct labor arrangements which influence picking styles and speeds. One informant used only his own labor and that of his family; his philosophy was to work slowly but surely and always (in his words) "cleanly." What he meant was that the picking should be carried out in such a fashion that the leaves and branches of each coffee tree were left intact, that only ripe beans should be picked (that is, those that were completely red), and that no ripe fruits should be left on the ground. The harvest began with the *primer jalón* (Sp., "first stretch"), in which the ripest beans were picked; by the time this was done, the remaining beans had ripened and the rest could be picked by repeating the process, tree by tree, in the *segundo jalón* (Sp., "second stretch"). He had a limited amount of

Figure 7.2. A boy harvesting coffee. The harvest season lasts from December to February, and much of the work is done by women and children. *(Photo by Gabriela Zamorano)*

coffee planted (just over one hectare), and this kind of control over his product was entirely possible. Working in this fashion, a skilled campesino can fill a 60-kg bag of ripe coffee beans daily—equivalent to approximately one *arroba* (11.5 kg or 25 lbs) of green coffee.

Campesinos engaged in *gozona* also tend to work "cleanly," as if working their own land. In *gozona* one campesino (who, for example, might own a *cafetal* in a low-lying area where the beans mature quickly) "borrows" the labor of another for several days and then repays the work days when the companion's coffee (located at a higher altitude) has matured. *Gozona* allows the campesino to harvest the crop speedily. The company of another campesino is also welcome, for the harvest can be monotonous if done in solitude. As in other settings, *gozona* reinforces social ties. In this sense, coffee farming is strikingly similar to sugarcane cultivation.

In coffee cultivation *trabajando en compañía* (Sp., "working in company with another") is essentially a form of sharecropping or more precisely shareharvesting. It is popular among native Taleans who live outside of the region (in Oaxaca City or Mexico City), who grant a resident campesino family the right to harvest a *cafetal* and keep one-half (or, increasingly, two-thirds) of the harvest. It is also customary for the campesino to prune and weed the *cafetal* if necessary. Landowners often grumble about campesinos keeping more than their share, so pains are taken to find someone who can be counted on to do the work. It is necessary that the work get done: if abandoned for a season or two, the *cafetal* becomes less productive as weeds and vines choke the trees and the fruits go unharvested. Working in company binds together native villagers living outside the region and resident Taleans.

In other cases coffee farmers with several hectares employ hired hands to pick large percentages of their crop. Many of them are paid by the bag (25 or 30 pesos per bag during the 1996–1997 season) rather than by the day, so a mad rush to pick the beans follows, in which entire branches are stripped of fruits, leaves, and everything else. Green fruits are mixed in with ripe ones, branches are broken, and the most out-of-the-way branches go unpicked. Working in this fashion, a field hand can fill two or three bags daily. *Trabajando cochino* (Sp., "working dirty") seems to be the logical result of working on someone else's land on a per-unit basis.

The different picking styles illustrate differences between the conceptualization of coffee by *ricos* and *humildes* in Talea. The rich generally place primary emphasis on getting the most money out of their *cafetales*, while the humble—the great majority—give priority to subsistence cropping and rely on the *cafetal* as a supplement for household purchases and also as something to be cared for (by working cleanly) for the maintenance of future generations.

Many Talean coffee growers noted a severe labor shortage plaguing the village, especially during harvest time. This situation was apparently a problem as early as twenty years ago, perhaps as a consequence of migration out of the village (Hirabayashi 1993:53). Today field hands are difficult to secure. Many Talean youths prefer to remain in the village, where there is plenty of work in construction, carpentry, and other trades. Most of those who do go to the fields are unwilling to stay for more than a day at a time or to venture far away from the village limits. Therefore, many of the laborers come from the neighboring villages of Yatoni, El Porvenir, Las Delicias, and Yojobi. Older Taleans, whether rich or poor, insist that laborers in general are treated much better compared to a generation ago, when the *patrón* reportedly limited the amount of food they were served. (The *patrón* is customarily responsible for preparing meals—the field hands provide their own tortillas, however.)

According to one Talean informant:

In the old days we were allowed only one bowl of beans to eat with our tortillas. Nowadays if the *patrón* doesn't give you meat on occasion, or eggs or rice or some other pasta, he's seen as a miser [*codo*], and he's going to lose his *mozos* [field hands] to some other *patrón*. Also, many *mozos* expect to be released early, as soon as the sun goes down, and this was not true before. Things have really turned around... now it's the *mozo* who sets the rules, not the *patrón*.[11]

The labor situation has become so acute, in fact, that several years ago one of Talea's most important growers is said to have imported labor from the Pacific coast of Oaxaca (perhaps contracted in Oaxaca City) for the harvest.

Sometimes campesinos are unable to harvest coffee on time, because of other commitments (for example, a delayed maize harvest) or a lack of field hands. In these cases the beans may begin to shrivel on the branch, making them impossible to pick without shaking them loose. In extreme cases plastic sheets must be spread below the branches to collect the precious beans. The harvest is an intricately timed process that must be done rapidly during a season punctuated by major fiestas and other obligatory activities. Add to these obligations the search for hired help and the cold weather and illnesses and one can easily understand why for many campesinos the months of December and January are the most stressful of the year.

Yields

How much coffee do campesinos harvest? A study of Mexican coffee reports that the 1982 national average was 12.3 *quintales* of green coffee per hectare (1 *quintal* equals 46 kg). According to the survey, Oaxaca has one of the lowest yields in the country (Nolasco 1985:137–138). Another source notes that "traditional" coffee cultivation can yield up to 12 *quintales* per hectare (Tyrtania 1992:213). Tyrtania claims that the regional average is approximately 8 *quintales* per hectare, but that Sierra *cafetales* can produce up to 20 in peak years.

Informants explained that it is difficult to estimate average yields because of variations in soil. In the words of one campesino: "Even within the same *cafetal*, there are good spots and bad spots. It's like a bowl of soup ... sometimes you get chunks of meat, nothing but good stuff, but other times you get mostly bone. That's the way it is with the soil." Weather patterns are another source of unpredictability.

Even more confusion follows from the fact that in a given *cafetal* the yield fluctuates every year. That is, if a "loaded" *cafetal* that is fully mature produces 30 *arrobas* in 1995, it will produce only about half of that amount in 1996, for like a person it needs time to "rest" or "recuperate." In 1997 it will resume full production of approximately 30 *arrobas*. Under ideal soil and weather conditions, one can expect to harvest 60 to 65 *arrobas* per *almud* (15–16 *quintales* per hectare) during a full production year, according to Talean growers.

There is apparently no direct correlation between productivity (yield) and the size of a *cafetal*. In other words, a large *cafetal* or plantation does not consistently yield more *quintales* of coffee per hectare than a small *cafetal*. In fact, the opposite is frequently true, since a small *cafetal* can often be worked much more intensively than a large one (Nolasco 1985:139). This is one reason why coffee is sometimes described as a "democratic crop" (Greenberg 1994:25). It does not show a tendency to favor large growers.

Depulping, Fermenting, Washing, and Drying Beans

At the end of each day, as dinner is being heated, the campesino removes the pulp from the coffee beans using a cast-iron machine called a *despulpadora* (Sp., "depulper"). The campesino uses a gourd to pour coffee beans into a receptacle at the top of the machine, from which they are drawn into a chamber where a rotating perforated copper drum pulls the skin and pulp from the seeds. The seeds fall into a bag, while the skin and pulp are released on the other side of the apparatus. The depulper may be rotated by hand or connected to an electric or gas motor; consequently, campesinos with more than two hectares of coffee land often purchase motors to make the work easier.

In years past, I was told, depulpers (like sugarcane mills) were made from wood by carpenters in Talea, and I had the opportunity to see one in the neighboring village of Yaeé. These apparently required much more physical effort to operate. The cast-iron models come in several different sizes, the smallest of which weighs approximately 30 kg. The cost in 1995 was N$700, worth approximately US$100 or twenty-eight days of labor.

The coffee beans, coated in the pulp's clear mucus, are set aside in the sun for two to three days so that the slime can ferment. They may be left in the plastic fertilizer bags for this purpose or may be placed in wooden tubs and covered with plastic sheets. This process is called *pizcando* (Sp., literally "harvesting," but the term also sounds like *pellizcando*, "pinching") because once the fermentation has eliminated the slime the beans make a grinding or crunching sound when rubbed together in the clenched fist of the campesino.

The INMECAFE, in its efforts to assist Talean coffee growers, issued plastic bins in the 1980s for the purpose of fermenting depulped coffee beans. The black bins, measuring about 50 cm by 75 cm by 100 cm, were put to good use, but in ways that the designers did not intend. According to one campesino:

One year they gave us the bins for fermenting coffee. But they weren't convenient— they were too big for the small amounts of coffee we harvest and besides that they were really strong boxes that could be put to better uses. Soon the campesinos began using them for other purposes. Chente used his to store firewood; Pedro filled his with green bananas and covered it with plastic so that they could ripen more quickly; Urbano poured sugarcane juice and pulque in his, covered it with a plastic sheet, and used it as a *tepachero*, to ferment his cane juice! But the biggest surprise of all was when the engineer came to visit the *cafetales* and saw one bin lined with blankets and a bundle of rags. Suddenly the rags began to move. A woman had put her baby in it—she used it as a crib for her child! The engineer just stared blankly at the bin. He was impressed!

At times the possibilities available to campesinos appear to have no limit.

Next the coffee is "washed." The beans are transferred to a specially made perforated tin (sometimes as large as a bathtub), and a water hose is dropped into the tank. A stirring paddle is used to loosen the remaining bits of pulp and other "trash" from the coffee beans. The effluence drains away through the perforations, the beans are transferred gourdful by gourdful to a smaller "tin" (a plastic tub with holes burned into the bottom of it with a heated nail), and then the process is repeated, using the arm and hand instead of the stirring paddle. After this second washing, the beans may be dried.

The clean coffee is sun-dried for two, three, or more days, depending on the weather. In years past, most campesinos laid straw mats across a patio and spread the coffee beans out upon them using a rake-like apparatus. Today a number of wealthier campesinos have constructed cement patios for this purpose. This eliminates the need for straw mats, though there are some who say that concrete is too hot and does not allow the coffee to dry evenly. Others with *cafetales* near the village transport their coffee to Talea and lay it out on the edges of one of the principal streets, on a cement rooftop (if they have a cinder-block house), or on a cement patio. The beans are spread out in the sun before breakfast and gathered before the sun sets. Campesinos test the beans' dryness by either biting a few samples or else pressing them with a thumbnail. If the gray heart of the bean (referred to as *café oro* [Sp., "gold coffee"], found inside the parchment) can be indented, it has not yet "reached its point." Once it is dry, there is little danger that the stored coffee will mildew. It also makes more sense to dry the coffee thoroughly before transporting it to Talea so that the beasts of burden will be able to carry as many beans as possible.

Most campesinos who grow coffee are very proud of its quality. If the coffee growers of the Sierra have a reputation for producing dirty coffee, I was told, this is unjustified; it is the *coyotes* (intermediaries) who add trash to skim off a higher profit. Rumors circulate about the dishonest practices of certain coffee buyers in Talea who are said to add dried pulp or water or both to green coffee before shipping it to Oaxaca City to increase the weight. By the early 1990s all of the buyers were allied with a Talean native who, according to informants, is one of the most powerful coffee men in the state of Oaxaca. He is said to own more than ten trucks and buys coffee over an area that extends from the Sierra to the Mixes to the Cañada and even to the Southern Sierra. One informant estimated that close to 2,000 bags (approximately 500,000 pounds) of green coffee are received each week at his warehouse in Oaxaca. Perhaps as a result, people from smaller Rincón villages sometimes stereotype Talea as a village inhabited by rich people who are mostly coffee buyers.

Figure 7.3. A Rincón family sun-drying coffee beans. Coffee is the primary source of cash income for most farmers working in the Rincón. *(Photo by Rafael Colin)*

The campesinos' opinion of retail coffees was summed up concisely in one phrase: "Nescafé no es café" (Nescafé is not coffee). They are correct. The retail coffee sold by Nestlé in Mexico consists of up to 30 percent additives (Hernández Navarro 1992).

Coffee for Household Consumption

Although the vast majority of Talean coffee is stored in bags for eventual sale, nearly all households keep a supply of toasted ground coffee. Women are the experts in coffee toasting. The "traditional" method involves toasting the beans on a clay griddle until the coffee "reaches its point." This can be determined by the color of the bean, which should turn a chocolate color. To ensure that the beans toast evenly, they use a tiny wooden paddle to stir and rotate the beans. The women then grind the beans either by hand or (more commonly) at an electric mill that specializes in grinding coffee to a fine, flour-like consistency.

Coffee has taken the place of other beverages which were more common in years past, most notably *atoles* and teas. In the Sierra the beverage is prepared with relatively little coffee (approximately two teaspoons per liter of water) and is heavily sweetened with *panela* (known as *piloncillo* in other parts of Mexico).[12] It tastes rather like the *café de olla* prepared in Mexico City and other parts of the nation. To prepare a good batch of coffee, one should drop a fist-sized lump of *panela* into a clay pitcher of water (approximately two liters), allow it to boil, and then drop four to five teaspoons of coffee into the container. After the water froths to the surface once or twice, the pitcher should be taken off the fire and the coffee grounds allowed to settle. Taleans drink their coffee very hot.[13]

One young Talean entrepreneur processes and packages coffee in the village. He roasts the coffee in an electric roaster, grinds it in another machine, and uses a third to package it in plastic bags, which feature the brand name "Café Taleano." He recently secured a contract with the CONASUPO stores to supply coffee to several dozen communities in the region. The coffee—consisting of toasted arabica beans free of fertilizers and pesticides—sold for eight pesos per half-kilo (just over US$1.20 at the time) when a similar quantity of mass-market coffee in the United States sold for more than three times the amount.

Sustainable Agriculture: The Case of Shade Coffee in Talea

In environmental terms, Talean coffee cultivation might be described as environmentally sustainable. *Cafetales* mimic the structure of a forest: *guajinicuiles* form an upper layer of foliage, coffee trees form a secondary layer, and smaller food plants and vines form a lower layer. As mentioned earlier, *guajinicuiles* offer protection from high winds and freezes, while their leaves cover the floor of the *cafetal*, providing a natural compost that inhibits weed growth and retains soil humidity. Like other pod-producing plants, the *guajinicuil* fixes nitrogen in the soil. Shade trees offer a number of other benefits as well: a "smoother" tasting (less bitter) coffee; protection from the coffee leaf spot fungus (*Phyllosticta coffeicola*) and other diseases; soil conservation; and a high degree of biodiversity (Nolasco 1985:113). In fact, recent studies have demonstrated that this system of "shade" coffee is almost as good as original forest in terms of biodiversity.

Some have suggested that shade coffee growing techniques in Meso-

Figure 7.4. Cafetales in Talea are a good example of "shade coffee" systems that replicate forests. The coffee trees (which appear as dark shrubs) are covered by a canopy of *guajinicuil* shade trees. Scattered throughout the *cafetal* are smaller food plants such as chayote vines, chile plants, and chives. *(Photo by Gabriela Zamorano)*

america may have developed from cacao cultivation methods invented long before the arrival of Columbus (Greenberg 1994). Indeed, the two crops are grown using very similar techniques in parts of southern Mexico, but it is difficult to say whether in fact shade coffee groves were deliberately modeled after cacao groves in the 1800s, mainly because of the scarcity of historical materials.[14]

By contrast, many large coffee plantation owners follow an agribusiness approach: they monocrop the trees in direct sunlight, which has serious environmental consequences: "sun" coffee requires heavy fertilization and fumigation and destroys the biodiversity characteristic of shade coffee plantations, creating an ecological wasteland. In extreme cases the soil can be ruined as a result of sun coffee plantations. One of the most widely publicized effects of the monocrop is the disappearance of several species of migratory birds from these regions (Greenberg 1994).

Talean campesinos also plant food crops either within the *cafetal*, as we have seen, or at the perimeter. In addition to chayotes, chiles, and chives, I saw many other food plants grown within *cafetales* and at the edges: oranges, limes, avocados, banana trees, cactus, maize, sugarcane, beans, and peas. Campesinos not only create forest-like spaces; they also intercrop, which has substantial advantages as far as soil conservation, erosion, and pest control are concerned (Richards 1985; Altieri 1987). Talean coffee-growing methods appear to be remarkably well suited to maintaining lands for future generations.

Talean Coffee Experiments: The Campesino as Scientist

Like agricultural scientists, Talean campesinos conduct experiments, formulate hypotheses, mold their results to theoretical frameworks, and disseminate their findings, from campesino to campesino and from parent to child. Their knowledge increases through a steady process of trial-and-error experimentation.

When coffee was first planted, the ancestors of today's Taleans apparently planted trees extremely close together (less than a *brazada* apart—"as if it were maize"), perhaps hoping to harvest more fruit this way. That caused problems, for the coffee trees could not extend their branches to receive sufficient sunlight. This was explained by campesinos in terms of an imbalance of "hot" and "cold." The crowded *cafetales* suffered from an excess of "cold," and it became impossible to harvest the beans because the tree branches could not be pulled down past the branches of adjacent trees. After the experiment failed, future generations spaced their trees at greater distances.

Campesinos also know that a *cafetal* needs humid soil and shade or the sun will "burn" it. They have found that one tree in particular, the *bocayo*, provides a good soil suitability test for *cafetales*. A *bocayo* sapling is planted in a prospective site; if it develops well, chances are that coffee trees will prosper there. In the winter shade trees "blanket" the coffee trees and thus provide protection from another kind of "burn"—frostbite.

Papaloapan Commission engineers and Mexican Coffee Institute agronomists arrived in the 1950s and 1960s with a host of new techniques and technologies, but were met with skepticism by most campesinos. After fifty years of learning how to grow arabica trees successfully under local soil and weather conditions, Talean growers were reluctant to begin using new seed varieties. Those who did planted only a few trees at a time, perhaps planting a single row of the *cafetal* as a test. The same procedure has been carried out with chemical fertilizers and pesticides, but with less frequency because of the high cost. The jury is still out on the new varieties; time will tell whether or not they are worth the trouble of replanting every fifteen years. Pesticides and fertilizers appear to have failed in the Sierra's *cafetales* in terms of effectiveness, cost, and (as we shall see below with respect to organic coffees) marketability.

The INMECAFE created other problems; in particular, it seems that in the 1970s its policies accelerated the process of social stratification. According to one anthropologist, growers were only extended loans if they implemented labor-intensive improvements (terracing, chemical fertilization). Only the wealthiest villagers could afford to do the work; when they began to realize increased production, they became even wealthier than the smaller growers. The INMECAFE "began to penalize people who brought in stained or badly dried, low-grade beans, by refusing to buy until later, by offering a lower price, or by refusing credit. Slowly the smaller and poorer producers were being edged out of coffee growing; this in turn undermined their ability to survive in the rural area" (Young 1982:172). In the end, many of the INMECAFE projects proved to be impractical and ineffective.

Even today various officials from the government and NGOs arrive with plans to improve coffee but are politely ignored by many campesinos. After one such visit, an informant confided:

Those kids come from Oaxaca City or Mexico City. *They don't know our lands.* We have poor soil. Our fields have only a thin layer of earth and below that there is rock. They come from lands with rich soil, and they have a different way of doing things. Pure theory, no practical experience.[15]

Many Talean campesinos see themselves as expert cultivators of the crop, for they have highly effective techniques for growing coffee on land that they and their ancestors have worked for centuries.

Hypotheses, experiments, observations, theories, dissemination of results — all of these elements are present in Talean coffee cultivation. From this perspective, the campesinos of the Northern Sierra are engaged in genuine scientific activity.

Coffee and "Tradition" in Talea

Over the course of the twentieth century many Taleans began to refer to coffee as a "traditional" crop. How did this occur? This is a fair question, for a great deal of research emphasizes the destructive tendencies of cash cropping in general and coffee cultivation in particular. Coffee has been described as a crop that can lead to social stratification in egalitarian communities (Hernández Díaz 1987), land conflicts, increased violence and homicide (Greenberg 1989), the deterioration of biodiversity (MacVean 1996), environmental degradation (Borgstrom 1967), a dependency on far-flung markets and transnational conglomerates (Early 1982), and the weakening of indigenous peoples' self-sufficiency (Paredes, Cobo, and Bartra 1997).

Yet most villagers see coffee favorably, and over the course of the twentieth century coffee became a vital part of what it means to be Talean. It has become normalized, a part of the local scene. Villagers refer to the arabica variety—first introduced at the beginning of this century—as "creole" or "native," as if it were indigenous to the area. Bumper stickers, T-shirts, and video cassettes sold in the village feature images of coffee trees and ripe red beans. The standard invitation to eat is "let's go drink coffee." The heavily sweetened beverage is consumed with every meal, and it is more common to wean infants with bottles of warm coffee than with bottles of milk. It is just as common for visitors to be given ground coffee as a gift as to be given black beans or *panela*. Coffee, in effect, has been appropriated by the Taleans. This seems counterintuitive, given the fact that with coffee the Taleans have connected themselves to unpredictable world economic markets. Why did this occur?

Coffee and Its Meanings

I have already discussed how coffee helped Taleans survive when their means of livelihood, mining, came to an end. Income derived from coffee helped them obtain staple foods which probably would have been impossible to cultivate on a small land base.

Taleans also like coffee because, relatively speaking, it is easy to cultivate. It does not require stooping all day (as when weeding a maize field) or working in the scorching sun (as when planting beans) or getting hands stuck with the painful hairlike barbs that cover sugarcane stalks. When the mosquitos are in full force, picking coffee can be inconvenient, but never exhausting and certainly not dangerous. Once the trees begin growing, they are nearly maintenance-free, for little weeding is necessary and branches only occasionally need pruning. Arabica trees, if cared for, can last fifty years or more. Picking coffee is one of the few jobs that even small children can do, and the harvest is a rare opportunity for entire families to spend time working together. Very young children occasionally work, and it is not considered exploitation:

My Magali is one clever little girl. She went with her mother yesterday to pick coffee in Victoria's *cafetal*, and she strapped a tiny little straw basket around her waist and started picking beans . . . and she's a good worker, too! Of course she could only reach the lowest branches, but she knew to pick only the red berries and she filled up her basket several times. Other kids spend all day playing in the *cafetales*, making trouble, but my daughter is a worker. And she's only four!

For men, the harvest is a nice change from the relative solitude that often characterizes the rest of the farming cycle. For women, it is an opportunity to be outdoors, away from the monotony of daily household chores. Much gossip, joking, singing, and flirting goes on in the *cafetales*.

Campesinos often talk about coffee trees as if they were savings accounts. Having the foresight to plant coffee after marriage ensures that one will have cash for the expenses that children bring:

Thank God that I woke up once I got married and planted my *almud* and a half [approximately one-third of a hectare] of coffee. As soon as the trees started giving fruit, Rosa, my oldest child, entered school. Where would I have gotten the money for her uniform? For notebooks? For pencils? Then the trees started giving more fruit, and Yadira and Panchito entered school. More uniforms, notebooks, pencils, materials for science projects, and everything else. And luckily by the time Rosa comes of age the trees will be at their peak, so we'll have the two or three thousand pesos [approximately US$350 to US$500] necessary to buy the chickens and the ingredients for the *mole* and for the wedding party. If it weren't for my *cafetal*, a single wedding could put me in debt.

Thus coffee helps pay for life-cycle events like births, weddings, and funerals and for school and medical expenses. Coffee trees, in fact, are intimately bound to the life cycle of individual campesinos and their children, and at times the growers talk about the struggle to cultivate a *cafetal* successfully as if it were a child:

It took me a hell of a lot of work and sacrifice to raise that little *cafetal*. Cleaning [weeding] it, pruning it, replanting saplings that didn't develop well, fertilizing it with rotting pulps. But it grew well, and my trouble paid off—now it hardly needs any care.

The striking thing about this is that the time needed for an arabica tree to mature and the time needed for a child to mature are nearly equal. Time and again, coffee is described as a means to a practical end: as a source of cash for educating children and putting them through school. It is important to note that school is no longer optional, for at the village level a *regidor de educación* (Sp., elected councilman in charge of education) serves as a truancy officer, enforcing what has become a general consensus among villagers: at least six years of elementary school education are necessary "to defend oneself" against unscrupulous villagers, politicians, and outsiders. Coffee, in short, is a subsistence strategy for maintaining oneself and one's household in the contemporary world, because without the crop formal schooling would become impossible.

Many campesinos have fond memories of working as children alongside their parents in young *cafetales* that they would later inherit. Consequently, *cafetales* and nearby ranch houses often hold sentimental value. Another informant commented on the interrelatedness of children and *cafetales*:

It's hard to have kids without a *cafetal*, and it's hard to have a *cafetal* without kids. These days, school is obligatory—it's not like the old days when nearly everyone including wives and children would live out on the ranches. So your kids are in school and you have to have the money for that. Even births cost money now that everyone calls on the doctor or Nati [the registered nurse] to deliver the baby. Martita's birth cost me 700 pesos [US$100] last year! But during the school break at the end of the year—and once they are done with school—you need your kids to help you out with the harvesting or the coffee trees won't get picked.

It is not uncommon for young people to be given a small *cafetal* as an advance inheritance after getting married. Single men or women who are likely to inherit significant amounts of coffee land are considerably more eligible than those who are not.

Some campesinos admitted that planting coffee trees is a much sounder investment than putting money in the bank. Even with its occasional price swings, coffee is still a safer bet than the Mexican peso. Coffee may rise and fall, they say, but the peso keeps sliding down. From mid-1994 to mid-1997 the value of the peso with respect to the dollar was cut in half.

Most campesinos refuse to sell all of their coffee at once (even if the price is high), because in Talea cash is generally a much more liquid substance than coffee. Once within the household, money is difficult to monitor and tends to slip away quickly; but a sack of coffee is easier to monitor, and selling it usually requires a short conference between husband and wife. Selling coffee in Talea is not like trading stocks and bonds. It is more like withdrawing money from a bank account.

Low coffee prices, like poor harvests, can be disastrous under certain circumstances. An ill child can drain cash resources very quickly:

My poor daughter. She wouldn't take milk from my breast; you see I was suffering from *susto* [literally "fright" — see Rubel 1964] and she wouldn't take my breast because my milk was bad, but she would drink powdered milk—NIDO [Nestlé's Mexican brand]. The problem was the price of coffee was so low [in 1993] that we had to exchange one *arroba* of coffee for one can [4 liters] of powdered milk. Thank God she developed well and she's healthy now, rosy cheeks and pretty and clever but *jesús!*—she was an *expensive* little girl!

When asked about possible alternatives to coffee, informants said that long ago they sold *panela* to earn cash. But *panela* does not always sell—and rarely sells in large quantities. Besides, the market is typically saturated by people from nearby villages (particularly from neighboring Yatoni) who sell the same product. With coffee, campesinos don't always get the price they would like, but at least they can sell as much as they want. The alternative to coffee, judging from the words of the Taleans, is not *panela*. It is Los Angeles.

How Coffee Became Traditional: Mixed Cash and Subsistence Cropping

Taking these perspectives into account, it is easier to understand why coffee is embraced as a "traditional" crop in Talea. Though still valued for its economic worth, coffee is important for other reasons as well. In the village coffee cultivation has occurred in a more or less egalitarian manner, on small plots distributed among many households, and thus replicates a pattern that has existed in the Northern Sierra for many years.

Coffee has not replaced maize farming. The saying "You can eat maize, but you can't eat coffee" has guided most of the Sierra's campesinos, including Taleans. Out of more than 300 households describing themselves as "campesino," more than three-quarters reported planting maize in 1995, and the majority of them planted more than half of what their households consumed. Most growers continued planting maize through the coffee boom and were prepared when the bottom dropped out of the market.[16] In this sense, coffee might be seen as one element in a multifaceted household maintenance strategy revolving around four crops: coffee, beans, maize, and sugarcane. Crop diversification allows farmers to survive fluctuations in the price of coffee or maize and bean crop failures.

To reiterate: coffee has not taken the place of maize farming. It has certainly changed Talea's physical appearance (having led to the construction of a road, cinder-block buildings, etc.) and during some periods has promoted migration out of the village. But it has also allowed many Taleans to provide food for their households more easily, not unlike the case of Totonac communities in the Northern Sierra of Puebla state:

> Coffee cultivation has reinforced the domestic mode of production because the two principal crops—maize and coffee—exist in a complementary relationship with respect to the agricultural cycle . . . Coffee cultivation has effectively countered the capitalist process of the proletarianization of the *campesinado* . . . In the Northern Sierra of Puebla, a contrary process exists. (Ruiz Lombardo 1991:17; my translation)

When I left Talea in 1997, Mexico's economic crisis was so severe that more than twice as many people were migrating into Talea as were migrating out. Most of these people were Taleans returning from Mexican cities (where unemployment was extremely high), and many were returning to work in the fields. The boom in coffee prices undoubtedly attracted them; in February 1997 one could earn up to 60 pesos (almost US$9) a day picking coffee in Talea, compared to a daily wage of 25 pesos (US$3.75) working as a day laborer in Mexico City. Nearly three years later, in late 1999, the trend had reversed. Low coffee prices triggered the migration of dozens (perhaps hundreds) of Taleans to Tijuana and Los Angeles.

Coffee also forms the basis for a uniquely Talean tradition vis-à-vis other Sierra villages. The Taleans have distinguished themselves from their neighbors over the course of the century by adopting coffee to supplement their subsistence base—to a greater degree than any other village except Tanetze. This has occurred in part because Talea, long considered "neither fish nor fowl" (Nader 1964) due to its post-Conquest origin, has been locked in rivalries with surrounding villages, not unlike communities throughout the state of Oaxaca (Dennis 1987). Coffee, after all, means money, and money means prestige; therefore, the village with the most coffee often has the most prestige.[17]

Talea's coffee tradition is a case of cultural creativity. At the turn of the century, faced with a situation in which their means of livelihood ended, people looked for a strategy for surviving on limited land. They found coffee. It was cultivated in egalitarian fashion, on small plots (probably improving on a more authoritarian work arrangement in the mines), and this practice continues today. By refusing to give up subsistence crops, they successfully survived price crashes. Most villagers think that the crop serves them well, and over the century they have evolved effective, environmentally sustainable farming methods and have even used coffee to define their village in relation to others.

Bitterness and Power: The Politics of Coffee in the Northern Sierra

The Taleans have created a "tradition" of coffee cultivation by adopting the crop and learning to grow it successfully in a geographic region they intimately understand. They have even incorporated coffee as part of their identity. But is it possible that, at a broader level, coffee cultivation—conceived as a set of knowledges embodied not only in techniques but also in social arrangements and institutions—might form a basis for wider movements that might potentially challenge the economic and political power of state and corporate entities? In short, can coffee cultivation form the basis of new cultural identities that might ultimately serve to promote political, economic, and social transformations in southern Mexico (Hernández Castillo and Nigh 1998)?

One thing is certainly clear: coffee has strong political implications. At least part of the discontent fueling the 1994 Chiapas uprising was associated with the hopelessness of small coffee growers and landless workers employed on large coffee plantations. Low coffee prices, the unscrupulous credit and purchasing practices of intermediaries, and abysmal labor conditions on plantations undoubtedly contributed to the events leading up to the revolt (Hernández Navarro 1994). Coffee cultivation was used hegemonically in the Rincón more than a century ago by caciques like Fidencio Hernández, but is it possible that coffee cultivation can be used counterhegemonically today?

The Collapse of INMECAFE and the 1989 Price Crash

INMECAFE reached its apogee from 1973 to 1989, when it began financing, purchasing, storing, and exporting coffee. Its building blocks were local groups of small coffee producers that received annual group loans, disbursed them among individual growers, and collected them (in green coffee) at the end of the cycle (Hernández Navarro and Célis Callejas

1994). In addition, INMECAFE played a vital role in guaranteeing a minimum price for small growers who might otherwise be cheated by intermediaries.

The percentage of Mexican coffee purchased by INMECAFE increased from less than 7 percent in 1971 to nearly 20 percent in 1973 and to nearly 50 percent in 1982 (Nolasco 1985:186). In the Sierra Juárez approximately 75 percent of the region's coffee was purchased by INMECAFE in the early 1980s (Tyrtania 1992:218). In 1982, however, coffee growers nationwide began mobilizing against INMECAFE, which was accused of offering low prices, delivering credits too late in the season, weighing coffee unfairly, downgrading the product, and, in some regions, conspiring with caciques. Protesters staged rallies at INMECAFE offices, demanding reforms and a more efficient organization. They also protested by cutting back on the amount of coffee delivered to the institute. In the early 1980s small growers began building grassroots cooperatives aimed at issuing credit, purchasing, processing, and marketing the crop (Hernández Navarro and Célis Callejas 1994:220-221). One of the earliest cooperatives was the Union of Indigenous Communities of the Isthmus Region (UCIRI), formed in 1981 in Santa María Guienagati, Oaxaca, with the help of progressive Catholic priests. The UCIRI specialized in organic coffee for the European market (Vásquez y de los Santos and Villagómez Velázquez 1993). Other regional cooperatives founded in Oaxaca during the 1980s included the multiethnic Union of Indigenous Communities of the Northern Isthmus Zone (UCIZONI) and the Union of Indigenous Communities-"One Hundred Years of Solitude" (UCI-100).

In 1989 several events occurred which profoundly changed the roles of INMECAFE and the new peasant cooperatives. President Carlos Salinas de Gortari announced the restructuring of INMECAFE, which relieved it of its most important functions.[18] A few months later the International Coffee Organization (ICO), a global coffee cartel, announced the failure of member nations to renew an agreement on export limits, which flooded the world market with coffee and led to a sharp drop in prices. This, coupled with the severe effects of Mexico's "structural adjustment" policies, led to the worst crisis in the coffee sector in nearly thirty years (Hernández Navarro and Célis Callejas 1994:217).

The regional coffee cooperatives mobilized rapidly. During the summer of 1989, twenty-six organizations formed the National Network of Coffee Producers or CNOC (Hernández Navarro 1992; Moguel 1992). Among the most important members were the UCIRI, the Coalition of Ejidos of the Costa Grande of Guerrero (Paredes and Cobo 1992), the Union of Unions of Chiapas (Harvey 1992), and the Union of Coffee Producers of Veracruz. By 1997 the CNOC included 107 organizations representing 70,000 small growers (Morales 1997), and eventually it secured contracts with European and North American buyers, including the English company Twin Trading, the Dutch company Max Havellar, and Aztec Harvests and Thanksgiving Coffee, both based in northern California (Hernández Navarro 1992:93; Moguel 1992:105).

To understand the rapid mobilization of these cooperatives, we must put them in proper context. Most importantly, the emerging networks included many indigenous people who had previously been affiliated in the early 1980s with the so-called Community Food Councils (CFCs) of the Mexican Food System (SAM), an ambitious but highly problematic project instituted to help Mexico achieve self-sufficiency in maize and beans by delivering subsidized inputs to farmers (Fox 1992). The CFCs were established by the government as community-level councils responsible for receiving and distributing hybrid seeds, fertilizers, crop insurance, and credit to small farmers. In the process indigenous people serving on the councils (often elected by their fellow villagers) learned how to gain access to state bureaucrats working within the system, who in turn might authorize them to use equipment such as trucks, processing machinery, warehouses, and so on. Once the SAM ended in the early 1980s, many ex-CFC members provided their knowledge, experience, and organizing skills to the emerging cooperatives and later the CNOC (Fox 1992; Moguel and Aranda 1992:168–169). Along similar lines—and just as important—were community groups in villages established in the 1970s to deliver credit from and collect coffee for INMECAFE. These so-called Unions of Small-Holding Producers of Coffee (UEPCs) consisted of people who served as liaisons between their home communities and INMECAFE. In this sense, they may be described as cultural brokers who maneuvered in the space between their villages and state bureaucracies like INMECAFE. They quickly applied their skills to the mobilization of cooperatives in the vacuum created by the withdrawal of INMECAFE in 1989.[19]

Oaxacan Cooperatives: CEPCO, UNOSJO, and MICHIZA

Early on, Oaxacan growers affiliated themselves with the CNOC through a statewide network called the State Network of Coffee Producers of Oaxaca or CEPCO (Moguel and Aranda 1992:181–182). CEPCO was founded in the summer of 1989 by a number of grassroots organizations including the

UCIRI, the UCIZONI, UCI-100, and the Pueblos Unidos of the Rincón of the Sierra Juárez (or Pueblos Unidos),[20] as many ex-CFC representatives and ex-UEPC members reestablished their institutional links with state bureaucrats and formed new ones with NGOs. CEPCO grew quickly: by 1994 more than 20,000 growers had joined, and by the early 1990s it "was the most consolidated, autonomous grassroots economic network in Oaxaca" (Moguel and Aranda 1992:187-188; my translation; Fox 1994:207).

In 1992 organizations from across the Northern Sierra formed the Union of Organizations of the Sierra Juárez of Oaxaca (UNOSJO), which was to become the most important regional grassroots alliance in the area. By 1996 the UNOSJO, based in Guelatao, included twenty-three organizations representing sixty communities and was involved in such diverse activities as road construction, forestry, staple foods production, and organic coffee production (Pascual López and Ortiz Medrano 1996). Another organic coffee group, the Mixe-Chinantec-Zapotec Organization (MICHIZA), has affiliates across the Northern Sierra as well. MICHIZA was started by progressive Catholic missionary priests in the early 1980s and includes approximately 300 members in the Rincón.

Although this effervescence in Mexico's coffee sector is an exciting event with broad implications for grassroots development, it has progressed unevenly and is far from complete. For example, in 1991 CEPCO estimated that it purchased only 8 percent of the coffee produced in the state of Oaxaca (Fox 1994). This is not to underestimate the impact of CEPCO; indeed, the organization has played a key role in establishing minimum prices since the collapse of INMECAFE. Perhaps more importantly, CEPCO has sensitized indigenous people to the possibilities of interregional alliances.

Even so, many are reluctant to join. They complain about the time that must be committed to these projects. Meetings are held on a weekly or biweekly basis, in the evenings, when many campesinos are in their ranch houses. For the coffee to be certified as organic, it is not enough for growers to refrain from using fertilizers and pesticides. They must also use organic compost and employ specific soil conservation techniques—which can only be done using labor-intensive processes that many campesinos cannot afford.

Other campesinos are reluctant to accept cash credit payments before the harvest. As mentioned earlier, cash is much more liquid than coffee, for it tends to drain away quickly once within the household. Furthermore, when growers accept cash advances they commit themselves to accept a price range whose upper limit may well be below the market price of the coffee at the time of the harvest (of course they also receive some assurance that the lower limit will only reach a certain minimum price during market downturns). This may explain why many campesinos involved in the organic coffee projects commit only a tiny fraction of their harvests. The programs are relatively new; as with new technologies, many campesinos are slowly experimenting on a limited basis to see whether the new arrangements are in their best interests.[21]

The Hazards of Organizing Coffee Cooperatives

As UNOSJO—one of the few Sierra-wide organizations in the history of the region—was being organized in 1991 and 1992, a number of mysterious events took place which, according to some informants, were meant to snuff out the multiethnic organization in the Sierra Juárez. Several key organizers were sent anonymous death threats through the mail. In May 1991 an engineer affiliated with the organization was mysteriously hit by an unidentified truck outside the village of Otatitlán one night. The engineer, who narrowly survived the incident and has since moved out of the state, was one of the most outspoken supporters of the embryonic UNOSJO and Pueblos Unidos. During the same year, Rosendo García Miguel, director of Pueblos Unidos, was assassinated under mysterious circumstances. According to a documentary video, the murderers were also from the Rincón, but it is not entirely clear whether they were influenced by actors from the "outside." Demonstrations calling for an investigation were largely ignored by state officials (Manzano 1993). More recently, anonymous and implausible accusations have been made on a Oaxaca City radio station linking UNOSJO to the Popular Revolutionary Army (EPR), a guerrilla army that has been active in Oaxaca and Guerrero since 1996. The accusations appear to be part of a smear campaign (Gaspar González 1997).

The new grassroots organizations seem to threaten at least some established interests, and in the case of UNOSJO it may well be that regional- or state-level political figures are fearful of the potential political power that a united Sierra Juárez might exercise. Historically the region has been characterized by strong opposition to penetration by outside forces (whether Aztec or Spanish) and fearless ferocity in battle (Garner 1988; Ruiz Cervantes 1988; Chance 1989). As recently as the 1990s Sierra villages "kidnapped" visiting state officials (including the governor) when demands were not met or when political promises were not kept, and government functionaries are often hesitant to enter the region (Fox and Aranda 1996: 59–60). This unwavering political will, coupled with the remarkable economic autonomy and long tradition of village democracy in Oaxaca (and the Sierra Juárez in particular), may make some members of the PRI uneasy.[22] The fact that the organizations are in some cases drawing together ethnolinguistic groups that have had antagonistic relations since the pre-Columbian era—as in the case of the Mixe and the Sierra Zapotec—might be especially troubling for some government and military officials. In the context of nationwide debates on regional autonomy for Mexico's indigenous peoples, the implications of the new cooperatives—many of which are explicitly devoted to the social and cultural invigoration of their communities—may be enough to concern the current political establishment. Efforts to promote autonomy and self-sufficiency, it seems, are increasingly seen as acts of subversion (Guadarrama Olivera 1996).

Much of the new literature on coffee in Mexico focuses on the impressive gains made by grassroots cooperatives, and they should give reason for optimism. But the contemporary coffee scene is sobering. The dismantling of INMECAFE did indeed open up a space for independent "civil society" groups operating below the level of the state (for example, CNOC and CEPCO). However, international forces—specifically the transnational corporations discussed earlier—have also been able to amplify their already extensive presence in the Mexican coffee sector. As stated earlier, in 1991 CEPCO estimated that it purchased only 8 percent of the coffee produced in the state of Oaxaca (Fox 1994). In the Rincón the percentage collected by cooperatives is even smaller. It is probably safe to assume that, in the absence of INMECAFE and in light of the previous history of transnational corporations in the coffee sector,[23] most of the remaining coffee went to intermediaries who delivered it to Mexican exporters affiliated with foreign firms. Indeed, two of the most vocal proponents of the new coffee organizations have offered this assessment of the changes following the 1989 INMECAFE withdrawal: "The largest portion [of the purchasing, storing, and marketing of coffee] . . . has been taken over by reemerging middlemen and by large transnational corporations which now operate directly in this area they once relegated to intermediaries." In short, the cooperatives find themselves "swimming with the sharks" (Hernández Navarro and Célis Callejas 1994:222).

Still, the importance of the cooperatives should not be underestimated. Organic coffee programs in particular may represent the best hope for an ecologically sound strategy in Mexico's rural communities, especially since the global market for such products is rapidly expanding (Jonathan Fox, personal communication, 1997). Generally speaking, the quality of

Mexican coffee is woefully poor (comparable to Brazilian "standard grade" coffee). Other countries such as Jamaica, Guatemala, and Colombia have put more effort into improving the quality of their coffee; consequently, Mexico's product has gotten steadily worse with respect to those of other Latin American countries. Attempts to produce quality organic coffee in Oaxaca, Chiapas, and Michoacán, though small, offer some hope, and the growing international market for specialty coffees may be a sign of things to come.

To summarize, coffee has been absorbed into the Talean agricultural system in the course of the last century by farmers with a proclivity for experimentation. In spite of generally favorable market conditions for the crop, in most cases it has supplemented, not replaced, subsistence crops, made possible in large part by the introduction of chemical fertilizers (for maize and beans) a generation ago. Coffee is generally grown on relatively small plots (less than four hectares), and growing methods appear to be ecologically sound. Most notably, coffee is grown in shaded groves that resemble the natural structure of forests. The crop is now considered "traditional" by many villagers and has become an important part of the village's identity vis-à-vis surrounding communities. Increasingly, as a partial result of involvement in coffee cooperatives, it is linked to emerging regional identities as well. Coffee has become appropriated by many rural people and directed toward ends that are unexpected and in some ways potentially counterhegemonic. It appears to offer a way for villages and regions to retain a level of economic autonomy in the modern world system.

CHAPTER 8

AGRICULTURE UNBOUND

Cultivating the Ground
between Science Traditions

ACTION IS URGENT. In the affluent industrialized countries, where food is plentiful and obesity is common, there is a frequent tendency to forget that today 800 million people go hungry... In 30 years it is estimated that the earth will have 8.5 billion inhabitants. If they are to be adequately fed, the world food production must more than double... By providing plants that are resistant to disease and pests, like Bt maize, gene technology can make an important contribution to ensuring the food supply for the growing world population.
Novartis, "Maize Is Maize" (2000)

Outside the United States, you'll find 96 percent of the world's population. Inside, you'll find the means to feed them.
TV commercial, Archer Daniels Midland corporation

Those who emphasize the dismal side of development may be underestimating the ingenuity and vitality of indigenous economic systems when given the opportunity to develop within their own logic of subsistence needs.
June Nash, "The Challenge of Trade Liberalization" (1994)

The Rincón Zapotec have incorporated new crops and techniques into their farming and dietary repertoires to improve upon a strong base of agricultural knowledge and practice. Through this process they have created a system which is neither static nor "underdeveloped," but rather modern, flexible, and dynamic. Indeed, the current arrangement of maize, bean, sugarcane, and coffee production in Talea might even be described as "postindustrial" to the extent that it emerged in its present form after the abandonment of silver mines early in the twentieth century. The ancestors of many of today's Taleans mined for nearly 200 years and probably came to understand how world markets worked from experience, since local mining was affected by radical fluctuations in global prices throughout the eighteenth and nineteenth centuries (Chance 1989). They became specialists in managing global economic change—they were "flexible workers" able to extract ore from a mine one year and to plant maize, beans, and sugarcane the next. It appears that Talean campesinos never lost their subsistence base; even when linked to world markets through cochineal, cotton, gold, and coffee production, they supported themselves by growing their own food. For the Spanish colonists residing in the area, this effectively served as a subsidy—the villagers had to feed themselves and care for their families and communities while they supplied tribute to the colonizers; for the Rincón Zapotec, the arrangement ensured a level of economic autonomy and independence from the Spanish.

Talea is an interesting case because it was apparently founded after the Conquest on a relatively small amount of land acquired from neighboring villages. These special historical circumstances probably motivated villagers to experiment with new agricultural techniques early on. For example, coffee production is likely to have emerged as a solution to the problem of scarce land, since it allowed farmers to buy the equivalent of several hectares of maize in exchange for cash derived from a single hectare of coffee. Part of this change might also be attributed to the village's diverse population (including miners drawn to the Rincón from outside the region), who may have brought with them creative solutions to practical problems and an overall attitude of acceptance toward new approaches. Since coffee was introduced in Talea, the village has changed more rapidly than any other in the Rincón.

Over time campesinos in Talea adopted a number of weights, measures, and implements, some from the prehispanic period and some from the colonial era, which are still in use today. Most are based on the dimensions of the user's body. This system affords campesinos a distinct advantage, since implements fashioned in this way are ergonomically designed; they correspond to the user's body, ensuring a custom fit. This makes rigorous farming work less punishing. Campesinos also creatively make extensive use of what we might call "scrap" materials, which might be thought of as an economic strategy for using mass-produced goods (often made of special materials that would be impossible to produce locally, such as plastic) at a nominal cost and sometimes at no cost at all. The production of such implements and artifacts might be described as a form of craftwork, if we define craft items as handmade goods used by the craftsperson or made for people in face-to-face contact with the producer.

Maize is the foundation of the alimentary base in Mexico, and most families in Talea continue to produce it for their own consumption. Archaeological evidence indicates that the crop was domesticated in Mesoamerica approximately 7,000 years ago. It was among the first crops taken to Europe after the arrival of the Spanish and spread rapidly across much of the Old World in the 1500s; in fact, it was probably a key element in the demographic explosion in Europe and Asia beginning in the sixteenth century. Native American production techniques have changed significantly since then; in particular, the adoption of steel implements, plows and draft animals, and more recently chemical fertilizers has revolutionized production. These techniques have been combined with much older ones that are now being taken seriously by contemporary cosmopolitan scientists. Most importantly, intercropping and other sustainable practices are increasingly seen as possible models for sustainable farming in the so-called First World. Campesinos, however, continue to cultivate maize using distinct conceptual notions of a personified, living earth; a recognition of the importance of reciprocity; an emphasis on the quality of their food; the necessity of maintaining a humoral balance in their diets, their land, and other spheres; and a long-term view of maintaining their households and lands.

Sugarcane was probably domesticated in New Guinea before its diffusion across Asia and Europe. It became one of the most important tropical crops produced in the European colonies during the era of empire — often using the labor of African slaves. The place of sugarcane in village life appears to contradict the usual pattern associated with the cultivation of the crop. Though historical evidence indicates that sugarcane is most frequently cultivated, harvested, and processed in authoritarian work arrangements (either for cash or under a slave labor arrangement), in Talea a mixed system has been adopted in which sugarcane work relies heavily on reciprocal exchanges of labor, equipment, and food gifts. It also is an important ritual food associated with the deceased. Taleans have continued to consume locally produced raw brown sugar (*panela*) even though it is more expensive than white sugar purchased at stores. This appears to be related to the villagers' distinct preference for high-quality products that are locally grown and processed: "knowing where it comes from" is one of the primary criteria used in judging food. Villagers also prefer *panela* because it is "hot" (in humoral terms) and thus less upsetting to the body than "cold" white sugar.

About a century ago Taleans began cultivating coffee on small parcels of land. Coffee was first discovered in Africa before its center of production shifted to the Middle East. European colonists brought it to the New World in the seventeenth century, and by the end of the nineteenth century Mexico had become an important exporter of the crop. Like sugarcane, coffee has been incorporated into the agricultural and dietary repertoire of Taleans, and villagers describe it as a "traditional" crop. Farmers have developed techniques tailored to the peculiarities of Talean soils. Coffee has allowed them to further diversify their farming activities and made it easier for most villagers to maintain their households economically. Significantly, coffee has not entirely replaced the production of subsistence crops; coffee production has fluctuated over the years as coffee prices have risen and fallen. Most growers refuse to give up maize and bean cultivation, probably because this would leave them vulnerable to a drop in international prices. Farmers generally view coffee as a long-term investment strategy, a means of securing the future of their children. The crop has even become a part of the villagers' identity; since Talea has produced more coffee (and converted more land to its production) than any other Rincón village, growing it is considered an important part of being Talean. Recently, farmers have formed coffee cooperatives to help market their product directly to buyers in the "First World," and new identities—often "indigenous" identities—are being created in the process (Hernández Castillo and Nigh 1998).

Viewed comprehensively, Zapotec farming appears to be a scientific and technological system in which the care of living matter—broadly construed as humans, animals, plants, earth, and water—is a means of achieving *mantenimiento* or maintenance of individuals, families, villages, and village lands. Consequently, a sophisticated agricultural strategy has emerged, allowing villagers to participate in global markets and a cash economy while continuing to grow a significant quantity of their food. Indeed, it appears that in the last century Talean campesinos have judiciously grafted cash cropping onto a subsistence economy in an effort to maintain household stability through crop diversification.

There are two pieces of evidence pointing to the effectiveness of such a flexible food system: its staying power and the recent adoption of some of its features and conceptual schemes by cosmopolitan science practitioners and laypeople in Europe, the United States, and other societies. My goal here is to polemically engage those who have established dualistic boundaries around agricultural systems: anthropologists describing local science traditions as if they had not been influenced by contemporary cosmopolitan science and vice versa; development experts cordoning off "modern" from "traditional" agriculture in order to promote the diffusion of factory farming; and representatives of agribusiness and biotechnology companies who use images of the "starving millions" of the "Third World" to justify the diffusion of their products and the expansion of corporate power.

The Place of Food and Farming: A Dialogue

It is probably not an overstatement to say that food and farming will be among the most critical issues facing the world within the next decade or two. Since the mid-1990s researchers have brought attention to the gravity of the world food situation (Brown 1996, 1997), shrinking crop land (Gardner 1997), and the security dangers posed by environmental and agricultural crises (Renner 1997). Troubling reports about world population appeared as the 6 billion mark was reached in the final months of 1999 (*New York Times* 1999). Recent reports from the Worldwatch Institute predict that, in light of expanding world population, stagnant grain harvests since 1990, the lowest levels of grain carryover stocks in more than forty years, unsustainable water and crop land use, and rising planetary temperatures due to the effects of greenhouse gases, global shortages of basic foodstuffs are an increasing probability. The trends are particularly alarming because from the 1940s until the end of the 1980s phenomenal annual increases in grain output were the rule.[1]

Against such a backdrop, alternative and sustainable subsistence strategies such as Rincón Zapotec farming may be more valuable as structural models for the future rather than as quaint survivals from the past. To the extent that they have incorporated crops, techniques, and knowledges from many corners of the world, Zapotec foodways are thoroughly modern; campesinos have intelligently weighed the advantages and disadvantages, adopting what works by their standards and rejecting what does not. Communities like Talea are especially important places to look for ideas about how to face a future of food scarcity because food is abundant in the region, most of it is produced locally, and farmers have succeeded in growing it sustainably year after year for millennia. If Worldwatch director Lester Brown is even partially correct in his predictions, then people in industrialized societies may soon find themselves eating as much of the rest of the world eats today, deriving most of their calories from basic grains rather than meat and other animal-based proteins. In this context, understanding the farming and foodways of people who rely heavily on locally produced foods might be of real importance and urgency.

But established norms will probably make it difficult to change course, because they have often created conceptual blind spots which render some questions invalid in the very process of defining problems. In these cases certain phenomena in industrialized society that alternative views or commonsense thinking might consider wasteful, destructive, or irrational can be explained in "rational" terms.

A relevant example might serve to illustrate. In 1997 the U.S. Department of Agriculture reported that 96 billion of the 356 billion pounds of food available for human consumption in the United States—more than 25 percent of the nation's total—were lost somewhere between farmers' fields and garbage disposals (Passell 1997). According to the report, 11.3 billion pounds of fruit, 15.9 billion pounds of vegetables, 8.2 billion pounds of meat, poultry, and fish, and 17.4 billion pounds of milk were lost or disposed of in 1995. The department took into account food that was lost when it was processed or shipped, when it was thrown away by restaurants, food services, and households, and when retailers discarded spoiled or out-of-date perishables. Though most people in the United States might find such wastefulness outrageous, a number of economists seem to espouse its rationality (Passell 1997:C1):

> In an era of abundance, one in which time and labor are more precious than the mass-produced goods of modern farms and factories, economists argue that prices should rule. From this perspective the decision to wrap and store the dinner leftovers . . . should depend on whether there is a better way to spend the time and effort. "There's an optimal level of food 'waste,' " insists Robert Frank of Cornell University, "just as there is an optimal level of dust in your house." Indeed, Robert Hahn of the American Enterprise Institute suggests . . . that waste is a sign of societal success rather than failure. "If there was ever a non-issue this is it," he argued. "The primary reason so much gets tossed is that America has the cheapest food in the world."

Some Clinton administration officials, most notably vice-president Albert Gore and secretary of agriculture Dan Glickman, disagreed with such views. Gore and Glickman proposed a "gleaning and food recovery" (Passell 1997:C1) initiative to recover some of the losses. The goal: to feed an extra half-million people in the United States.

In this case, some experts are able to look past the startling fact that more than one-fourth of the nation's food supply is lost by making assumptions that make such waste seem rational and even desirable—a sign of "success rather than failure." The questions about malnutrition in the United States, about the high environmental costs of producing "the cheapest food in the world," and about whether landfills can sustain this quantity of waste are avoided, since they occupy logical spaces outside the field of vision.[2]

The example demonstrates another point: that experts and laypeople do not always share the same assumptions; nor do experts always agree. Indeed, in this case most people in the United States (including Gore and Glickman) probably hold a view more like that of the Rincón Zapotec with regard to waste, even if they do not always put it into practice. It may be that soon these views will win out.

There are several fields in which "alternative" concepts and practices are making headway in U.S. society, entering areas formerly monopolized by the dominant discourses of established "experts." Often they are informed by the knowledges, practices, and conceptual frameworks of societies practicing local sciences, though they may undergo cultural translation as attempts are made to explain their effectiveness. For example, U.S. physicians who explain acupuncture's effectiveness in terms of "endorphin release" are qualitatively no different than Zapotec farmers who explain the beneficial effects of chemical fertilizers in terms of hot/cold classificatory systems—each group attempts to classify effective methods into its respective established theoretical terms. Furthermore, a curious phenomenon may manifest itself in some cases: cosmopolitan scientists experimenting with local science methods sometimes appear to act more in line with prescriptions implied by Zapotec concepts than with their own. Under these circumstances, the boundary between "the cosmopolitan" and "the local" begins to blur.

From one perspective it appears that a kind of "convergent evolution" between the Zapotec and (certain) cosmopolitan scientists is occurring, the converse of the evolutionary process characterizing ancient Zapotec and Mixtec societies in Oaxaca, which began diverging geographically and culturally approximately 6,000 years ago (Flannery and Marcus 1983).

Although it is not their central focus, Kent Flannery and Joyce Marcus (1983:3) discuss how convergent evolution may occur: "Alternatively, after a long separation, two societies may converge either as the result of a similar adaptation or of the acculturative effects of a third society." Looking at cultural change in a contemporary context, we might consider how cosmopolitan and local agricultural systems are, in a sense, converging as crops, technologies, and conceptual categories move rapidly across the globe.

Though such cross-cultural borrowing may be occurring with increasing regularity as the globalization of ideas, artifacts, and people accelerates, diffusion has long been a feature of agricultural systems—and, more generally, knowledge systems—everywhere. Some local knowledges and practices in the contemporary United States are being absorbed by certain conventional sciences.

The Resurfacing of Submerged Knowledges in the United States

Rincón Zapotec farming is not a closed system; indeed, some of the most important contemporary crops were introduced relatively recently in historical terms. Furthermore, certain technologies, such as chemical fertilizers, were introduced less than a generation ago by campesinos pragmatically able to fit them into their conceptual schemes. But technological transfer is a two-way street; even as the Zapotec experiment with artifacts and methods developed for factory farms, people in industrialized societies have begun to borrow certain methods based on tried and true techniques used in rural Mexican villages and other societies with strong local science traditions.

A brief disclaimer: I do not want to suggest that the following shifts have become standard practice among cosmopolitan scientists (much less among U.S. factory farmers or agribusiness firms). Like many activities occurring outside of "normal science" (Kuhn 1962:10), these changes have often been initiated by "rogue figures outside the establishment . . . [operating] from the margins, where the intellectual point of departure and operating assumptions . . . are substantially different" (Scott 1998:279). Sometimes laypeople excluded from established expert discourses influence the acceptance of new views as well.

Sustainable Agriculture in the United States

In the 1960s and 1970s a number of critics and activists in the United States—both cosmopolitan scientists and laypeople—began to make the case for a more sustainable approach to farming and livestock production. Sustainable agriculture, according to its proponents, was a response "to the oldest faults of agriculture: its self-destructive tendency to wash away and sterilize the soil, to encourage deserts to develop where people have been raising their food, and to drive the best and the brightest (or at least the most ambitious) people off the land" (Jackson, Berry, and Colman 1984:ix).

The move toward ecologically sound farming occurred at a number of levels. At the grassroots level activists began employing "new" practices on experimental farms and communes in the United States. Many were highly educated young people seeking to build alternative communities that were more viable than the country's cities, and some became pioneers in the nascent organic farming movement.

As these experiments were being realized, critics started writing books aimed at convincing the general public of the need to reform factory farming, certain large-scale agricultural techniques, and unhealthy U.S. diets. For example, the Food First Institute began publishing a series of popular books in the early 1970s to address themes related to agriculture, politics, and sustainability (Lappé and Collins 1977). Texas agriculture commissioner Jim Hightower published a series of scathing critiques in the 1970s that dealt with the manipulation of taxpayer-financed land-grant colleges by corporations (1972) and the monopoly practices of multinational food conglomerates (1975). Other influential texts included *The Unsettling of America* (Berry 1977) and *The Poverty of Power* (Commoner 1976). Even research by the U.S. Department of Agriculture—often criticized for its close ties to agribusiness—described the limitations and consequences of factory farms and monocropping (USDA 1973). Other studies such as Rachel Carson's *Silent Spring* (1962), E. F. Schumacher's *Small Is Beautiful* (1973), and the Club of Rome's *The Limits to Growth* (Meadows 1972) also had strong impacts on the drive toward sustainable agriculture. The books dealt with such issues as world poverty, corporate power, nutrition, democracy, and the effects of pesticides. They attracted such a large audience that it would be difficult to overestimate their impact in the United States.

Sustainable agriculture and organic farming have also received impetus from the work of biologists, entomologists, geographers, and other cosmopolitan scientists inspired by the effectiveness of local science traditions. The work of Robert Chambers (1983), Paul Richards (1985), Miguel Altieri (1987), Gene Wilken (1987), and many others has opened the way for "new" agricultural techniques, including integrated pest management, intercropping, and organic composting—cutting-edge practices that have rapidly gained acceptance among contemporary cosmopolitan scientists and increasing numbers of farmers. For example, *Agroecology: The Science of Sustainable Agriculture* (Altieri 1987) has been accurately described as an attempt to reintroduce "indigenous, organic farming techniques and the preservation of small farms" as part of an "old-is-new science of agriculture" (Bierma 1997:7). Altieri argues that the crop diversification strategy employed by "traditional" farmers to ensure household survival (what the Rincón Zapotec call *mantenimiento*) holds a lesson for everyone, particularly those living in societies where monocropping is the norm (Altieri, quoted in Bierma 1997:9):

If you homogenize—at least in ecology—you become a highly vulnerable society. In order to be a resilient country, your agriculture needs to be diverse biologically and culturally. Because it is the traditional knowledge that will illuminate for us how to make agriculture sustainable in each environment.

Similarly, the author of *Indigenous Agricultural Revolution* (Richards 1985) argues that the quickest route to agricultural improvements in the so-called Third World will probably come from methods practiced in ecologically specific niches in regions that have historically been considered "underdeveloped" or "marginal" by development agents—not from specialized high-tech research laboratories. Appealing to the sensibilities of "good scientists" in the preface to *Good Farmers*, Wilken (1987:x) implies that learning about "traditional resource management" might serve as a basis for duplicating such techniques in other places: "To Western readers some of the tools and techniques may seem quaint, even primitive. But as good scientists we eschew such ethnocentric evaluations and ask instead: 'Do these methods work?' 'Can we learn from them?' 'Could they be used elsewhere?' " The answer seems to be a resounding "yes" if we look at the growing interest in organic farming (USDA 1980; Coleman 1989; NRC 1989; Bender 1994), multicropping (Papendick, Sánchez, and Triplett 1976), coverage in popular scientific magazines (Lockeretz, Shearer, and Kohl 1981; Reganold, Papendick, and Parr 1990), and the creation of scientific journals such as the *American Journal of Alternative Agriculture* and *Agro-ecosystems*.

Indeed, ecological approaches represent a kind of rupture; they have called certain dominant paradigms into question and have attempted to construct analytical models representing alternative frameworks for viewing the world. The growing literature on "ecological economics" or "environmental economics" may serve to reconcile alternative paradigms with those conventionally employed by economists (Martínez Alier 1987; Daly and Cobb 1989; Constanza 1991). Scholarship from the so-called Third World, particularly works from South Asian scholars in the wake of the Green Revolution, has had a particularly strong impact on critiques of agricultural development (Shiva 1991).

In a different vein, ecological notions of a "living earth" (Lovelock 1979, 1995; Ericson 1988; Nisbet 1991) and a "hungry planet" (Borgstrom 1967) that threatens the long-term survival of a careless human species explicitly suggest a personification of "nature" remarkably similar to Zapotec ideas that take for granted the vitality of animate, personified forests, mountains, soils, rivers, and plants. It is probably not an exaggeration to say that these ideas are part of how some ecologists and laypeople have conceived such phenomena as global warming, pollution, and deforestation—symptoms of planetary "pain" in need of "healing" (Lovelock 1991).[3]

Organic Farming: Addressing U.S. Concerns about Food Quality

Closely related to ecological perspectives are critiques of dietary practices in industrialized societies, particularly in the United States. An interest in diet emerged in part from environmental and consumers' rights activists concerned about the use of pesticides and insecticides, fertilizer runoff, and other harmful elements in U.S. food production. Later a second wave of reports about the carcinogenic and deleterious cardiovascular effects of diets heavily laden with cholesterol and fats—characteristic of the contemporary diets of industrialized societies—added fuel to the fire.

Diet for a Small Planet (Lappé 1971), a pragmatic guide for preparing nutritious meals, sold more than a million copies. Many of the book's recipes are based on cooking in "traditional" societies, including the Mesoamerican "trinity" of maize, beans, and squash. In *Food First* (Lappé and Collins 1977), Frances Moore Lappé overturns conventional notions about the so-called Third World by demonstrating that allegedly "impoverished" diets consisting of minimal amounts of animal proteins in fact provide more than adequate amounts of vitamins, proteins, and carbohydrates— and make more efficient use of farming land. As in the case of the Rincón Zapotec, people in subsistence farming societies are often efficient consumers of food in the sense that they obtain many more calories per unit of land than their counterparts in industrialized societies, who almost inevitably use a large proportion of farmland to grow grains for animal feed.

Reports from established medical research journals and the U.S. Senate Committee on Nutrition and Human Needs (1978) drew special attention to the health risks implicated in U.S. dietary patterns. The Senate committee report in particular created a stir by reporting that in 1977 six of the ten leading causes of death in the United States were linked to diet. High levels of fat and low levels of complex carbohydrates in the diet were found to lead to heart attacks, strokes, arteriosclerosis, cancer, cirrhosis of the liver, and diabetes, which together accounted for half of all deaths in the United States. Leticia Brewster and Michael Jacobson, in a critical analysis of *The Changing American Diet* (1983:4) from 1910 to 1976, document the tremendous increase in meat, processed food, and sugar consumption over the course of the twentieth century and summarize the health implications of their findings: "Our 'new' diet, in combination with our sedentary lifestyle, has contributed to what must be considered a national epidemic of obesity, diabetes, heart disease, stroke and tooth decay . . . These problems cost Americans tens of billions of dollars a year in direct medical expenses and indirect losses to the economy."[4] By 1999 studies indicated that nearly 60 percent of the adults in the United States were overweight and that nearly 20 percent were obese (Schultz and Fischman 1999).[5]

Such dietary concerns helped promote organic farming in the 1990s. From 1990 to 1997 organic food sales increased by 20 percent annually and reached a record level of $2.5 billion in sales in 1997 (Adelson 1997). This may partially explain recent efforts by agribusiness and food-processing conglomerates to persuade the U.S. Congress to relax organic food certification standards; it seems that such firms want to cash in on this rapidly growing market (Schmelzer 1998). These efforts were unsuccessful, due in large part to protests from concerned citizens.

How is this relevant to Zapotec farming and foodways? Significantly, it signals a partial shift away from dominant ideas about agriculture and food in industrialized societies (specifically the United States), from an emphasis on maximizing quantity to an emphasis on quality. The evidence indicates that, like the Rincón Zapotec, a growing number of consumers are willing to sacrifice cheap food in exchange for products that are healthier and free of harmful chemicals, and a growing number of U.S. farmers are willing to provide them with such foods.

Related to this are the health consequences of relying upon the long food chains which characterize the U.S. food system, linking consumers to distant producers through a winding circuit of food-processing plants and transportation systems spanning the globe (Fox 1997). Recent high-profile cases of *E. coli* outbreaks and food contamination in processing plants, agricultural fields, and other points in the food chain have prompted the

U.S. Food and Drug Administration (FDA) to reform routine inspection procedures for testing food safety (Janofsky 1997). The incidents might partly explain the rapidly increasing popularity of farmers' markets and co-operative farming arrangements—in which urban consumers deal directly with the growers and their families, establishing personal relationships. It is further indication that more U.S. consumers—like their Zapotec counterparts—want to know where their food comes from.

This shift provides us with evidence that the prevailing agricultural goal of yield maximization (which makes heavy pesticide and herbicide use, genetically modified seeds, large doses of chemical fertilizers, hormone injections for livestock, long food chains, etc., appear rational) may be reaching its limits in the United States. Blind as it is to certain demand-side concepts of "quality," the supply-side model has been implicitly called into question by shifting consumption patterns in U.S. society sparked at the grassroots. In the process it has largely been interpreted or translated into the terms of two cosmopolitan sciences, nutrition and modern biomedicine.

"Alternative" Medicine in the United States

Perhaps nowhere is the popular acceptance of "alternative" and local knowledges in the United States more visible than in the field of medicine. According to a recent report, one-third of the U.S. population made use of alternative medical treatments, in some cases against the advice of their doctors (Shweder 1997). Even the medical establishment seems to be coming around. In late 1997, for example, the Office of Alternative Medicine of the National Institutes of Health issued a report recommending that American physicians integrate acupuncture into standard medical practice. An independent twelve-member panel of researchers found that acupuncture was an effective therapy for certain medical conditions, especially those involving nausea and pain, and that treatments were remarkably safe, with fewer side effects than many established therapies.

Many U.S. physicians have been reluctant to recognize the legitimacy of acupuncture, probably because the explanations offered by Chinese specialists are based on theoretical concepts of opposing forces, *yin* and *yang*, which when out of balance disrupt the natural flow of "energy" (*qi*) in the body. A wider acceptance of acupuncture by the U.S. medical establishment will probably rely upon "translation" into the terms of modern biomedicine: "there is considerable evidence that acupuncture causes a release of natural pain-relieving substances like *endorphins*, as well as *messenger chemicals and hormones* in the *nervous system*. Further, it . . . appears able to alter *immune functions*" (Brody 1997; emphasis added). It should be mentioned that, apart from acupuncture, Chinese traditional medicine has enriched European medical traditions with countless remedies, ideas, and practices over hundreds of years and Chinese experts were among those who created modern biomedicine.

Another recent project has attempted to translate Mayan herbal remedies into the terms of modern biomedicine and has found that a high percentage of the plant species show antimicrobial activity in laboratory tests (Berlin et al. 1996). Even though Mayan categories for classifying plants and illnesses differ sharply from those used by cosmopolitan botanists and physicians, the remedies seem to be effective when viewed from the framework of either system.

With respect to the hot/cold paradigm used by Taleans and many other Latin Americans, there is renewed interest on the part of modern biomedical researchers who abandoned humoral therapies centuries ago. A recent scientific mission organized by Stephen Koslow of the U.S. National Institute of Mental Health (NIMH) took a party of U.S. psychologists and pharmacologists to India to examine Ayurvedic medicine (a south Asian version of humoral therapy) and its effects (Shweder 1997:6): "It is the bet of some neuro-pharmacologists at NIMH that Ayurvedic practitioners know something about barks, roots, leaves, and other botanical provisions for human beings that they can no longer afford to overlook."

These examples demonstrate how knowledge — in this case "alternative" medical practices and theories — can travel back and forth across societies over centuries. Similar cross-cultural technological movements have been explored by other scholars (see, for example, Bernal 1987; Weatherford 1988; Pacey 1990), but the examples reviewed here show that it is possible for entire cosmological views or conceptual frameworks — including, for example, the notion of a living earth — to diffuse in a mass society within a relatively short period.

The Modernization of Mexican Agriculture in the Era of "Free" Trade

Yet in spite of the resurgence of "alternative" knowledges in the United States, Europe, and other societies, creative solutions to agricultural problems are often relegated to second-class status in practice; with few exceptions, agricultural development models based on factory farming continue to be exported into countries with strong local science traditions and rich genetic banks. A two-way process is occurring in which some cosmopolitan scientists are learning and writing about specialized, highly effective local science traditions and technologies, while international development agents continue exporting development models to the so-called Third World—sometimes with the cooperation of national elites trained at the Massachusetts Institute of Technology, Harvard, Berkeley, and other foreign institutions.

The conventional wisdom among development economists models successful farming in terms of factory farming. "Traditional" agriculture is described as less productive and efficient than factory farming systems that make use of tractors and mechanized implements, chemical fertilizers and pesticides, and hybrid seeds. An introductory textbook on development economics succinctly draws the distinction (Todaro 1989:298):

World agriculture, in fact, comprises two distinct types of farming: (1) *the highly efficient agriculture of the developed countries*, where substantial productive capacity and high output per worker permits a very small number of farmers to feed entire nations; and (2) *the inefficient and low-productivity agriculture of developing countries*, where in many instances the agricultural sector can barely sustain the farm population, let alone the burgeoning urban population, even at a minimum level of subsistence.

Both progressive development economists and conservative neoclassical economists generally take this binary (and social evolutionary) view of world agriculture as a given.

In spite of the practical problems present in such models, described eloquently in many critical studies from Asia (Rahnema 1986; Shiva 1991), Latin America (Wilken 1987; Rodríguez 1989; Wright 1990), and Africa (Richards 1985), evolutionary thinking persists, perhaps as strongly today (at least in the area of food production) as at any time in the past century. In Mexico this process began in earnest in 1988 as part of Carlos Salinas de Gortari's economic liberalization policies. Beginning in 1994, Mexico's commitment to NAFTA (which among other things permits highly subsidized U.S. corn to enter the Mexican market) and its unilateral reform of constitutional Article 27 in 1992 (which allows the privatization of *ejidos*) opened the way for the expansion and consolidation of Mexican factory farms. Consequently, transnational corporations such as Archer Daniels Midland, Ralston Purina, Anderson Clayton, and others have entered directly into new joint ventures with Mexican conglomerates.

It is somewhat ironic that these and other transnational corporations, in conjunction with Mexican development agents, are promoting and propagating factory farming—a system that has proven to be ecologically and socially destructive in the U.S. Midwest and other areas—in the very regions where innovative farmers developed maize thousands of years ago (Carson 1962; McWilliams 1971; Berry 1977).[6] The durability, security, and ecological sustainability of 7,000 years of maize farming may soon be overturned by forces that view such techniques and knowledges as obsolete. Factory farming, the late Paul Feyerabend (1978) might tell us, is being aggressively pushed by proselytizers in increasingly greater areas of the globe with a missionary zeal that is religiously intolerant of alternative methods of growing food, even when they have demonstrated clear advantages over methods that have not been tested in the specialized ecological niches which sustain many other societies. But the real issue may have more to do with raw economic power than with science: a large number of transnational corporations make factory farming possible—including seed companies, food conglomerates, equipment manufacturers, chemical companies, and biotechnology firms—and for these entities a tremendous amount of political and economic power is at stake.

From Green Revolution to Gene Revolution

Genetically modified (GM) crops are a case in point. As in the case of Green Revolution crops, GM crops have been portrayed by biotech companies (and more than a few scientists) as *the* solution to world hunger—although little is known about their long-term effects.

One of the most common GM crops is Bt corn, named after *Bacillus thuringiensis*, a microorganism that produces a protein that destroys the intestinal walls of a pest known as the corn borer. (The larvae of the corn borer feed on various parts of the maize plant.) Bt corn is genetically modified by having part of the *B. thuringiensis* DNA inserted into its own. In this way it is able to produce the protein that kills the corn borer. By reducing crop losses associated with this pest, Monsanto and other biotech companies claim that they can help farmers increase yields.

Companies who have developed GM crops (including Monsanto, Novartis, DuPont, Dow, AgrEvo, and others) have aggressively promoted GM seed sales, in part to recoup the vast sums of money spent in acquiring seed companies and conducting the scientific research necessary to produce the seeds. This has led to one of the most rapid transformations ever in the U.S. food supply. Bt corn—which was only introduced in 1995—

by 1999 accounted for approximately 40 percent of the corn harvested in the United States. Approximately 50 percent of the 1999 soybean crop was GM.

Yet very little is known about the long-term health and environmental effects of GM crops. Although the USDA, the FDA, and the Environmental Protection Agency (EPA) have consistently justified their approval of GM foods, there are doubts. In 1999 journals such as *Nature* and *Lancet* published reports about the negative health effects of Bt corn pollen on monarch butterflies (Ewen and Pusztai 1999) and GM potatoes on rats (Losey, Rayor, and Carter 1999), respectively.

Another issue has to do with the looming threat of "genetic contamination" from GM maize pollen. In 1999 European researchers discovered that pollen from GM maize plots had traveled more than three miles away, much farther than the 200-meter buffer zones established in some European countries. Some U.S. seed companies are apparently urging farmers not to guarantee the genetic purity of their crops to buyers until they genetically test grain.

A range of possible environmental effects may be associated with GM crops. For example, there is concern that they may cross with wild relatives to create new "superweeds." This is an issue because many of the GM crops are genetically designed to withstand higher pesticide dosages (Roush and Shelton 1997). A good example is Monsanto's line of "Roundup Ready" seeds, designed specifically to endure high dosages of Roundup brand pesticide—made by none other than Monsanto. Others note that GM crops that target insects may lead to the evolution of "superpests" as the surviving bugs reproduce and multiply through natural selection processes. Still others are concerned that useful insects may be destroyed through exposure to Bt corn pollen (Yoon 1999a, 1999b). In the United States these concerns became so great that in February 2000 the EPA set limits on the amount of GM corn that may be planted by farmers (Yoon 2000).

The threat of genetic contamination in Mexico and Central America—the center of origin of maize and the area with the most varieties of the plant (more than 20,000)—is a growing concern. Since the implementation of NAFTA in 1994, increasing amounts of U.S. maize have been imported by Mexico, and in the last four years this has come to include GM maize. In transporting the grain, critics note, it is easy for GM kernels to escape into local ecosystems; they may drop from a rail car or a truck and take root in rural areas. Given maize's tendency to disperse its pollen easily, it would be relatively simple for Bt genes to spread via open field pollination processes and thereby contaminate the milpas that hold the world's richest genetic reserves of maize. Although Mexico's Ministry of the Environment banned the importation of GM maize in 1999, there are doubts about how thoroughly the restriction is being enforced (Tricks and Mandel-Campbell 1999).

Scientists, Farmers, and Citizens: Views on GM crops

On the first day of the new millennium the Mexican daily newspaper *Reforma* featured a commentary on food in the twenty-first century (Borlaug 2000:4A). The piece, entitled "Biotechnology against Hunger," was written by microbiologist Norman Borlaug, one of the principal architects of the Green Revolution and winner of the 1970 Nobel Peace Prize. He argues that the Green Revolution would probably have ended world hunger by the end of the twentieth century except for the fact that

the population monster continues growing. During the 1990s alone, world population increased by nearly 1 billion . . . clearly the most fundamental challenge of the future is to produce and equitably distribute a food supply adequate for this very crowded planet . . . I think we have the agricultural technology to feed the 8.3 billion people anticipated in the next quarter century . . . [But] the extremists in the ecological movement of the wealthy nations seem to be doing all that they can to hold back scientific progress . . . [by] slowing the application of new technology, including transgenic technologies, biotechnologies, and more conventional agricultural science methods.

Borlaug's tone is dark, and he masterfully combines moralistic appeals with images of a Malthusian doomsday. The style is strongly reminiscent of the Green Revolution literature of fifty years ago:

Agricultural scientists and decision makers have the moral obligation to explain to political, educational, and religious leaders the magnitude and seriousness of the problems of arable land, food, and population . . . If we cannot do this in a direct way, we will be contributing to the chaos of countless millions of deaths from starvation.

These kinds of arguments—which rely on images of "countless millions" of starving people—are virtually identical to those made by biotechnology companies who produce genetically modified crops (Monsanto Company 1998; Novartis 2000). This line of argument was pressed into service in an extraordinarily forceful fashion beginning in October and November 1999, as a massive public relations campaign was launched by biotech industry alliances. The members have financed scientific research, lobbied legislators, regulators, and farm organizations, and used newspapers, TV commercials, and the Internet to disseminate information about the alleged benefits of GM products (Barboza 1999a).

Monsanto has perhaps been the most creative in its public relations efforts; on October 1, 1999, the company unveiled a new exhibit at Walt Disney World's Epcot Center that, in the company's own words, "offers a fascinating look at life sciences and new technologies designed to address the food, health, and nutrition challenges of the next millennium." Monsanto officials are hopeful that the exhibit, called "Beautiful Science," will attract many of the 13.5 million people who visit Epcot each year. The company refers to the exhibit as a "public information program" designed to "communicate the benefits of science and technology to the general public" (Monsanto Company 1999).

The actions of the biotech industry are in many ways a reaction to a surge of grassroots opposition to GM food on the part of organizations representing consumers, environmentalists, and farmers. In 1999 a number of events occurred which focused attention on the topic of agricultural biotechnology. First, in October the European Union clarified its rules on the labeling of GM foods (*Wall Street Journal* 1999). Specifically, only foods containing less than 1 percent GM material are to be labeled "GM free." This had an immediate impact on U.S. farmers, most of whom had not been separating GM crops from non-GM crops; they now found themselves facing the prospect of losing a good part of their European sales. A backlash on the part of U.S. farmers against GM seeds was evident: on average, they planned to plant 16 percent fewer acres of GM corn in 2000 than they did in 1999 (ACGA 2000).

Another kind of backlash was made apparent on December 15, 1999, when a group of farmers filed a class action lawsuit against Monsanto, claiming that it misled growers about the possible health effects and marketability of its GM corn and soybeans. The group also charged Monsanto with helping to form an international cartel to control the world corn and soybean seed market (Barboza 1999b). With stock prices plummeting and investors upset at his failure to judge consumer attitudes, Monsanto CEO Robert Shapiro announced that the company would merge with Pharmacia & Upjohn (a pharmaceutical firm), and rumors circulated about the imminent sale of its agricultural unit.

Although some observers have claimed that critiques of GM foods are based on illogical European "food fears" (Fumento 2000) or the radicalism of environmental extremists (Borlaug 2000), there is recent evidence that ordinary U.S. citizens have also become concerned. Forty-nine members of the U.S. Congress sent letters to the FDA requesting that mandatory food labels be put on GM foods (BioDemocracy Campaign 2000). In a recent survey 70 percent of U.S. consumers said that they wanted more extensive labeling of GM foods and 40 percent said that they wanted more stringent regulation of agricultural biotechnology (*AgraFood Biotech* 1999:11).

From an anthropological perspective one of the most striking things about the public relations campaign launched by the biotech industry in 1999 (and earlier) is its strategic use of science and scientists to accomplish quite specific political and economic goals. Increasingly, the companies have funded individual academic scientists and even academic departments (Press and Washburn 2000). Here we see how science may become politicized as those who control funding may seek to determine research questions and claim part of the results as their own.

The influence of scientific research funding by biotech companies was well illustrated in another event which occurred in late 1999: a November 30 press conference in which Republican Senator Kit Bond of Missouri (home of Monsanto) appeared with a number of prominent scientists at the World Trade Organization meetings in Seattle to discuss the benefits of transgenic crops. With great fanfare, Bond released a letter of support for the use of biotechnology signed by more than 300 scientists from the public sector, the private sector, and academic institutions (Kit Bond News 1999). Many of the academically based scientists who signed the letter had received research funds from biotech companies. Bond went to great lengths to portray critics of biotechnology as "anti-science," even though some of the most outspoken critics have been scientists at the cutting edge of their respective fields, as in the case of Dr. Martha Crouch.

Crouch, a professor of biology at Indiana University, Bloomington, directed a lab whose research got the cover story in *Plant Cell*, a leading journal in the biotech field. Soon afterward she wrote an editorial for the same publication announcing that such scientific research was unethical and that she would no longer conduct it (Crouch 1990). In a recent interview she explained her decision (Crouch 2000:15):

I thought it was inevitable that biotechnology would strengthen industrial agriculture... Modernization of agriculture around the world... has resulted in a greater gap between rich and poor... The research I had done indicated that hunger had increased and more people had starved rather than fewer as a result of the Green Revolution.

Crouch began to see biological science in its social context and realized that in spite of its power it has limitations. Most importantly, its increasingly close connections with the biotechnology industry have, in Crouch's view, compromised the freedom of its researchers. This gets to the heart of the issue of the autonomy of science. The kinds of research questions that are funded (and realized) may become those which coincide with the objectives of corporate sponsors. Though these questions may adhere to our working definition of science (as a search for truth about the world), they search for a kind of knowledge that is often geared to a specific aim: profit maximization.

Crouch's insight into the mechanistic, reductionistic worldview of some biotech scientists is especially revealing (Crouch 2000:15):

[I began to realize that biotech] would result in an agriculture that treats the fields like a factory and the plants . . . as little photosynthesis machines that would crank out a certain amount of product that you could patent or own. And when you genetically engineered the plants, you would think of them in terms of turning on and off switches that would increase or decrease specific products and allow you to do these specific manipulations. You are basically treating the agricultural environment as if it was a factory where you are making televisions or VCRs.

Crouch is describing a decontextualized science, one that does not analyze the wide range of possible effects that GM crops might cause over time. Questions about long-term environmental and health impacts are absent, in part because of the breakneck speed with which the research is being done and the even faster pace at which the products of the research—including GM crops—are being introduced into ecosystems. Crouch ends by arguing that in the long run it may be wiser for agricultural biotechnology researchers to reflect more about their work and to look for more localized solutions to the problems of agriculture and nutrition.

Valuing Local Knowledge

The mechanistic, accelerated approach of the agricultural biotech scientists described by Crouch is radically different from that of Rincón Zapotec farmers. So is the scientific style. When Rincón campesinos obtain a new variety of seed (whether maize, coffee, or another crop) or a new technology (such as chemical fertilizer), experiments are done at a relatively slow pace and in a limited area—perhaps one row at the edge of a milpa, for example. After several years of experimentation, often carried out informally by farmers planting at different altitudes, in different soil types, or at different times of the year, information is exchanged (again by informal channels: chatting while hiking to work, at the cantina, or at a fiesta). A dialogue begins to take shape about the characteristics of the plant or technology, its apparent advantages and disadvantages, and how it affects taste, texture, the soil, other plants, or the human body. Approximately three-fourths of Talean households are campesino households, so many people are local science experts; many opinions and hypotheses are set forth. Because the food chain is short, environmental and health effects tend to be observed quickly and corrected. This is often a matter of survival for Rincón farmers.

Practice and experience are vital in the improvement of Zapotec food and farming systems. Talean farmers were keenly aware of the fact that farming and food practices frequently gave them clear advantages over academically trained specialists. Agronomists and agricultural engineers, from the perspective of many campesinos, are unsuccessful largely because of the fact that they are often theorists, but not practitioners. I was reminded by informants time and again that most specialists have never had to survive by eating plants that they themselves have grown. Few have had to live with the consequences of neglecting a work animal or livestock. Campesinos, by contrast, are both theorists and practitioners, I was told, and are thus able to correct (or at least improve upon) flawed elements in their system of farming and food knowledge.

There are good reasons for preserving and promoting farming systems like those of the Talean campesinos. Small farms have served as banks for much of the genetic material that has been used to create new hybrids; in light of the continued loss of plant species following "the Columbian exchange," this is a remarkably important goal in and of itself, given the threat of genetic standardization (Harlan 1975:251-258; Nabhan 1992: 146-147, 150-151). As geographer Carl Sauer noted in 1941 on the eve of the Green Revolution (quoted in Wright 1984:140):

A good aggressive bunch of American agronomists and plant breeders could ruin the native resources for good by pushing their American stocks ... Mexican agriculture cannot be pointed toward standardization on a few commercial types without upsetting native economy and culture hopelessly. The example of Iowa is about the most dangerous of all for Mexico.

An increasing number of scholars have argued that the cluster of inventions collectively known as the Industrial Revolution was developed by eighteenth-century artisans and engineers, not theoretical scientists (Adas 1989:77). Indeed, scientific revolutions sometimes represent little more than the codification of knowledge developed decades or centuries earlier by pragmatic people who did not have access to printing presses. In the case of agriculture Richards (1985:71) notes that in West Africa "traditional" farms have historically served as agricultural research stations and that frequently the farmer, not the agronomist, has been the source of scientific revolutions:

Innovations normally associated with the so-called "agricultural revolution" of the eighteenth century in large measure had been developed and put into "running order" at an earlier period by ordinary farmers without access to (or need of) means of publicizing their experiments and innovations. The main sense . . . in which there was an "agricultural revolution" in the eighteenth century was that this was the point at which many of the agricultural issues and innovations of the previous two centuries came into public and literary focus.

In Talea a similar phenomenon can be observed; in particular, the best campesinos are rarely even in the village—they are typically found on relatively remote farms, an hour or two away by foot. Still, their relative isolation does not diminish their importance for the future of contemporary cosmopolitan science (Wilken 1987:9):

A long record of successful farming, often under difficult conditions, attests to their ["traditional" farmers'] effectiveness. Are they efficient? . . . [That is] a difficult question; the answer depends on one's objectives. Is production paramount? Income? Employment? Ecological stability? Security? If the last two are indices of overall success, then traditional systems should receive high marks. Some of the practices . . . have been in use for centuries, even millennia, and have been the means by which their farmer/practitioners survived and even prospered. The fact that they still are in use is strong evidence for a social and ecological stability that modern, mechanized systems could well envy.

Mexican farming may depend on the knowledge of campesinos who have developed sustainable subsistence methods that have served them for thousands of years, often on arid, marginal lands. They have largely succeeded in preserving a wide range of native maize varieties and developed techniques that have promoted species biodiversity. This might represent one step toward countering the effects of genetic homogenization that characterized the Green Revolution (and now, potentially, the Gene Revolution).

In other words, the very future of Mexican agriculture might depend on the knowledge of campesinos like those working in Talea.

Science, Democracy, and Zapotec Foodways in a New Millennium

What would the Rincón Zapotec say about such events? It may be helpful to step back for a moment and consider how they might conceive of "development" and "progress." According to anthropologists Alvaro González and María Teresa Prado (1992:14–15), there is linguistic evidence indicating that people in the Sierra might view "development" processes with skepticism, since equivalent terms for "development" and "progress" do not exist in Zapotec, Chinantec, or Mixe languages. The terms can only be approximated with words that mean "to rise up," "to grow," "to advance," "to prosper," or "to realize." Most of these concepts have meaning in reference to cyclical notions of time.

In Talea campesinos employ the two Spanish terms to denote distinct processes. "Development" (*desarrollo*) is a term used to describe the growth, prosperity, or precociousness of people, plants, animals, or minds: things with a life cycle short enough to observe within a human lifetime. In this sense it is a consequence of successful "maintenance," because, with adequate provisioning of food and care, children, milpas, mules, cattle, trees, and other plants will grow properly and quickly. This notion of development, linked as it is to living things (with humans as the point of reference), is strikingly different from ideas of economic development and development projects commonly used today in the United States.

"Progress" (Sp. *progreso*) and (to a lesser degree) "advancement" (Sp. *avance*), in contrast, have more to do with notions of "modern" economic development—these terms are associated with paved roads, electricity, and the trappings of modern urban life. Incidentally, many Taleans often use the term "progress" to describe and distinguish their village favorably from other Rincón villages, indicating that Talea's tradition is one of change. The term's generally positive connotation clashes with the negative effects associated with some new technologies, and there appears to be a conceptual disjuncture with respect to "progress" that has not yet been resolved.

The term "science" (Sp. *ciencia*) is also used by Talean farmers and is associated with airplanes, tractors, and other large machines produced in distant cities. A certain mystification of science seems apparent; perhaps as a result of decades of formal schooling with standardized textbooks espousing nationalist modernization, a culturally specific notion of science has emerged, not unlike the one commonly seen in the United States. Radios, televisions, and the Spanish version of *Reader's Digest* have further congealed such ideas about science in the minds of many villagers.[7] Nevertheless, we should not forget that local critiques of contemporary cosmopolitan sciences are abundant, as in the case of campesinos who describe visiting agronomists as well-intentioned theorists who are ineffective at dealing with practical problems.

Historically, Talean farmers have not exported staple crops out of the Sierra; nor have they worked *ejido* plots. As a result, the Article 27 reforms have not directly affected them. Still, if they were asked to assess the factory farming model being imported from the United States into Mexico, they would probably be critical, since the concepts underlying Zapotec farming and foodways are markedly different from the conventional factory farming categories. Talean farmers might claim that such a change represents too big a tradeoff: food quality might suffer drastically; reciprocal institutions (*gozona*) and more or less egalitarian social structures might be weakened; the earth and forests would likely be upset and vent their wrath on future generations; and *mantenimiento* might become a difficult if not impossible task in monocropped fields drenched in "hot" pesticides, herbicides, and fertilizers. The genetic modification of maize might be viewed as morally wrong and perhaps even foolhardy, since maize has a soul, a memory, and a willingness to seek revenge. Factory farming, in short, might possibly undermine the fundamental principles underpinning a flexible, dynamic Zapotec science with millennial roots.

This may be one of the things that most distinguishes Rincón Zapotec farming (and many other local subsistence strategies) from dominant agricultural practices in U.S. society: the time variable. As Nader reminds us (1998), industrialized societies represent a mere drop in the bucket in the two-million-year history of humankind. Agriculture has been practiced only a little longer, but has generally afforded humans the opportunity to correct themselves before their activities became too destructive in environmental terms. Only relatively recently, however, has a reflective view begun to emerge among some thoughtful people as certain "anomalies" (Kuhn 1962)—such as unprecedented environmental damage and a series of relatively new health epidemics affecting industrial societies (Proctor 1995)—demand fresh approaches in both practical and theoretical terms (see, for example, Steingraber 1997). It is in this context that local science techniques and concepts—once considered "inefficient" and "unproductive" by many in industrialized societies, but perhaps better attuned to longer wavelengths (or perhaps cycles) of time and therefore better able to avoid or confront long-term problems successfully—have become relevant to the survival of people living in industrialized societies.

Rincón Zapotec farmers and their families earn their livelihood by tilling the earth, sowing seed, and harvesting, preserving, and preparing crops. They also experiment with new crops, methods, and technologies, use trial-and-error experimentation, and communicate their findings among colleagues and to younger generations. In these respects they very much resemble contemporary cosmopolitan scientists, though institutional differences are obvious. Perhaps the only qualitative difference is that each group proceeds from a distinct set of conceptual categories. Most Zapotec scientists tend to acknowledge the differences, adopt what seems to work according to their own categories, and reject what does not. Many contemporary cosmopolitan scientists—at least practitioners of "normal science"—appear much more reluctant to do so, as Kuhn (1962) has illustrated. However, there are those who do—as in the case of agroecologists and other scientists working at the cutting edge of their disciplines—and they are frequently responsible for successfully pushing the boundaries of cosmopolitan science.[8] Even so, their work is often not accepted by other cosmopolitan scientists, particularly those unwilling to recognize the possibilities of new approaches to thinking.

Why might this be so? When did this insistence on a single science occur? Feyerabend (1978:79), relying on source material from the ancient Greeks, posits that the defense of "one view, one procedure, one way of thinking and acting to the exclusion of everything else" extends thousands of years into the past. More recently, even religion and "ancient science" were shut out by the "new scientific philosophy" in Europe, and Feyerabend (1978:102) implies that such a view was bound up with political events:

In the 16th and 17th centuries there was a fair competition (more or less) between ancient Western science and philosophy and the new scientific philosophy; there was never any fair competition between this entire complex of ideas and the myths, religions, procedures of non-Western societies. These myths, these religions, these procedures have disappeared or deteriorated not because science was better, but because *the apostles of science were the more determined conquerors*, because they *materially suppressed* the bearers of alternative cultures. There was no research. There was no "objective" comparison of methods and achievements. There was colonization and suppression of the views of the tribes and nations colonized ... Again the superiority of science is the result not of research, or argument, it is the result of political, institutional, and even military pressures.

Cosmopolitan science is not autonomous or context-free. In some cases there is strong evidence that politics and big money lead the way. When agricultural research, for example, is carried out at land-grant colleges financed by taxpayers, one must ask why research projects favor large agribusiness concerns by allowing them to define research problems. In effect, a certain kind of agricultural research and development is subsidized by U.S. citizens. Under the banner of science, such endeavors mask the influence of corporations on the U.S. Congress and their domination of governmental agencies responsible for making key agricultural policy decisions (Hightower 1972). In the words of Laura Nader (1996:9):

Science is not free of culture; rather, it is full of it . . . Politicization of science is unavoidable, [because] behavior is affected by those who control funding and who often determine the research questions . . . Denial of a contextualized science, or the assertion that science is autonomous, strikes at the scientific endeavor, defined as a process of free inquiry.

Yet the need for cosmopolitan sciences to broaden their scope appears to be increasingly urgent in a number of fields as environmental and social limits are reached (witness the case of greenhouse gas emissions, the growing consensus on global warming, and the ecological limitations of plantation farming techniques in tropical regions)—a situation which has some arguing for a return to "machete technology" (Yoon 1998).

A model for science in the new millennium might benefit greatly from a recognition that science traditions have crossed boundaries for many centuries and that cross-cultural movements of technology and ideas can be tremendously fruitful. Anthropologist Shiv Visvanathan (1998:41) describes how a "celebration of science" emerged in the pre-Nehru era in India, selectively borrowing from cosmopolitan science and technology. It was a period of hope and optimism; there were those who thought that "a scientific India . . . would humanize the aggressive West" as it emerged as a part of the "Swadeshi (local, indigenous, native) movement" in the first few decades of the twentieth century (Visvanathan 1998:41). Exciting experiments were carried out, including efforts to examine childhood and nature in a new way, the testing of various systems of medicine such as homeopathy, the development of farming using environmentally sound techniques, and the attempt to incorporate local systems of technology and architecture in large cities such as New Delhi. Underlying this creativity was a recognition of the plurality of science traditions and a desire to forge democratic science policy on a national scale. Others have outlined the potential benefits that might accrue from a democratic approach (such as the establishment of "science shops" in Holland) and have even discussed how the United States might move in such a direction (Winner 1987; Sclove 1995).

The consequences of *not* following a democratic approach—or of working within too narrow a time frame without considering the range of possible long-term effects—appear in newspaper headlines with increasing regularity. The problem of nuclear waste, for example, has staggering safety and security implications: an accident could have grave environmental and public health consequences for literally centuries. Yet there was no democratic debate on nuclear energy—or on factory farming or on the introduction of genetically modified crops. A reexamination of science policy—including a consideration of local traditions and theoretical frameworks like those employed by the Rincón Zapotec and other peoples who have survived thousands of years under adverse conditions—may be something that we can no longer afford to postpone.

There is a fertile ground between science traditions that has barely begun to be cultivated by scholars, scientists, and interested citizens. This analysis of Rincón Zapotec farming is an attempt to stretch the conventional boundaries; specifically, I have argued that the boundaries of knowledge have overlapped and blurred to such a degree that the links between knowledge systems simply cannot be ignored or minimized by anthropologists interested in agriculture. In this study I have attempted to build upon the foundation laid by those doing critical connecting work between science traditions. In exploring Rincón Zapotec farming and foodways, it becomes clear that the divisions between local and cosmopolitan sciences, or an imagined "West" and "the rest," are inadequate for thinking about agriculture today. In a pragmatic way the farmers have not concerned themselves with such distinctions; indeed, even relatively new technologies and crops have rapidly been normalized, perhaps because they have seldom been incompatible with the conceptual bases underlying Rincón agriculture.

Obviously it is quite impossible to accurately describe Zapotec farming as "traditional" or "modern" or, as one ethnographer put it, "Indian or Spanish" (Parsons 1936). Even "syncretism" is too facile a term. It is perhaps more appropriate to describe the agricultural enterprise in the Rincón as Ricard (1966 [1933]:3) once described Mexico's religious sphere, "fused and sometimes juxtaposed," because Zapotec agriculture includes crops and techniques with European, Asian, African, and Oceanic origins. Farming systems in industrialized societies like the United States have drawn (and continue to draw) liberally from "Other" traditions as well. The science of the Zapotec, upon closer analysis, becomes inseparable from cosmopolitan science, and cosmopolitan science inseparable from that of the Zapotec. The question before us all (Zapotec and cosmopolitan scientists alike) is about boundaries and knowledge: how we might intelligently, consciously, and democratically improve our science practices—that is, how we might do better science—by tapping the plurality of influences available to us today.

APPENDIX A

PRONUNCIATION OF RINCÓN ZAPOTEC TERMS

Rincón or Nexitzo Zapotec can be written and pronounced using a modified Spanish alphabet. All pronunciations are identical to those in Mexican Spanish, but there are several additional sounds in Zapotec:

1. There are six vowels in Nexitzo Zapotec, rather than five. The sixth vowel is similar to the schwa sound in English (e.g., the *e* in *cider*), but is pronounced with a tightly constricted glottis. It may be denoted with the letter *h* (which in Spanish is mute), as in the word *bhni*, man.

2. All vowels have two variants, a "normal," unabbreviated version and a more abrupt version that is abbreviated with a glottal stop. The glottal stop may be denoted with an apostrophe following the abbreviated vowel. For example:

gui'   fire, light
za'a   sweet corn
yu'u   house
dxi'a good

3. There are four consonant sounds that do not have equivalents in Spanish but do have equivalents in English. These may be denoted as the following:

(a) *dx*, pronounced like English [j] (e.g., *jar*)
(b) *sh*, pronounced like English [sh] (e.g., *shy*)
(c) *z*, pronounced like English [z] (e.g., *zip*)
(d) *zh*, pronounced like English [zh] (e.g., *pleasure*)

For example:

bedx	bull
padiush	greetings
zroj	hot (piquant)
zhua	maize

This is a modified version of a system developed by Cado Pérez, a Talean who organized and taught a class on written and spoken Zapotec during the 1980s. He has also translated several items from Spanish to Rincón Zapotec.

APPENDIX B

TALEAN FOOD PLANTS

ENGLISH	SCIENTIFIC NAME	SPANISH[1]	ZAPOTEC
I. Maize			
maize	*Zea mays*	maíz	zhua
"	"	" grande	zhua zhn
"	"	" chico	zhua cuídi
"	"	" de tres meses	zhua huén
"	"	" negro	zhua gasj
"	"	" blanco	zhua chiguích
"	"	" amarillo	zhua gach
"	"	" colorado	zhua zhib
"	"	" pinto	zhua pint
II. Beans			
bean	*P. vulgaris*	frijol	za
black bean	"	" delgado	za las
white bean	"	" blanco	za chiguích
bean	"	" zatope	za dúpi
green bean	"	" de milpa	za leyhl
"	"	" za láya	za láya
"	"	" cuarentena	za cuan
haba bean	"	" haba	za rrab
pea	*Pisum* sp.	chícharo	za lbérj

265

ENGLISH	SCIENTIFIC NAME	SPANISH[1]	ZAPOTEC
III. Squash			
squash	*Cucurbita* spp.	calabaza	yútu
"	*Cucurbita* sp.	" támala	yútu gu
"	"	" chompa	yútu chína
"	"	" huiche	yútu yag
"	"	" largucha	yútu nicách
"	"	chilacayota	yútu wedx
IV. Chiles			
chile	*Capsicum* spp.	chile	guí'na
"	*C. pubescens*	" bolero	guí'na bolér
"	*C. frutescens*	" cimarrón	guí'na cimarón
"	*C. annuum*	" piquín	guí'na piquín
"	"	" parado	gui'na zui'
"	"	" jalapeño	guí'na jalapén
"	"	" guajillo	chilcózle
V. Coffee			
coffee	*Coffea* spp.	café	café
"	*C. arabica*	café arabe	arabe, criollo
"	"	caturra amarilla	caturramarilla
"	"	caturra roja	caturra roja
"	"	mundonovo	mundonovo
"	"	garnica	garnica
"	"	vegetal	vegetal
"	"	borbón	borbón
VI. Sugarcane			
sugarcane	*Sacharum officinarum*	caña de azúcar	yhtj
"	*Sacharum* sp.	" blanca	yhtj chiguích
"	"	" negra	yhtj gasj
"	"	" de castilla	yhtj shtil
"	"	" de campeche	yhtj campéch

ENGLISH	SCIENTIFIC NAME	SPANISH [1]	ZAPOTEC
"	"	" jaba	yhtj jaba
"	"	" quijote	yhtj quijote
"	"	" morada	yhtj moradu
"	"	" de brasil	yhtj brasil

VII. Vegetables

ENGLISH	SCIENTIFIC NAME	SPANISH	ZAPOTEC
chayote	*Sechium edule*	chayote	cuan yhtzi
tomatillo	*Physalis ixocarpa*	miltomate	beshigísh
radish	*Raphanus sativus*	rábano	rábano
cabbage	*Brassica oleracea*	col	colsh
collard green	"	acelga	acelga
lettuce	*Lactuca sativa*	lechuga	lechuga
potato	*Solanum tuberosum*	papa	tormít
tomato	*Lycopersicon esculentum*	tomate	bésh
wild tomato	"	tomate delgado	beshlás

VIII. Fruits

ENGLISH	SCIENTIFIC NAME	SPANISH	ZAPOTEC
banana	*Musa* spp.	plátano	yhla
"	*Musa* sp.	" chaparro	yhla chapárr
"	"	" manzanita	yhla manzán
"	"	" burro	yhla búrr
"	"	" castilla	yhla shtíl
"	"	" de la india	yhla delíndiu
"	"	" guineo	yhla gasj
"	"	" morado	yhla morádu
"	"	" siendoboque	yhla siendbók
"	"	" de beyácu	yhla bduá'
orange	*Citrus* spp.	naranja	huí
"	*Citrus* sp.	" china	huí chína
"	*C. aurantium*	" agria	huí zi
"	*C. sinensis*	" dulce	huí criollo
tangerine	*C. reticulata*	mandarina	mandarína

ENGLISH	SCIENTIFIC NAME	SPANISH[1]	ZAPOTEC
grapefruit	*C. paradisi*	toronja	toronja
lime	*C. aurantifolia*	limón	limón
sweet lime	*C. limetioides*	lima limón	limalimón
sweet lemon	*C. limetta*	lima chichita	huí zídx
apple	*Malus* sp.	manzana	manzán
peach	*Prunus persica*	durazno	tras
thornapple	*Crataegus pubescens*	tejocote	tejocóte
blackberry	*Rubus aderotrichus*	zarzamora	beská'
white zapote	*Casimiroa edulis*	zapote amarillo	ladzh
black zapote	*Dispyrus ebenster*	zapote negro	lá'u
mamey	*Calocarpum mammosam*	mamey	lazhún
pomegranate	*Punica granatum*	granada	bza'a
pineapple	*Ananas comosus*	piña	duadxína
mango	*Mangifera indice*	mango	mángu
guava	*Psidium* sp.	guayaba	huiyáj
papaya	*Carica papaya*	papaya	papaí
plum	*Prunus americana*	ciruela	yadx
loquat	*Eriobotrya japonica*	níspero	nísperu
custard apple	*Annona* sp.	anona	lagúchi
watermelon	*Citrullus lunatus*	sandía	sandié
yellow mombin	*Spondias mombin*	ovo	yadxko'
fig	*Ricinus communis*	higo	hígo
cantaloupe	*Cucumis melo*	melón	melón

IX. Quelites (Greens)

squash leaf	*Cucurbita* spp.	guía de calabaza	cuan sétu
"	*Cucurbita* sp.	támala	cuan setugú
"	"	chompa	cuan setuchína
"	"	huiche	cuan setuyág
"	"	chilacayota	cuan setuhuédx
chayote leaf	*Sechium edule*	quelite chayote	cuan naph
coraltree leaf	*Erythrina* sp.	" de zompantle	cuan ptútsu

Talean Food Plants   269

ENGLISH	SCIENTIFIC NAME	SPANISH[1]	ZAPOTEC
bean leaf	*P. vulgaris*	" de frijol	cuan za
"	"	" frijol negro	cuan za las
"	"	" de haba	cuan zarráb
"	"	" de za láya	cuan za láya
"	"	" de za tope	cuan za dúpi
pea leaf	*Pisum* sp.	" de chícharo	cuan zalbérj
N/A	[unidentified]	" de conejo	cuan conej
N/A	[unidentified]	" de berro	cuan bérru
N/A	[unidentified]	" de cuche	cuan iquiáj
N/A	[unidentified]	" de la virgen	cuan shunádx
N/A	[unidentified]	oreja de toro	cuan guídnag bedx
N/A	[unidentified]	piojito	cuan béchi
N/A	[unidentified]	huele de noche	cuan zhu'
N/A	[unidentified]	N/A	cuan besh
N/A	[unidentified]	N/A	cuan guíticuch

X. Cactus and Century Plants

cactus	*Opuntia* sp.	nopal de huevo	biédxít
"	"	" castilla	biéshtíl
"	"	" de zorra	biébuïza
century plant	*Agave* sp.	maguey (pulque)	duá nup
"	"	maguey (flor)	duá bla'

XI. Flavorings

yerba buena	*Satureja douglasii*	yerba buena	zhuiyáj
yerba santa	*Piper auritum*	yerba santa	ladxuá
epazote	*Chenopodium ambrosioides*	epazote	bht
onion	*Allium cepa*	cebolla	cebolla
chives	*A. schoenoprasum*	cebollina	layh
garlic	*A. sativum*	ajo	ájo
parsley	*Petroselinum crispum*	perejil	perejil
cilantro	*Coriandum sativum*	cilantro	cilandru
pennyroyal	*Hedeoma* sp.	poleo	zhuiyájzh

ENGLISH	SCIENTIFIC NAME	SPANISH[1]	ZAPOTEC
oregano	*Origanum vulgare*	orégano	gueréganu
thyme	*Thymus vulgaris*	tomillo	tumíllu
aguacatillo	*Persea* sp.	aguacatillo	yeshuló'
allspice	*Pimenta dioica*	pimienta	pimiént
chipil	*Crotalaria longirostrata*	chipil	chepil
bay leaf	*Laurus nobilis*	laurel	yéshu bgín
cinnamon	*Cinnamomum zeylanicum*	canela	canél
cocolmecatl	*Smilex* sp.	cocolmeca	yhtzi buíraj
cacao	*Theobroma cacao*	cacao	zuí'a
lemon grass	*Cymbopogon citratus*	té limón	té limón
spiny cilantro[2]	[unidentified]	cilantro espina	cilandru yhtzi

XII. Food Plants Collected in the Forest

wild blueberry	[unidentified]	N/A	rahuás
pitaya palm	[unidentified]	pitaya	pitaya
mushroom	Basidiomycetes class	hongo	be'i
"	*Boletus* sp.	lengua de toro	bé'iludxibédx
"	*Amanita cesarea*	yema de huevo	bé'i gúlagdxit
"	*Ramaia* sp.	cresto de gallo	bé'iludxbuïraj
"	[unidentified]	hongo de elote	bé'iyïtza

XIII. Other Food Plants

avocado	*Persea americana*	aguacate	yéshu
"	"	" delgada	yéshugá
"	"	" gruesa	yéshuchúga
"	"	aguacatillo	yéshulo'
"	"	chupón	yéshudúdxi
yam	*Dioscorea* sp.	camote	guyág
"	"	camote morado	guyúlu
chayote yam	*Sechium edule*	chayocamote	guyán
pacay	*Inga guinicuil*	guajinicuil	yajtúlu
tepejilote	*Chamaedorea tepejilote*	tepejilote	yútzu
pecan	*Carya illinoensis*	nuez decuilapa	nuezcuilápa

ENGLISH	SCIENTIFIC NAME	SPANISH[1]	ZAPOTEC
corn smut	*Ustilago maydis*	huitlacoche	béhuaj
wheat	*Triticum* sp.	trigo	trigo
N/A	[unidentified]	gallito	yagyéchu

[1] Spanish names may not coincide with the names given to the plants in other parts of Mexico or Latin America; I have included the Spanish names used by the Taleans.

[2] "Spiny cilantro" is an unidentified plant with a taste similar to cilantro, yet is unrelated.

APPENDIX C

TALEAN LIVESTOCK AND GAME ANIMALS

ENGLISH	SCIENTIFIC NAME	SPANISH[1]	ZAPOTEC
I. Livestock and Domesticated Animals			
bull, cow	*Bos* sp.	toro, vaca	bedx
chicken	*Gallus* sp.	gallo, gallina	bhráj
pig	*Sus scrofa*	cerdo	cuch
goat	*Capra* sp.	cabra	chivo
turkey	*Meleagris gallopavo*	guajolote	brus
rabbit	*Leporidae* sp.	conejo	coneju
honey bee	*Apis mellifera*	abeja	buíz
sheep	*Ovis aries*	borrego	bhcuzhríla
II. Game Animals			
wild hen	*Gallus* sp.	gallineta	bdxí'a
deer	*Odocoileus* sp.	venado	bdxinaghshi
quail	*Geotrygon* sp.	codorniz	butó'
armadillo	*Dasypus* sp.	armadillo	bugúpi
grasshopper	*Sphenarium histro*	chapulín	yajza'
brocket deer	*Mazama* sp.	mazate	bdxinaghshi
peccary	*Tayassu tajacu*	jabalí	cuchighshi
weasel	*Mustela frenata*	comadreja	bunél
small weasel	*Mustela frenata*	" chica	bunelbíz
coati	*Nasua narica*	tejón	bdxídzu
wild rabbit	*Osyetolofagus cuniculus*	conejo	conéju

Talean Livestock and Game Animals 273

ENGLISH	SCIENTIFIC NAME	SPANISH[1]	ZAPOTEC
paca	*Agouti paca*	tepescuintle	biajág
black crab	Decapoda order	cangrejo negro	buatijér
wasp larva	*Vespa* sp.	panal de castilla	pch'h
"	"	" de tierra	biyáj
"	"	" de campana	bechidh
"	"	" de torta	bziú' rhn
"	"	" de piedra	bechidh guiáj
"	"	bolsa del cielo	bzudyubá
wild turkey	*Meleagris gallopavo*	pavo	brus
chachalaca	*Ortalis vetula*	chachalaca	chachalaca
dove	*Zenaida* sp.	paloma solitaria	palóm
"	"	paloma montés	palomguí'a
anteater	*Myrmecophaga* sp.	oso hormiguero	osormiguéro
frog	*Rana* sp.	rana	bludzuguín
porcupine	*Erethizon* sp.	puerco espino	bédxightzi
iguana[2]	*Iguana iguana*	iguana	iguán
opossum[2]	*Didelphis marsupalis*	tlacuache	cháje'eza

[1] Spanish names may not coincide with the names given to the animals in other parts of Mexico or Latin America; I have included the Spanish names used by the Taleans.
[2] The iguana and the opossum are no longer eaten in Talea.

APPENDIX D

SELECTED AVERAGE CROP YIELDS

Milpa system (intercropped maize field) [1]
maize: 4–5 *fanegas* (566–708 liters) per *almud* (0.25 ha) [2]
beans: 12 *almudes* (71 liters) per *almud* (0.25 ha)
squash: 70–80 squashes per *almud* (0.25 ha)

Frijolar system (intercropped bean field) [3]
maize: 7–8 *almudes* (41–47 liters) per *almud* (0.25 ha)
beans: 1 *fanega* (132 liters) per *almud* (0.25 ha)

Sugarcane: [4] 125–150 *pancles* (875–1,050 kg) per *almud* (0.25 ha)

Coffee: 3–4 kg of dry coffee beans per tree

Chiles
chile bolero: 160 chiles per plant
chile cimarrón: 200–300 chiles per plant

Vine plants
chayote: 150–180 chayotes per plant
squash: 15–20 squashes per plant
watermelon: 10–12 watermelons per plant
melon: 10–12 melons per plant

Fruit trees [5]
banana: 200–300 fruits per tree
orange: 200–300 fruits per tree
tangerine: 200–300 fruits per tree

lime: 100–150 fruits per tree
lemon: 100–150 fruits per tree
lemon-lime: 100–150 fruits per tree
grapefruit: 70–80 fruits per tree
apple: 150–200 fruits per tree
peach: 150–200 fruits per tree
hawthorn: 120–144 liters of fruit
blackberry: 12–14 liters of fruit
annona: 100–150 fruits per tree
zapote: 150–200 fruits per tree
mamey: 150–200 fruits per tree
pomegranate: 170–180 fruits per tree
mango: 400–500 fruits per tree
guava: 400–500 fruits per tree
papaya: 140–150 fruits per tree
plum: 120–144 liters of fruit
loquat: 120–144 liters of fruit
fig: 120–144 liters of fruit
yellow mombin: 120–160 liters of fruit

Other trees

avocado: 150–200 fruits per tree
allspice: 3 *arrobas* (34.5 kg) dry peppers per tree

Animals[6]

cattle: 18–22 *arrobas* (207–253 kg) after 10 years
chicken: 4–5 kg after 6–8 months
turkey: 5–7 kg after 6–8 months
rabbit: 3.5–4.5 kg after 5–6 months

---

[1] Interviewees were asked: "How much maize [or beans or squash] can a campesino expect to harvest from a milpa of one *almud*?" One *almud* is an area equal to the amount of land needed to plant an *almud* (approximately 5.9 liters) of maize seed. It is equal to approximately 0.25 hectares. See Chapter 3 for a description of weights and measures.

[2] Average yield information varies from quite precise data (for example, in the case of milpa systems, *frijolar* systems, and coffee trees) to rough approximations (for example, in the case of some fruit trees). Refer to Appendix B for scientific names.

[3] Interviewees were asked: "What quantity of beans [or maize] can a campesino expect to harvest from a bean field of one *almud*?"

[4] Sugarcane fields, like bean fields, are measured in terms of *almudes* of maize (0.25 ha).

[5] Interviewees were asked: "How much fruit might an average mature tree bear in a year?"

[6] Interviewees were asked: "How large does this animal grow when mature, and how much time passes before it reaches that size?"

APPENDIX E

RECIPES

Recipe 1. Chayote Root Yam in Yellow Sauce

(Spanish name: *Amarillo de chayocamote*)

(Zapotec name: *Gínduguyán*)

Ingredients

  2 guajillo chiles, toasted
  5 cloves of garlic
  6 small tomatoes, roasted (or boiled) and peeled
  1 fist-sized lump (approx. 0.2 kg) of maize meal
  2 chayote root yams (20–25 cm long), diced
  3–5 pinches of salt (according to taste)
  4 tender leaves of yerba santa (*Piper auritum*)

Directions

1. Prepare sauce by grinding (on a metate grinding stone) guajillo chiles, garlic, and tomatoes into a thick liquid. Pour into bowl and set aside.

2. Place maize meal in another bowl. Stir in a small amount of water until it forms a batter the consistency of pancake batter. Set aside.

3. In a clay cooking pan, boil the water. Add chayote root yams and salt.

4. When chayote root yams are tender, lower heat (simmer) and add sauce and maize batter. Stir occasionally.

5. Allow sauce to thicken. Add yerba santa leaves and, if needed, salt. After 5 minutes remove from fire. Serve with fresh corn tortillas.

Recipe 2. Beans with Sweet Corn and Pennyroyal

(Spanish name: *Frijol con elote*)

(Zapotec name: *Lan zá'a*)

Ingredients

- 2 cups of black beans
- 1 clove of garlic, chopped
- 4 small leaves of epazote (*Chenopodium ambrosioides*)
- 1/2 handful of fresh chives, finely chopped
- 1/2 handful of salt
- 1 1/3 cups of sweet corn
- 1 chile seco (hot red tree pepper), ground
- 1 clove of garlic, ground
- 3 tablespoons of *panela* (unrefined brown sugar)
- 1/2 teaspoon chopped *poleo* leaves

Directions

1. Prepare beans: put beans, garlic, epazote, chives, and salt into a clay pot with 2 liters of water. Cover and boil at high heat until beans are tender (approximately 2 hours), adding water if necessary.

2. Grind (on a metate stone or hand grinder) or blend (in a blender) sweet corn into a paste. Place in a bowl and slowly stir in water to make a batter the consistency of pancake batter.

3. Combine beans and corn batter in a clay cooking pan. Add ground chile and garlic.

4. Bring to a low boil, stirring frequently. Stir in *panela* and *poleo*. When mixture thickens, remove from fire. Add salt to taste. Serve with fresh tortillas.

Note: The *poleo* plant is the pennyroyal (*Hedeoma pulgioides*), a member of the mint family (Labiatae). Those wishing to prepare this dish might substitute fresh peppermint (*Mentha piperita*) for pennyroyal.

NOTES

1. The Conceptual Bases of Zapotec Farming and Foodways

1. In Mexico the term "campesino" literally means "country person" or "person who works in the fields." I have chosen to use this term rather than "peasant" because many Talean campesinos do not meet any formal definition of "peasant." The anthropological interest in "peasants" began in the 1930s with the study of "folk society" (Redfield 1930). Since that time various efforts have been made at defining "peasants" and the communities in which they live (Kroeber 1948; Wolf 1955, 1966; Foster 1967). The term has recently come under fire from anthropologists, particularly Michael Kearney (1996), who argues that it has outlived its usefulness.

2. I have chosen the terms "local science" and "cosmopolitan science" to refer to two different approaches to knowledge. Local science is situated in a specific place; it "is particular, by definition; it can be acquired only by local practice and experience ... the holder of such knowledge typically has a passionate interest in a particular outcome" (Scott 1998:317-318). As examples, we might think of the farming knowledge of Zapotec campesinos, the hunting knowledge of the Cree, the medical knowledge of Chinese acupuncturists or of orthopedic surgeons who specialize in knees, or the practical knowledge of ship pilots. James Scott (1998:6-7) has expressed a similar idea with the term *métis:* "practical knowledge that can only come from experience." The emphasis here is on the nature of knowledge as something constructed from *practice* and *experience*, as in the case of Chinese medical specialists (Farquhar 1994) and West African subsistence farmers (Nyerges 1996). Local knowledge is typically dynamic and flexible enough to use in dealing with a range of unforeseen circumstances and is not usually codified.

I use the term "cosmopolitan science" to describe bodies of knowledge which are truly cosmopolitan or international in scope, in the sense that (1) they draw upon science traditions from many societies around the world (Chinese, Indian, European, Mesoamerican, etc.) and (2) they are practiced in many different countries, by people from many different cultural and ethnic groups.

Cosmopolitan science represents a sharp and self-conscious rupture with Western classical scientific traditions (astrology, biblical creationism, Hippocratic medicine, etc.). It includes such fields as modern international biomedicine, nuclear physics, agronomy, molecular cell biology, agroecology, and ethnobotany. Cosmopolitan sciences are often quite precise and powerful because their practitioners radically restrict the field of scientific inquiry to only a few variables. Theories are codified in a set of universal rules or general scientific laws which only change when they are superseded by others.

These terms, though not without problems, are much better than the alternatives. "Western" versus "non-Western" science has, in blunt terms, racist connotations. What is often incorrectly referred to as modern Western science is in fact derived from many science traditions around the world. "Traditional" versus "modern" science is a problematic distinction because "tradition" implies stasis, which is not true in the case of local knowledges. Similarly, "indigenous" knowledge systems are rarely completely indigenous.

3. In order to draw a distinction between *Homo economicus* and the subsistence farmer, several terms have been coined that describe the latter forms of culture/agriculture as "traditional" (Wilken 1987), "indigenous" (Richards 1985), "peasant" (Foster 1967), "sustainable" (Altieri 1987), or subject to a noncapitalist "moral economy" (Scott 1976).

4. See Malinowski (1935) and Netting (1993).

5. See also Norman (1974), Belshaw (1979), and Toledo (1980).

6. See also IDS (1979), Richards (1979), Bandyopadhyay and Shiva (1981), Kidd and Colletta (1982), Posey (1984), and Warren (1984).

7. These bear some resemblance to anthropological notions of "world view" (Geertz 1973), "paradigms" (Kuhn 1962), "themes" (Geertz 1959; Opler 1959), or "cultural patterns" (Kroeber 1917; Benedict 1934).

8. It seems likely that hot/cold food classifications, for example, came from the outside (see note 14 below); thus it would clearly be a prescriptive guide. The *mantenimiento* and reciprocity concepts, however, are probably logical, inevitable deductions that have been made over the course of many centuries.

9. This process has been documented in other parts of the world where shifts have been made from subsistence food production to cash cropping (Gudeman 1978; Pratt 1994).

10. Among other things, this is likely to depend on international coffee prices, maize and bean subsidies, migration, and other external factors.

11. The *matelacihua* appears across Mesoamerica. The term comes from Nahuatl.

12. There are contrary trends among segments of the U.S. population, as we shall see in the concluding chapter. Specifically, the market for organic foods is growing rapidly in the United States.

13. There may be other criteria as well. Oranges are a "hot" color, while limes are a "cool" color.

14. Specifically, most of the soil, water, air, and forest categories are demonstrably indigenous to the Americas, while the food categories come from the Old World.

15. I am grateful to Eugene Anderson for helping me outline the central ideas in this section.

2. Locating Talea

1. Laura Nader (personal communication, 1997) notes that a generation ago "Sierra Juárez" referred exclusively to the area lying immediately to the west of the Rincón (between

the Rincón and the Valley of Oaxaca). The terms "Sierra Madre Oriental," "Sierra Madre del Sur," "Sierra Zapoteca," and "Sierra Zapoteca-Mixe" have been used historically to refer to the mountain system as a whole.

2. According to Tyrtania (1992:22), the region appears in turn-of-the-century documents as the Cordillera del Margen (literally, "mountain range of the margin").

3. Many informants told me that the last owners of the Santa Gertrudis mines (forty-five minutes' walk below Talea) were an Italian family named Tomacelli. They were undoubtedly a powerful family when the construction of the church began, and it is reasonable to think that they may have brought the famed painting from Italy and perhaps played some role in initiating the project as well.

4. It seems that the intentions of the Spanish were eventually overturned and appropriated in many indigenous communities. By the 1920s the Mesoamerican indigenous village was being described by Mexican anthropologists of the *indigenista* variety (who sought the cultural integration of the country's indigenous people into national life) as the biggest obstacle to the modern development of Mexico. Such views dominated Mexican anthropology through the 1970s (see Gamio 1916; Caso and Aguirre Beltrán 1960; and de la Fuente 1960).

5. The structures also show how the civil and religious sides of life were linked: *tequios* were, according to older informants, required to build the Catholic church.

6. Not all young people frequent the village streets after dark. In particular, most campesina girls—those belonging to the *gente humilde* (humble people)—are not allowed to leave the house unchaperoned. It should be added that not all young campesino men congregate around the courts either. Though it is fairly common for some to shower, change into clean, ironed clothes, and wolf down dinner in order to *cotorrear* (chat, gossip, "hang out") or *cazar* ("hunt") girls, many are too tired for the social scene or else stay out in the fields. There are clearly distinct patterns based on socioeconomic differences in the village.

7. Ruins of what was probably a community settled by the ancestors of the Taleans exist today at this site. It consists of a level area marked off by long stone walls ranging from approximately one to three meters in height, arranged at right angles. According to some Taleans, this is where a church was going to be built centuries ago, but the project was abandoned due to a lack of water. Today Sudo' (a Zapotec term that means "where the temple was") is mostly covered by coffee and pimiento trees.

8. These fields have been abandoned due to the introduction of chemical fertilizers on the one hand and increased migration on the other. Fertilizers effectively doubled the maize yield, which eased land pressure to such a great degree that the fields began to be seen as too distant (more than two hours' walk) to work conveniently. Outmigration, made easier by improvements to the road to Oaxaca City, also helped to keep land pressure from rising.

9. *Bienes comunales* should not be confused with *ejidos*. The latter were institutionalized in many other parts of Mexico following the 1910 Revolution and were codified through much of the century in Article 27 of the Mexican Constitution, which was radically reformed in 1992 (Randall 1996). The former, unlike *ejidos* (which are essentially individually worked plots owned formally by the community), are lands which contain resources (above all, firewood) which may be exploited by community members at little or no charge.

10. Old World diseases devastated Native Americans, who had not previously been exposed to the disease pathogens and therefore had no immunity to them (Crosby 1972).

11. This system of obligatory community service is referred to as the cargo system or civil-religious hierarchy in the anthropological literature. See Carrasco (1961), Wolf (1966), De-Walt (1975), Rus and Wasserstrom (1980), and Chance and Taylor (1985).

12. For most of the colonial period, only males from elite indigenous families were allowed to vote.

13. The revolts occurred in Totontepec in 1748, Latani in 1783, Cacalotepec in 1779, and Yahuive in 1812.

14. Zapotec resistance continues today and is strikingly evident in the semipermanent demonstrations at the Oaxacan state government building. In 1998 and 1999 the most persistent protests were organized by the Zapotec from a group of villages known as the Loxichas, in the Southern Sierra. The villagers have protested the incarceration of dozens of Loxichans accused of being leftist guerrillas associated with the Popular Revolutionary Army or EPR. For a review of Zapotec resistance movements in the Isthmus of Tehuantepec, see Campbell et al. (1993).

15. The document, entitled "Memoria y probanza de la fundación del Pueblo de Talea, hoy Villa de Castro," is written in Zapotec and also in Spanish.

16. The other documents refer to the towns of Solaga, Otatitlán, Yatzachi, and Juquila.

17. One of these men is Juan de Salinas, *alcalde mayor* of Villa Alta from 1555 to 1560. The other is Fray Jordán de Santa Catalina (Chance 1989:30-31).

18. García is regarded by many Taleans as the most important village leader in recent history, a visionary who spent much of his money on projects, including an electric generator. García used the current generated by a river 600 meters below Talea to illuminate public spaces such as streets and the municipal palace. But electricity also served an entrepreneurial function: it powered an electric mill (the first of its kind in the region) for grinding hulled maize into tortilla dough—for which García charged a small fee.

19. Although the commission's work improved conditions in many communities, bringing electricity, roads, and potable water to a relatively inaccessible part of the country (Poleman 1964), the damming of the Papaloapan River resulted in the displacement of tens of thousands of Mazatec and Chinantec people—up to 20,000 according to some estimates—in the late 1960s and early 1970s (Bartolomé and Barabas 1974; Partridge, Brown, and Nugent 1982). Many of them fell ill and even died after being relocated in lands far outside the region and were denied adequate support following the relocation.

20. For a description of typical small industries in the various Zapotec regions in the 1940s, see Alba and Cristerna (1949b).

21. According to Laura Nader (personal communication, 1996), Talea had telephone and telegraph service prior to the Mexican Revolution of 1910, though it was discontinued during the Revolution. Talea has also had telegraph service since the 1940s.

22. Many Talean merchants grumble about the growing markets on the other side of the mountain, and a handful make the one-day trip to the larger of the two markets (Yaeé). The road from Maravillas to Yaeé (linking Juquila Vijanos, Reforma, Tanetze, Yaviche, Yagallo, Lachichina, and Yaeé), completed in the early 1980s, has been a major factor in the Talean market's decline. By 1997 construction was underway to extend the road from Yaeé to Lalopa. Once completed, the road will link those on the east side of the mountain (Talea, Las Delicias, Porvenir, Yatoni, Otatitlán) to those on the west, forming a circuit that will include nearly all Rincón villages. The history of this road is marked by conflict and confrontation between the state government and Pueblos Unidos, an organization of nine (now twelve) Rincón villages that pressured state officials with protests in Oaxaca City and on one occasion by holding road construction crews and machinery hostage. Pueblos Unidos continues to exist as a cooperative organization providing daily bus service from Yaeé to Oaxaca City.

Talean informants described the market of the late 1950s with nostalgia: it was so crowded that the buzz of the people in the village center could be heard from the edges of Talea. Pack animals came from Solaga and Zoogocho, so heavily loaded with merchandise from the Valley of Oaxaca that they could be heard grunting as they climbed the steep trail from Santa Gertrudis. Laura Nader's film *To Make the Balance* (1966) includes footage that supports these descriptions.

23. Chance's categories include three of the six Zapotec dialects outlined by Nader (1964: 204): Serrano, Nexitzo, and Villalteco. Nader's categories are based on the work of Morris Swadesh (1949).

24. In other words, language and culture are not always congruent. As Nader (1969:331) noted, "linguistic areas do not necessarily coincide with cultural areas. A gross description of cultural areas could state that the people of the northern Sierra have more in common with each other than they do with the valley Zapotec, that the Chopan Zapotec may have more in common with the Chinantec than with the Sierra Zapotec, that the valley Zapotec may have more in common with valley Mestizo towns than either have with the Tehuantepec Zapotec, and so on."

25. In some Sierra villages migrants living in other cities or countries are obligated to pay fiesta expenses upon entering the village to enjoy the festivities. For example, informants from Yaeé reported that during the opening days of the annual fiesta the village police wait for incoming vehicles in order to collect the mandatory tax. I was also told that in certain Cajonos villages young men are so scarce (due to migration) that they are summoned from as far away as California to serve cargos. They are given the choice of returning to fulfill their duties, providing a relative to serve as a substitute, or paying a fee to hire a replacement (approximately N$900 or US$120 per month at the time my research was conducted).

26. The Sierra was not always a tranquil place. As recently as the 1910s revolutionary *caudillos* (military strongmen) such as Isaac Ibarra and Fidencio Hernández led groups of mountain Zapotec soldiers to the state capital, and federal troops swept the Sierra on several occasions looking for the armed groups (Pérez García 1956; Garner 1988).

27. For profiles of the EPR and early summaries of its activities, see Moore (1996), Preston (1996), and Gatsiopoulos (1997).

3. The Craft of the Campesino

1. However, one important piece of evidence should make us wary of generalizing the slash-and-burn hypothesis for the entire Rincón: high population estimates. Although most villages had fewer than 1,000 residents in the mid-1500s, two of the largest Rincón villages, Tanetze and Yaeé, had more. Tanetze, with an estimated population of nearly 3,300 (see Chance 1989), was by far the largest Rincón village. (Even Talea's current population is well below this level.) It may be that such population density was caused by "congregation" policies in the colonial period (see Chapter 2). Also, it is possible that in that period Tanetze's lands covered a considerably larger area than they do today. Nonetheless, it seems likely that the limits of the village's land were being pressed. It may be that even before the Conquest the people of Tanetze had adapted by moving toward a system more resembling fallow cultivation.

Slash-and-burn and fallow systems should be seen as ideal types, the extreme ends of a continuum. One could easily imagine intermediate systems in which, for example, four patches of land are each cultivated for two years and then allowed to rest for eight; or where three patches of land are each cultivated for one year and allowed to rest for two; and so on. These would seem to lie between the ideal types.

2. If the population had remained constant, it seems plausible that hoe cultivation might have been adopted since it is a more intensive activity than plow agriculture. That is, hoe cultivation can produce up to twice as much more than plow agriculture for a given unit of land (Wolf 1959:198).

3. The willingness of many Taleans to experiment with new techniques and crops may thus be related to an early need for surviving on limited land. Talea appears to be markedly different from Tzintzuntzan, Michoacán, where limited crop land led not to experimentation, but rather to a "conservative" outlook in which new ideas were often rejected. The dismal view of a miserly world was dubbed "the image of limited good" by Foster (1967).

4. Not everyone in Talea uses these measurements. Most masons and carpenters, for example, use the metric system and employ tape measures and meter sticks. The metric system, according to Foster (1960), was adopted in Spain around 1850, and it seems likely that it spread to many parts of Latin America shortly thereafter.

5. The *dedo* in Mitla, however, is different from the Talean measure: the former denotes the length of the index finger, while the latter denotes its width.

6. Alba and Cristerna (1949a:473) report that in the Sierra five liters equals an *almud* in most villages, except San Mateo Cajonos, Betaza, and Yalálag, where it equals half an *arroba* (5.75 kg). They say that in the Rincón village of Cacalotepec a four-liter *almud* is used.

7. The dimensions of the Talean *almud* box are approximately 21.6 cm square by 12.7 cm deep or 5,920 cubic cm (5.92 liters). Assuming that the weight of a liter of dry maize is equivalent to 0.71 kg, the Talean *almud* of maize weighs approximately 4.2 kg. This accords with data collected by Malinowski and de la Fuente (1982 [1940]:177) in the Valley of Oaxaca.

8. There appears to be a general consensus in the village that at least some formal education—particularly reading, writing, and basic math—is necessary so that villagers can "defend themselves" from more educated villagers and the occasional charlatans and government officials who visit the community. Not all agree that junior high school is a good thing. According to a key informant, both boys and girls often learn to be "lazy people" in junior high; that is, they are often reluctant to take up physical work and are generally not as skilled at farming as their campesino counterparts. Furthermore, in a conversation with a group of campesinos, it was argued by some that a handful of pregnancies among junior high school girls and a scuffle or two among a few junior high school boys were evidence that "education is something learned in the home, not in school." Indeed, the term "studies" (Sp. *estudios*) is used to refer to what is learned formally in schools; "education" (Sp. *educación*) refers to moral upbringing: a person's ability to interact respectfully and responsibly with peers and superiors.

9. By July 2000 this appeared to be a very real threat. Dozens of young Talean men and women were leaving the Sierra to migrate to the United States. Depressed coffee prices had devastated the local economy; according to informants, young villagers were leaving in droves in pursuit of economic opportunities.

4. "Maize Has a Soul"

1. Not all Green Revolution scientists were so ignorant, however. Some of its principal architects freely admitted that they could not significantly increase maize yields in Mexico

because local farmers had improved the crop about as much as was biologically possible. I thank Eugene Anderson for making this point.

2. The Mixe of San Pablo Chiltepec use the word *pu:ck* or "navel" to refer to this part of the kernel (Lipp 1991:18). The nucleus gets eaten by the ubiquitous granary weevil because, in the words of the campesinos, the "force" of the plant is concentrated there.

3. Older informants in Talea give accounts similar to those of villagers from Yojobi (Castellanos 1988). I did not have the opportunity to talk about the Yojobi conflict with villagers from Tabaa.

4. According to a number of campesinos, one of the gravest offenses that a woman can commit is to let a bucket of tortilla dough spoil. Husbands have been known to punish their wives for this. For the case of maize, at least, Foster's "image of limited good" (1967) seems to hold up.

5. Maize was obviously a dominant element in the ancient Mesoamerican civilizations, but our primary concern in this section is with maize as a vehicle for communion with the deities in the Northern Sierra of Oaxaca.

In the Aztec creation myth humans were created five times, each time more perfectly evolved than the last. *Teosinte*—called *cencocopi* and *acicintli* by the Aztec—was the food of the humans living in the eras of the Third and Fourth Suns. Not until the era of the Fifth Sun (the era immediately preceding the arrival of the Spaniards) did maize, a more perfect grain, become the principal Aztec food. Another narrative relates the story of Quetzalcoatl, the Aztec god who like Prometheus tried to "civilize" humans—by giving them maize. The Mayan creation myth *Popul Vuh* relates the story of how, after several failed attempts at human creation, a successful formula was found by mixing the blood of the gods with maize dough (Warman 1988:48). For the Zapotec of Loxicha in the Sierra Sur of Oaxaca, the dualistic deity Mzian has a male half, Mdi, who represents lightning and rain and a female half, Mbaz, who represents the earth; Ndubdo, maize, is their child (Weitlaner 1965:561).

6. According to one source, San Isidro Labrador plays a much more important role in rural Latin American societies than in Spain (Foster 1960:116-117).

7. Crosses had sacred connotations in ancient Mesoamerica. See Parsons (1936:230), Ricard (1966 [1933]:31), and Romero Frizzi (1996:65-71).

8. It has been reported that the Mixe used miniature images to petition the "mountain spirits" for such things as oxen (Beals 1945:85-86). De la Fuente (1949:266), writing about the Cajonos Zapotec of Yalálag several years before the discovery of the Green Cross, noted that cross-shaped trees held a special place among supernatural things: "The trees which demand reverence are the living crosses (Sp. *cruces vivas*), whose power is in some respects similar to that of holy [Christian] crosses."

9. There is certainly an element of animal sacrifice; during the 1997 fiesta, four bulls were slaughtered in order to feed the visiting pilgrims. Though I witnessed no other animal sacrifices during the May celebration, an off-season visit revealed that the lagoon is used in much the same way as the sites of old: two chicken heads were floating in the middle of the pool, eggshells and chicken feathers littered one end of the bank, two loaves of bread were lying under the water, and a stone platform—obviously some kind of an altar—was covered with packages of cigarettes and decaying flower petals.

10. By no means are all indigenous peoples "stewards of the earth" or "closer to nature." As C. Jay Ou (1996) has shown in an important essay on the Monitored Retrievable Storage plan for storing nuclear waste on U.S. Indian reservations, such imagery can be manipulated in a hegemonic fashion by power holders. Although many indigenous peoples in the Ameri-

cas—including the Rincón Zapotec, the Cree of St. James Bay, Canada (Scott 1996), and others—do seem to have a profound sense of environmental stewardship, others do not. In Chiapas, for example, indigenous peoples have to a certain degree carried out part of the deforestation that has characterized the recent environmental tragedy there. In many cases they have been forced into this situation by a population boom and by unscrupulous ranchers who have taken over choice farming land previously belonging to indigenous communities (see Collier 1995). Clive Ponting (1991) has reviewed a number of cases in which ancient civilizations (including great Mesoamerican indigenous societies) also appear to have contributed to the collapse of ecosystems.

11. In the past even village fiestas were realized in this way, with each family contributing a designated quantity of maize, beans, and *panela* for the fiesta. This custom continues in smaller villages. In Talea today fiesta taxes are charged in cash to each head of a household.

12. Among the Zapotec, turkeys are symbolically connected with weddings. According to informants, in "the old days" the parents of a prospective bride would actually set a minimum price (in turkeys) that suitors had to meet. In Talea the turkeys no longer function in this way. They serve as symbols, and four to eight are typically given to the bride's family by the groom's family.

13. In recent years—and probably as a response to migration—distinctly urban elements have become a part of weddings, including ornate white wedding dresses, rice-throwing, stacked wedding cakes, and, in a few cases, wedding rings. Other elements from Mexican national culture have also been introduced: the *arras* (Sp., thirteen coins symbolizing the provision of plenty to the household), the *lazo* (Sp., two large rosaries connected to form a lasso which binds the couple together), and the Niño (Sp., the Christ child, which is the centerpiece of household altars in Talea).

14. One of the more interesting versions is an apparently successful crossing of a rare relative of *teosinte* (*Zea diploperennis*) and a variety of *Tripsacum*. The hybrid, patented by geneticist Mary Eubanks, produced ears of corn remarkably similar to the fossilized remains found at archaeological sites. For a summary of debates over maize origins, see Dold (1997).

15. There is nothing inherently attractive about agriculture; Sahlins (1972) has amply demonstrated that the transformation to sedentary agriculture represents a significant loss of leisure time for hunter-gatherers. The reasons for the change are probably related to increasing land pressure, since agriculture allows a much higher population density than hunting and gathering.

16. As a corollary we might consider the U.S. "agricultural revolution" as seen from the perspective of its victims and those who have not enjoyed its fruits. As Warman (1988:211-213; my translation) notes: "The history of North American agriculture is a process of accelerated capital accumulation. It is also a history of inequality, exclusion, and submission. Each step in the process has created its marginalized groups. Those groups survive. There are American Indians, the first to be dispossessed—valiantly resisting yet segregated by their identity. There are the rural poor in the economically depressed countryside. In 1965 there were 14 million rural poor in the United States who were never reached by poverty eradication programs. There are the urban poor, many of them blacks who were once farm workers or sharecroppers, who abandoned or else were expelled from the southeastern states after 1945. There are the migrant farm workers (many of them 'illegals') who tolerate low wages and unstable work patterns, who with their labor make nonmechanized agricultural tasks profitable for U.S. farmers. There are the malnourished, who since 1965 have been attended to with a

food subsidy program [food stamps], which is ironic in a country that even if it so desired could not eat all of the food it produces."

17. We should remember that the Green Revolution was supported by funds from the Rockefeller and Ford Foundations. It is no accident that the "technological packages" relied heavily on oil and machinery.

18. In Mexico many of the larger farms are located in states north and west of Mexico City. They are frequently geared toward the production of cash crops like tomatoes and asparagus, sometimes under direct contract with large U.S. buyers, including transnational corporations and major supermarket chains.

19. To be sure, the *ejidos* faced serious problems by 1990 (de Janvry et al. 1995:72–74) that, according to proponents of Article 27 reforms, could only be addressed by a complete overhaul of the system (Warman 1994). Others convincingly argue that the reforms will devastate small producers and the environment and are not significantly decreasing the state's political-economic presence but simply disguising it (Bartra 1995; Toledo 1995). The debate on whether the *ejidos* were more or less productive than private-sector farms continues (see Thiesenhusen 1995 for a review).

20. According to one source, in the 1960s the average annual household production of maize among the Sierra Zapotec was enough to last just under six months (Berg 1968).

21. The actual values were three pesos for a day's work, six pesos for an *almud* of maize in the early 1960s; and twenty-five nuevos pesos for a day's work, ten to twelve nuevos pesos for an *almud* of maize in 1996.

5. From Milpa to Tortilla

1. Property taxes are paid once a year to the municipal authorities. The *impuesto predial* in 1996 was either seventeen or thirty-four pesos depending upon the size of the property; the money stays within the municipal coffers, unlike the situation many years ago when state officials assessed and charged property taxes. The changes have come within the last ten years or so.

2. Older informants describe a number of institutions whose members owned parcels of land in common. Talea's religious brother/sisterhoods (Sp. *hermandades*) and musical groups, the *banda* and the *orquesta*, each owned plots of land on which staple foods were planted. Harvests were used to feed the groups' conductors, who were dedicated to musical pursuits full-time. The church also owned land, with the fruits used to feed the priest. John Chance's research (1989) indicates that the practice is rooted in the colonial period when *cofradías* (Sp., brotherhoods) owned land in common and used the proceeds to finance religious festivals. Today less than 4 percent of Talea's land pertains to *barrios*.

3. In Talea I was given the following rules of thumb. Large yellow maize should be planted in early March in *tierra fría* (at approximately 1,600 m altitude) and in late April (April 25, ideally) near the *tierra caliente* of Santa Gertrudis (at approximately 1,000 m). Large white maize should be planted in late March (around March 28–30) in *tierra fría* and in late April (around April 23–25) in *tierra caliente*. See also Tyrtania (1992:155–162).

4. According to a number of informants, in some of the Rincón villages on the other side of the mountain—Yagallo and Lachichina, for example—women care for oxen. Taleans find this bizarre. Intervillage farming variations are common, sometimes due to environmental differences, sometimes not.

5. Plowmen charged approximately N$80 per day (more than three times the daily wage) in the 1995-1996 season.

6. The *chompa* is valued for its seeds, which are prepared as *pepitas* (roasted seeds eaten as snack food or ground with chiles to make a tasty paste). The *támala* is prized for its greens and its flowers. The *huiche* has tender "meat," which is prepared in a tomato and chile sauce, or diced and dropped into the pot of beans, or boiled with *panela* to brew a refreshing drink in the summertime.

7. If allowed to roam freely through the milpa, the oxen wander, choosing the best stalks and leaving behind others.

8. On windy days burning is postponed so that fires do not spread. In addition, all material to be burned is kept away from the milpa's edges to avoid setting fire to nearby trees. Uncontrolled fires are disastrous during the dry season.

9. De la Fuente (1949:79), for example, reports a spacing distance of 2 paces (approximately 2 *varas*) in Yalálag, while Lipp (1991:19) reports a distance of 1-2 meters in the Mixe region.

10. The work is scheduled according to altitude. For example, if a campesino has a milpa lying at a higher altitude and his father-in-law has one located 50 m below it, they might trade several days of labor so that the milpas are planted one after another, first at the higher altitude (since this maize takes longer to mature) then at the lower.

11. Another variety of maize, said to have come from the Cajonos Zapotec village of Zoochila, matures in three months. It is uncommon in Talea.

12. Later I discovered that four-year-olds are not only able to distinguish squash sprouts from other sprouts; they are also able to distinguish different kinds of squash sprouts by looking for leaf patterns.

13. "Three-month" (small) maize is typically planted in counterseasonal plots just after the rainy season ends. The plots are grown using a two-field fallow system, just as the seasonal plots are, but the two-year periods begin and end in the "off-season." A handful of campesinos have simple irrigation systems (powered by gravity pressure) to help the maize develop. On the whole, counterseasonal cultivation is still quite rare because few campesinos have the time to spare during the coffee harvest.

14. Some Latin American farmers classify fertilizer as "cold," not "hot" as in Talea. Organic fertilizers—specifically animal or green manures—are considered "hot" in these areas; to achieve a balance, organic materials are added to chemically fertilized soil. This has the beneficial effect of maintaining "soil body and tilth" (Wilken 1987:94). According to Nelson Graburn (personal communication, 1998), farmers in parts of rural England in the 1940s also described fertilizers as "hot" and carefully avoided "burning" crops, just as the Taleans do.

15. Husking practices vary greatly across Latin America. The Mixe husk maize in the milpa, as it is picked (Lipp 1991). Similar procedures are reported for coastal villages in Peru (Hatch 1974). But some Cajonos Zapotec villagers are said to leave individual ears unhusked until just before they are consumed.

16. The fungus is consumed by some North Americans, however: in some trendy U.S. restaurants it is apparently sold as "Mexican truffles" or "caviar Azteca" (Wiegner 1995). Throughout much of Mexico it goes by the Nahuatl name *huitlacoche*.

17. The poles should be made of a strong, light wood. Wood from a tree known as the *rahuás* (scientific name unknown) in Zapotec, found in the forests above the village, is well suited. So is wood from orange trees. De la Fuente (1949:81) reported that threshing was "an ancient technique and in disuse." This was obviously not the case in the Rincón.

18. Boiling maize in a lime solution has important nutritional effects: it releases niacin (vitamin B3), which is lacking in the traditional Mesoamerican diet. It took Europeans more than 400 years to discover that pellagra was caused by a vitamin B deficiency related to the consumption of maize, which was commonly eaten without legumes and green vegetables in parts of the Old World. South Europeans in particular suffered from pellagra, which was probably complicated by the fact that cornmeal was prepared in the form of *polenta* (in Italy) and *mamaliga* (in Romania), without boiling the kernels in a lime solution (Warman 1988: 149-167). Alkaline cooking enhances the protein quality of the corn by beneficially altering the relative amounts of different essential amino acids and by making niacin more available for absorption by the body (Katz, Hediger, and Valleroy 1974). It also increases the calcium content of corn (Ortiz de Montellano 1990:102).

19. The triumvirate might be more accurately described as a quadrivirate, if chile is included. Chile peppers provide a high percentage of vitamins and minerals in the traditional Mexican diet. For much of the year, they are the only source of vitamin C.

20. In the Rincón village of Yagavila 950 kg per hectare was reported as the average yield during the early 1980s, when no chemical fertilizers were in use (Tyrtania 1992:171). This seems to support the claims of Talean campesinos regarding fertilizers (that is, that they effectively doubled maize yields).

21. The question of whether the full moon affects the resistance of wood to termite infestation is an open one. There have been no studies done that I am aware of seeking to confirm the efficacy of the practice. However, if moonlight affects the behavior of termites—for example, if they are less active when the moon is waning—it is not beyond the realm of possibility to think that there may be some connection, because during this critical period freshly cut wood goes through a "hardening" process (assuming that it is cut during a full-moon period) as its moisture evaporates. For an insightful analysis of the effects of lunar phases on the behavior of fish and Micronesian islanders, see Johannes (1981:32-40).

6. Sweetness and Reciprocity

1. I only saw sugar used in large amounts during certain functions such as reroofing parties, when it was used to sweeten buckets of Kool-Aid or Tang for workers.

2. It is significant that food products that come from outside the region, particularly maize, beans, and white sugar (but also poultry, beef, mezcal, and other items), are valued less than crops grown locally.

3. Parsons (1936:207, 531) notes that, according to official Catholic doctrine, the celebration is a petition for God to release the souls of the deceased from purgatory. However, the Valley Zapotec of Mitla have interpreted it as a chance to achieve communion with the dead—hence the offering of ceremonial foods, flowers at the cemetery, etc. Parsons claims that they are keeping alive the ancestral death cults of the prehispanic period through a process of religious syncretism. El Guindi agrees (1986:123): "After death, adults go to another world, which is modeled after this one . . . After being in the world of the dead . . . they can obtain permission to come back once a year to the house of their families to visit the community of the living. This visit is celebrated by the entire community, with each household preparing for a three-day celebration known as 'All Saints.' Life in the Zapotec conceptualization is, therefore, cyclical, not linear, and death is the journey from this world to the world of the dead."

Foster (1960:346) found evidence of similar traditions in parts of Spain; there still existed "the belief that the souls of the deceased return to the earth to share these foods [bread and wine]." He also reports that in northern and central Spain it was common for young people to go from house to house on the night of November 1, saying prayers for the deceased and collecting offerings for the church. The structure of Todos Santos in southern Mexico appears to be remarkably similar to that in rural Spain, though the particular foods offered to the deceased may differ.

4. Mintz (1985:184), for example, notes that "the artistic and ritual association between sugar and death is not a Mexican monopoly; in much of Europe, candied funeral treats are popular." See also Brandes (1997, 1998).

7. The Invention of "Traditional" Agriculture

1. Even after the Cárdenas reforms and confiscation of German properties during World War II, large plantations often remained intact. They simply passed into the hands of Mexican plantation owners.

2. General Hernández, according to Pérez García (1956), was born in Ixtlán in 1832. He received some formal schooling and was then given an important position by Don Miguel Castro, owner of the Santa Gertrudis mines near Talea. Hernández affiliated himself with the Liberal party at the age of twenty-three and by 1860 was a soldier in the Reform Wars. He participated in the War of the French Intervention and afterward dedicated himself to agriculture and small-scale commercial activities. Hernández corresponded with President Benito Juárez and was occasionally called upon to act as "political chief and commander of the National Guard of the District because he was considered *the* public official of the region" (Pérez García 1956:133). During this time, he is said to have introduced coffee in the Rincón. Archival sources indicate that Hernández's "coffee decree" was issued in January 1875 (Esparza 1988:301). Hernández died in 1881 of hepatitis (Pérez García 1956:131-133).

3. Jiménez (1995) convincingly argues that such changes—which profoundly affected life in the United States but also in coffee-producing villages like Talea—were intimately related to the myriad effects of the rise of U.S. industrial capitalism. The creation of an urbanized, more homogeneous working class meant a standardized market; new mass-marketing channels (especially supermarkets) led to more efficient distribution; emergent food conglomerates began standardizing tastes and coffee blends for mass consumption; and consumer tastes were shaped by novel advertising techniques and an assertive trade group, the National Coffee Association. Technological developments also played a role: improved transportation led to better and cheaper access to coffee in the nation's interior, while new food production and processing techniques (such as vacuum packing) made mass marketing possible.

4. Although it is difficult to be sure, it seems that the transplant method was the preferred method in the Rincón until the arrival of the Papaloapan Commission engineers and INMECAFE agronomists. One informant told me that the first coffee trees in Talea were saplings brought from villages on "the other side" of the mountain, specifically Santa Cruz Yagavila and Teotlaxco, and that he first saw a nursery of coffee saplings in the late 1950s or early 1960s, when agronomists established a shaded nursery at a site near Santa Gertrudis.

5. There are a number of areas in which the need for shade trees is precluded, such as certain steep pockets of land that are veiled from the sun nearly year round by the shadow of the mountainsides above.

6. One kind of *guajinicuil* has a tough pod, but another has a softer pod that contains a cottony substance that surrounds its seeds. It has a sweet taste (rather like cotton candy in flavor and texture) and is often taken home as a treat for children during its season.

7. The same thing holds true for insecticides. Since 1989 there has apparently been a marked reluctance to invest money in such nonessential inputs.

8. Although not articulated by informants, another benefit is described in a detailed study on Mexican coffee which suggests that widely spaced plants can help prevent the spread of certain plagues and diseases to which crowded *cafetales* are vulnerable (Nolasco 1985:136).

9. It is generally noted that "the engineers" (Papaloapan Commission employees, INMECAFE employees, or both) brought the new varieties, which apart from the *caturras* include *mundo novo, bourbón, garnica,* and *vegetal*. It is interesting to note that the Taleans with significant quantities of the new varieties are often those who own the most coffee land.

10. The categories are used by the Taleans themselves, and other anthropologists conducting work in the Rincón have employed similar terms (Young 1976; Tyrtania 1992; Hirabayashi 1993).

11. Laura Nader (personal communication, 1997) notes that *mozos* used to be paid in advance by their *patrones* in the late 1950s and early 1960s. This practice occurs occasionally in Talea, when a relatively prosperous coffee grower or a merchant with coffee holdings extends a cash loan to a villager needing money. The debt is repaid during the coffee harvest season, with wage labor. Unlike the situation in some other parts of southern Mexico, where debt peonage is the rule, this is the exception in Talea.

12. A small number of households use white sugar (processed far from the Rincón) instead of *panela*, probably because it is cheaper for those who do not grow their own sugarcane. Most Taleans I spoke with, though, were repulsed by the thought of drinking coffee with white sugar because "it doesn't fill one's body the way *panela* does." The consistency of the coffee is noticeably thicker when *panela* is used. At least one Talean recently diagnosed as diabetic purchases aspartame (Nutrasweet, the artificial sweetener phenylalinine) in Oaxaca for his coffee. Many Taleans have heard that North Americans drink their coffee without sugar; they find this appalling because they do not think that bitter coffee has an agreeable taste.

13. In Talea, as in many other parts of the world, soft drinks are making inroads. At one gathering I attended (it was a "new" kind of occasion not commonly celebrated in the Rincón: a child's birthday party) coffee was not offered to the guests; all were served Coca-Cola and Fanta orange soda with their lunches. The young hosts see themselves as "progressive" (they are both employees and never work in the fields), have both lived in cities, and took advantage of a special promotion being offered by the region's Coca-Cola distributor: two 2-liter bottles were offered at the special price of N$20 (approximately US$2.50) with a five-gallon plastic bucket as a bonus. Soft drinks, beer, and brandy are high-status items; when important visitors arrive at a household, children are often quietly dispatched to purchase Coca-Colas and Coronas for the guests. This is a complete reversal of the situation in the contemporary United States, where soft drinks are viewed as a "working-class" or low-status beverage while gourmet coffees carry an aura of sophistication and distinction (Roseberry 1996).

14. The case of cacao is especially interesting. A recent report describes how large firms have joined together with environmental groups in order to promote small-scale cultivation. "Sun" plantations (which are more vulnerable to diseases and pests) have failed on a massive scale, according to the report, with blights affecting global production so severely that short-

ages of chocolate may be five to ten years away. Smallholder production is now being touted as the wave of the future (Yoon 1998).

15. The statement, made by a campesino in his fifties, is quite an accurate comparative analysis of soils. The Jalapa, Veracruz, region has some of the deepest levels of rich soil (which are most amenable to coffee cultivation) to be found in any coffee-producing region of the country, while many Oaxacan regions have much thinner soils with bedrock not far below the surface (Nolasco 1985:95, 99). The campesinos use many rules of thumb such as these to guide them in their cultivation of coffee, just as they have done for maize, beans, and sugarcane.

16. Many Talean campesinos take international price crashes more or less in stride. For example, most informants I spoke with saw the 1989 crisis not as a sinister First World plot but as an unlucky but temporary condition—rather like a natural disaster—even though intellectually most understood that far-flung forces lay behind the events. From experience, they know that droughts and hurricanes periodically damage maize yields; rootworms sometimes ruin the bean crop; and every thirty to forty years market crashes sap coffee's value. These events are not necessarily reasons to stop growing corn, beans, or coffee.

17. Displays of prestige are made especially clear at the annual January fiesta, when Taleans ignite a massive tower of fireworks which can cost more than 15,000 pesos (US$2,100)—made possible by taxes derived from coffee income. It was once suggested to me by an outsider who has spent a number of years living in the Sierra that the fireworks function to make villagers from across the region envious.

18. This may be seen as part of the larger process of Mexico's economic globalization which culminated in Salinas de Gortari's neoliberal market reforms and participation in NAFTA and the World Trade Organization.

19. They also received help from others. For example, some Catholic missionaries have been instrumental in helping the cooperatives find overseas markets for their organic coffee. NGOs such as the Centro de Apoyo al Movimiento Popular de Oaxaca, A.C. (CAMPO) have also provided technical assistance.

20. Pueblos Unidos was formed by approximately ten Rincón Zapotec villages in the early 1980s to pressure the state government to construct a road connecting the Rincón villages of Juquila, Tanetze, and Yaeé. (Interestingly, Talea—by far the largest village in the Rincón—was never invited to join the organization, probably because the road would pose a threat to many Talean merchants who depended on customers from these villages.) The Pueblos Unidos eventually acquired coffee-bean roasting machines. For a time their product, "Café Padiush" (*padiush* means "greetings" in Rincón Zapotec), was vacuum packed in beautifully designed bags. By the mid-1990s, however, the roasters were out of use due to the fact that insufficient quantities of coffee were being delivered to the cooperative.

21. Interestingly, most Taleans who have joined the new programs have relatively large coffee holdings—typically more than three hectares. This seems to follow a pattern characteristic throughout many coffee-producing regions in Mexico (Jonathan Fox, personal communication, 1997). If we combine this with the fact that many of the poorest (money-poor) villages in the Rincón also tend to be quite active in the new organizations, this case seems to resemble Cancian's argument (1972) about risk-taking: the wealthiest and the poorest are the most likely to try new techniques.

22. See Nader (1990) and Fox and Aranda (1996) for descriptions of Oaxaca's municipal tradition based on the cargo system and the extensive use of municipal committees, which

is remarkably democratic in comparison to other Mexican states. According to one of Fox's sources, less than 10 percent of Oaxaca's 470 municipalities are controlled by political bosses (Fox and Aranda 1996). It remains to be seen whether this system of autonomous municipal government will be sustained in the wake of the EPR and the subsequent militarization of the state. It is important to note that the Mexican government is not monolithic; nor is the PRI. Certain agencies and officials have had members who have provided support—moral, technical, and economic—to independent organizations, including indigenous cooperatives.

23. The mechanism of transnational corporate control in the purchasing sphere has historically been realized with the cooperation of Mexican intermediaries who act as agents of the firms (Nolasco 1985:245–246; my translation): "Out of 186 Mexican exporters (responsible for 70% of all exports), seven are based in the Mexican offices of the Volkhart Brothers corporation, which buys 14% of all Mexican exports. In other words, a small number of transnational firms operate through Mexican export agents who buy coffee in their own names (as if they had no affiliation with the firms) and then ship it abroad from key points in the coffee-producing regions."

According to Nolasco's calculations, in the late 1970s six firms purchased nearly 60 percent of all private sector (non-INMECAFE) exports, and twenty purchased nearly 80 percent. The coffee sector, she concludes, is a transnational oligopoly held in check only by the efforts of the INMECAFE. Now that there is no INMECAFE, one is left to assume that the cooperatives are the only institutions standing in the way of a total transnational oligopoly.

8. Agriculture Unbound

1. Reports contradicting the Worldwatch predictions have also been published in recent years (Mitchell, Ingco, and Duncan 1997), but these analyses are typically limited to the 1980s (when increases were still occurring), without extending into the 1990s. Worldwatch researchers claim that the most alarming trends occurred in that decade and only truly became visible in the mid-1990s. Others have cheerfully pinned their hopes on the future and argue that new plant biotechnologies will be developed, spurring a new Green Revolution certain to boost global yields once again.

2. It is possible that more than 25 percent of Mexico's food is also lost, but to pests and poor storage rather than to sheer wastefulness. I thank Eugene Anderson for bringing this to my attention.

3. This concept is by no means new in the "West." The living earth idea was characteristic of Celtic society and continues to exist among some Celts.

4. The possibility that the meat-centered U.S. diet might have functioned as an ideological tool used by the cattle industry and multinational corporations became plausible in the mid-1990s, when the "four food groups" concept inculcated in the minds of young students for much of the twentieth century was replaced with the "food guide pyramid" by the U.S. Departments of Agriculture and Health and Human Services (Burros 1992). The shift is worthy of a full anthropological analysis.

5. It would be misleading to suggest that only people in the United States have dietary problems and that all people in the so-called Third World have healthy eating habits. Indeed, people living in regions undergoing economic "development" often begin eating much like their counterparts in industrialized societies, as they encounter the same food suppliers (including transnational conglomerates like Pepsico, Ralston Purina, and McDonald's). Re-

cent articles have documented this process in Mexico City (Pelto 1987), Japan (Magee 1996), Zimbabwe (Murray 1997), and Hungary (McKinsey 1997). Needless to say, a variety of diet-related health problems (especially obesity, diabetes, cancer, and heart disease) now plague these societies.

6. Perhaps the main reason factory farming has succeeded in the United States is that it has been given a host of special legal and political favors (Bovard 1989). I thank Eugene Anderson for drawing my attention to this point.

7. *Gente de razón* (Sp., literally "people of reason" or "rational people") is a term that is still used by older campesinos in Talea as a gloss for well-dressed, highly educated people, usually descended from Spaniards. It is most typically used in accounts of events that took place generations ago. It is not surprising that the *gente de razón* are opposed to the *indios*— both were legal categories that the Spanish instituted as part of a colony-wide caste system during the colonial era. The term *gente de razón* is used quite unself-consciously in the village, and the Taleans in no way seemed to "believe" the implications of the terminology they were using. That is, I was not left with a sense that the Taleans think that people of European descent are in fact more reasonable or rational than themselves, even though their occasional use of the term might lead one to think so. Friedlander (1975) has written a magnificent book about the ideological construction of such terms in the colonial era and the functions they served.

8. It is important to recognize that part of their creativity often stems from the fact that they are working from different pools of evidence than colleagues engaged in the puzzle solving of normal scientists (Kuhn 1962). For example, in the case of agroecologists it is significant that Altieri and others are collecting their data from agricultural systems in lands pertaining to indigenous communities, with farming traditions that are hundreds or thousands of years old.

REFERENCES

Adas, Michael
1989 *Machines as the Measure of Men: Science, Technology, and Ideologies of Western Dominance.* Ithaca: Cornell University Press.
Adelson, Andrea
1997 "Organic Clothes on Backs, Not Minds." *New York Times*, November 6, A18.
AgraFood Biotech
1999 "Survey Conducted by Strategy One, Edelmann Public Relations." *AgraFood Biotech* 14:11.
Alba, Carlos H., and Jesús Cristerna
1949a "La agricultura entre los Zapotecos." In *Los Zapotecos*, edited by Lucio Mendieta y Núñez, 449–496. Mexico City: UNAM.
1949b "Las industrias zapotecas." In *Los Zapotecos*, edited by Lucio Mendieta y Núñez, 497–600. Mexico City: UNAM.
Alcina Franch, José
1972 "Los dioses del panteón zapoteco." *Anales de Antropología* 9:9–43.
Altieri, Miguel A.
1987 *Agroecology: The Science of Sustainable Agriculture.* 2nd ed. Boulder, Colo.: Westview Press.
Alvares, Claude
1992 *Science, Development, and Violence: The Revolt against Modernity.* Delhi: Oxford University Press.
American Corn Growers Association (ACGA)
2000 "Corn Growers Complete Survey on Farmer Planting Intentions for Upcoming Planting Season." See http://www.acga.org/news/.
Anderson, Eugene N.
1967 "The Ethnoichthyology of the Hong Kong Boat People." Ph.D. dissertation, University of California at Berkeley.

1996   *Ecologies of the Heart: Emotion, Belief, and the Environment.* New York: Oxford University Press.
Ball, Howard
1986   *Justice Downwind: America's Atomic Testing Program in the 1950s.* New York: Oxford University Press.
Balling, Robert C., Jr., and Randall S. Cerveny
1995   "Influence of Lunar Phase on Daily Global Temperatures." *Science* 267:1481-1483.
Bandyopadhyay, J., and Vandana Shiva
1981   "Alternatives for India: Western or Indigenous Science?" *Science for the People* 13:22-28.
Barabas, Alicia
1985   *Utopias indias.* Mexico City: Grijalbo.
Barboza, David
1999a "Biotech Companies Take on Critics of Gene-Altered Food." *New York Times*, November 12, A1.
1999b "Monsanto Sued over Use of Biotechnology in Developing Seeds." *New York Times*, December 15, C1.
Barry, Tom
1995   *Zapata's Revenge: Free Trade and the Farm Crisis in Mexico.* Boston: South End Press.
Bartolomé, Miguel, and Alicia Barabas
1974   "Hydraulic Development and Ethnocide: The Mazatec and Chinantec People of Oaxaca, Mexico." *Critique of Anthropology* 1(1):74-91.
1990   *La presa Cerro de Oro y el Ingeniero el Gran Dios: Relocalización y etnocidio chinanteco en México.* Mexico City: Consejo Nacional para la Cultura y las Artes.
Bartra, Armando
1995   "A Persistent Rural Leviathan." In *Reforming Mexico's Agrarian Reform*, edited by Laura Randall, 173-184. Armonk, N.Y.: M. E. Sharpe.
Beadle, George
1939   "Teosinte and the Origin of Maize." *Journal of Heredity* 30:245-247.
Beals, Ralph
1945   *The Ethnology of the Western Mixe.* Berkeley: University of California Press.
1975   *The Peasant Marketing System of Oaxaca, Mexico.* Berkeley and Los Angeles: University of California Press.
Belshaw, Deryke
1979   "Taking Indigenous Technology Seriously: The Case of Intercropping Techniques in East Africa." *IDS Bulletin* 10(2):24-27.
Bender, Jim
1994   *Future Harvest: Pesticide-free Farming.* Lincoln: University of Nebraska Press.
Benedict, Ruth
1934   *Patterns of Culture.* New York: Houghton Mifflin.
Berg, Richard L.
1968   "The Impact of the Modern Economy on the Traditional Economy in Zoogocho, Oaxaca, Mexico." Ph.D. dissertation, University of California at Los Angeles.
1976   "The Zoogocho Plaza System in the Sierra Zapoteca of Villa Alta." In *Markets in Oaxaca*, edited by Scott Cook and Martin Diskin, 81-106. Austin: University of Texas Press.

Berkes, Fikret
1977 "Fishery Resource Use in a Subarctic Indian Community." *Human Ecology* 5(4): 289-307.
Berlin, Brent
1974 *Principles of Tzeltal Plant Classification*. New York: Academic Press.
Berlin, E. A., B. Berlin, X. Lozoya, M. Meckes, J. Tortoriello, and M. L. Villareal
1996 "The Scientific Basis of Gastrointestinal Herbal Medicine among the Highland Maya of Chiapas, Mexico." In *Naked Science: Anthropological Inquiry into Boundaries, Power, and Knowledge*, edited by Laura Nader, 43-68. New York and London: Routledge.
Bernal, Martin
1987 *Black Athena: The Afroasiatic Roots of Classical Civilization*. New Brunswick: Rutgers University Press.
Berry, Wendell
1977 *The Unsettling of America: Culture and Agriculture*. San Francisco: Sierra Club Books.
Bierma, Paige
1997 "Cuba's Food Revolution." *Monthly* (Emeryville, Calif.) 27(10):7-9.
Biodemocracy Campaign/Organic Consumers Association
2000 "Forty-nine Members of the US Congress Send Letter to FDA Demanding Mandatory Labeling of Genetically Engineered Foods." See http://www.purefood.org/ge/cong49label.cfm.
Blaut, J. M.
1993 *The Colonizer's Model of the World: Geographical Diffusionism and Eurocentric History*. New York: Guilford Press.
Bonfil Batalla, Guillermo
1962 *Diagnóstico sobre el hambre en Sudzal, Yucatán*. Mexico City: INAH.
1983 *Culturas populares y política popular*. Mexico City: SEP-Museo de Culturas Populares.
Borgstrom, Georg
1967 *The Hungry Planet*. New York: Collier Books.
Borlaug, Norman
2000 "La biotecnología contra el hambre." *Reforma* (Mexico City), January 1, 4A.
Bovard, James
1989 *The Farm Fiasco*. San Francisco: ICS Press.
Brandes, Stanley
1997 "Sugar, Colonialism, and Death: On the Origins of Mexico's Day of the Dead." *Comparative Studies in Society and History* 39:270-299.
1998 "The Day of the Dead, Halloween, and the Quest for Mexican National Identity." *Journal of American Folklore* 111(442):359-380.
Braudel, Fernand
1967 *Capitalism and Material Life, 1400-1800*. New York: Harper Colophon Books.
Brewster, Leticia, and Michael F. Jacobson
1983 *The Changing American Diet*. Washington, D.C.: Center for Science in the Public Interest.
Brody, Jane E.
1997 "U.S. Panel on Acupuncture Calls for Wider Acceptance." *New York Times*, November 6, A10.

Brokensha, David L., D. M. Warren, and Oswald Werner (editors)
1980 *Indigenous Knowledge Systems and Development.* Lanham, Md.: University Press of America.
Brown, Lester R.
1996 *Tough Choices: Facing the Challenge of Food Scarcity.* New York: Norton.
1997 "Facing the Prospect of Food Scarcity." In *State of the World 1997*, edited by Lester R. Brown, 23-41. New York: Norton.
Burros, Marian
1992 "U.S. Department of Agriculture to Rank Food with Symbol of Pyramid." *New York Times*, April 28, B1.
Butt Colson, Audrey, and Cesareo de Armellada
1976 "An Amerindian Derivation for Latin Creole Illnesses and Their Treatment." *Social Science and Medicine* 17:1229-1248.
Buttel, Frederick H., Martin Kenney, and Jack Kloppenburg
1985 "From Green Revolution to Biorevolution: Some Observations on the Changing Technological Bases of Economic Transformation in the Third World." *Economic Development and Cultural Change* 34(1):31-55.
Bye, Robert A.
1981 "Quelites: Ethnoecology of Edible Greens, Past, Present and Future." *Journal of Ethnobiology* 1(1):109-123.
Campbell, Howard, Leigh Binford, Miguel Bartolomé, and Alicia Barabas (editors)
1993 *Zapotec Struggles: Histories, Politics, and Representations from Juchitán, Oaxaca.* Washington, D.C., and London: Smithsonian Institution Press.
Cancian, Frank
1972 *Change and Uncertainty in a Peasant Economy: The Maya Corn Farmers of Zinacantán.* Stanford: Stanford University Press.
1992 *The Decline of Community in Zinacantán, 1960-1987.* Stanford: Stanford University Press.
Carmagnani, Marcelo
1988 *El regreso de los dioses: El proceso de reconstitución de la identidad étnica en Oaxaca, siglos XVII y XVIII.* Mexico City: Fondo de Cultura Económica.
Carrasco, Pedro
1961 "The Civil-Religious Hierarchy in Mesoamerican Communities: Pre-Spanish Background and Colonial Development." *American Anthropologist* 63:483-497.
Carson, Rachel
1962 *Silent Spring.* New York: Houghton Mifflin.
Caso, Alfonso, and Gonzalo Aguirre Beltrán
1960 "Applied Anthropology in Mexico." In *Middle American Anthropology*, vol. 2, edited by Gordon R. Willey, Evon Z. Vogt, and Angel Palerm, 54-62. Washington, D.C.: Pan American Union.
Castellanos M., Javier
1988 *El maíz en Yojovi, Villa Alta, Oaxaca.* Mexico City: Primer Lugar.
Chambers, Robert
1983 *Rural Development: Putting the Last First.* London: Freeman.
Chance, John K.
1989 *The Conquest of the Sierra: Spaniards and Indians in Colonial New Spain.* Norman: University of Oklahoma Press.

Chance, John K., and William B. Taylor
1985   "Cofradías and Cargos: An Historical Perspective on the Mesoamerican Civil-Religious Hierarchy." *American Ethnologist* 12:1-26.
Chevalier, François
1963   *Land and Society in Colonial Mexico: The Great Hacienda*. Translated by Alvin Eustis. Berkeley: University of California Press.
Childe, V. Gordon
1951   *Man Makes Himself*. New York: New American Library.
Cockroft, James D.
1990   *Mexico: Class Formation, Capital Accumulation, and the State*. New York: Monthly Review Press.
Cohen, Jeffrey
1999   *Cooperation and Community: Economy and Society in Oaxaca*. Austin: University of Texas Press.
Coleman, Eliot
1989   *The New Organic Grower: A Master's Manual of Tools and Techniques for the Home and Market Gardener*. Chelsea, Vt.: Chelsea Green.
Collier, George A.
1975   *Fields of the Tzotzil: The Ecological Bases of Tradition in Highland Chiapas*. Austin: University of Texas Press.
Collier, George A. (with Elizabeth Lowery Quaratiello)
1995   *Basta! Land and the Zapatista Rebellion in Chiapas*. Oakland, Calif.: Food First.
Colson, Elizabeth
1973   "Tranquility for the Decision-Maker." In *Cultural Illness and Health*, edited by Laura Nader and Thomas W. Maretzki, 89-96. Washington, D.C.: American Anthropological Association.
Commoner, Barry
1976   *The Poverty of Power*. New York: Knopf.
Conklin, Harold C.
1954   "The Relation of Hanunóo Culture to the Plant World." Ph.D. dissertation, Yale University.
Corro, Salvador
1996   "Operativos militares en casi todo el país: Retenes, vuelos de reconocimiento y patrullajes para aplicar 'toda la fuerza del estado' al EPR." *Proceso* (Mexico City) 1036, September 8, 7-13.
Costanza, Robert (editor)
1991   *Ecological Economics: The Science and Management of Sustainability*. New York: Columbia Press.
Crosby, Alfred
1972   *The Columbian Exchange: Biological and Cultural Consequences of 1492*. Boulder, Colo.: Greenwood Press.
1986   *Ecological Imperialism: The Biological Expansion of Europe, 900-1900*. Cambridge: Cambridge University Press.
Crouch, Martha
1990   "Debating the Responsibilities of Plant Scientists in the Decade of the Environment." *Plant Cell* 2:275-277.
2000   "Interview with Martha Crouch." *Corporate Crime Reporter*, January 3, 10-16.

Daly, Herman, and J. Cobb
1989   *For the Common Good.* Boston: Beacon Press.
Davis, Shelton H.
1977   *Victims of the Miracle: Development and the Indians of Brazil.* Cambridge: Cambridge University Press.
de Janvry, Alain, Elisabeth Sadoulet, Benjamin Davis, and Gustavo Gordillo de Anda
1995   "Ejido Sector Reforms: From Land Reform to Rural Development." In *Reforming Mexico's Agrarian Reform,* edited by Laura Randall, 71-106. Armonk, N.Y.: M. E. Sharpe.
de la Fuente, Julio
1947   "Los Zapotecos de Choapan, Oaxaca." *Anales del Instituto Nacional de Antropología e Historia* 2:143-206.
1949   *Yalálag: Una villa zapoteca serrana.* Mexico City: Museo Nacional de Antropología.
1960   "Manuel Gamio." *Acción Indigenista* 86:1-4.
1962   *Relaciones interétnicas.* Mexico City: INI.
1964   "Ceremonias de lluvia entre los zapotecos." In *Educación, antropología, y desarrollo de la comunidad* by Julio de la Fuente, 183-190. Mexico City: INI.
Dennis, Philip
1987   *Intervillage Conflict in Oaxaca.* New Brunswick: Rutgers University Press.
Devitt, Paul
1977   "Notes on Poverty-Oriented Rural Development." In *Extension, Planning, and the Poor,* 15-29. Agricultural Administration Unit Occasional Paper 2. London: Overseas Development Institute.
DeWalt, Billie R.
1975   "Changes in the Cargo System of Mesoamerica." *Anthropological Quarterly* 48:87-105.
DeWalt, Billie R., and Martha Rees
1994   *The End of Agrarian Reform in Mexico.* San Diego: UCSD Center for U.S.-Mexican Studies.
Díaz, Bernal
1963   *The Conquest of New Spain.* Translated by J. M. Cohen. New York: Penguin Books.
Dold, Catherine
1997   "The Corn War: Competing Theories on the Origin of Corn." *Discover* 18(12):108-114.
Downs, P.
1984   "Agriculture: Learning from Ecology." *Technology Review* (July):70-71.
Dubinskas, Frank (editor)
1988   *Making Time: Ethnographies of High Technology Organizations.* Philadelphia: Temple University Press.
Durkheim, Emile, and Marcel Mauss
1963   *Primitive Classification.* Translated by Rodney Needham. Chicago: University of
[1903]   Chicago Press.
Early, Daniel K.
1982   *Café: Dependencia y efectos.* Mexico City: INI.
El Guindi, Fadwa (with Abel Hernández Jiménez)
1986   *The Myth of Ritual: A Native's Ethnography of Zapotec Life-Crisis Rituals.* Tucson: University of Arizona Press.

Ensminger, Paul A.
1996 "A Novel Method of Weed Control." *Biology Digest* 22:11-18.
Erickson, Jon
1988 *The Living Earth: The Coevolution of the Planet and Life*. Blue Ridge Summit, Penn.: Tab Books.
Erlich, Reese
1994 "Heavy Frosts Hurt Brazil's Smaller Coffee Farmers." *Christian Science Monitor*, August 19, 8.
Escobar, Arturo
1995 *Encountering Development: The Making and Unmaking of the Third World*. Princeton: Princeton University Press.
Esparza, Manuel
1988 "Los proyectos de los Liberales en Oaxaca (1856-1910)." In *Historia de la cuestión agraria mexicana—Estado de Oaxaca (Prehispánico-1925)*, edited by Leticia Reina, 269-330. Mexico City: Juan Pablos Editor.
1994 *Relaciones geográficas de Oaxaca, 1777-1778*. Oaxaca, Mexico: CIESAS.
Evans-Pritchard, E. E.
1976 *Witchcraft, Oracles, and Magic among the Azande*. Oxford: Clarendon Press.
[1937]
Ewen, Stanley W. B., and Arpad Pusztai
1999 "Effect of Diets Containing Genetically Modified Potatoes Expressing *Galanthus nivalis lectin* on Rat Small Intestine." *Lancet* 354:1353.
Farquhar, Judith
1994 *Knowing Practice: The Clinical Encounter of Chinese Medicine*. Boulder, Colo.: Westview Press.
Faust, Betty B.
1998 *Mexican Rural Development and the Plumed Serpent*. Westport, Conn.: Bergin and Garvey.
Ferguson, James
1990 *The Anti-Politics Machine: "Development," Depoliticization, and Bureaucratic Power in Lesotho*. Minneapolis: University of Minnesota Press.
Feshbach, Murray, and Alfred Friendly, Jr.
1992 *Ecocide in the USSR*. New York: Basic Books.
Feyerabend, Paul
1978 *Science in a Free Society*. London: NLB Press.
Flannery, Kent V.
1999 "Los orígenes de la agricultura en Oaxaca." *Cuadernos del Sur* (Oaxaca, Mexico) 14:5-14.
Flannery, Kent, and Joyce Marcus
1983 *The Cloud People: The Divergent Evolution of the Zapotec and Mixtec Civilizations*. New York: Academic Press.
Fleck, Ludwik
1979 *Genesis and Development of a Scientific Fact*. Translated by Fred Bradley and Thaddeus J. Trenn and edited by Thaddeus J. Trenn and Robert K. Merton. Chicago:
[1935] University of Chicago Press.

Foster, George M.
1948   *Empire's Children: The People of Tzintzuntzan.* Washington, D.C.: Smithsonian Institution.
1960   *Cultura y conquista: La herencia española de América.* Translated by Carlo Antonio Castro. Xalapa, Mexico: Universidad Veracruzana.
1967   *Tzintzuntzan: Mexican Peasants in a Changing World.* New York: Little, Brown, and Company.
1994   *Hippocrates' Latin American Legacy: Humoral Medicine in the New World.* Langhorne, Penn.: Gordon and Breach.
Fox, Jonathan
1992   *The Politics of Food in Mexico: State Power and Social Mobilization.* Ithaca: Cornell University Press.
1994   "Targeting the Poorest: The Role of the National Indigenous Institute in Mexico's Solidarity Program." In *Transforming State-Society Relations in Mexico: The National Solidarity Strategy,* edited by Wayne A. Cornelius, Ann L. Craig, and Jonathan Fox, 179-216. San Diego: UCSD Center for U.S.-Mexican Studies.
Fox, Jonathan, and Josefina Aranda
1996   *Decentralization and Rural Development in Mexico: Community Participation in Oaxaca's Municipal Funds Program.* San Diego: UCSD Center for U.S.-Mexican Studies.
Fox, Nicols
1997   *Spoiled: The Dangerous Truth about a Food Chain Gone Haywire.* New York: Basic Books.
Fradkin, Philip
1989   *Fallout: An American Nuclear Tragedy.* Tucson: University of Arizona Press.
Frazer, James
1911-   *The Golden Bough: A Study in Magic and Religion.* London: Macmillan.
1915
Freeman, M. M. R., and L. N. Carbyn (editors)
1988   *Traditional Knowledge and Renewable Resource Management in Northern Regions.* Boreal Institute for Northern Studies Occasional Publication No. 23. Edmonton, Canada: University of Alberta.
Friedlander, Judith
1975   *Being Indian in Hueyapan: A Study of Forced Identity in Contemporary Mexico.* New York: St. Martin's Press.
Frye, David
1996   *Indians into Mexicans: History and Identity in a Mexican Town.* Austin: University of Texas Press.
Fumento, Michael
2000   "Why Europe Fears Biotech Food." *Wall Street Journal,* January 14, A14.
Galinat, Walton C.
1992   "Maize: Gift from America's First Peoples." In *Chilies to Chocolate: Food the Americas Gave the World,* edited by Nelson Foster and Linda S. Cordell, 47-60. Tucson: University of Arizona Press.
Gamio, Manuel
1916   *Forjando patria.* Mexico City: Porrúa Hermanos.

Gardner, Gary
1997 "Preserving Global Cropland." In *State of the World 1997*, edited by Lester Brown, 42–59. New York: Norton.
Garner, Paul H.
1988 *La Revolución en la provincia: Soberanía estatal y caudillismo en las montañas de Oaxaca, 1910–1920*. Translated by Mercedes Pizarro. Mexico City: Fondo de Cultura Económica.
Gaspar González, Aleyda
1997 "Persiste el hostigamiento a UNOSJO." *Contrapunto* (Oaxaca, Mexico) 2(63), March 1, 5.
Gatsiopoulos, Georgina
1997 "The EPR: Mexico's 'Other' Guerrillas." *NACLA Report on the Americas* 30(4):33.
Gay, José Antonio
1950 *Historia de Oaxaca*. 2 vols. Mexico City: Talleres Verano.
[1881]
Geertz, Clifford
1959 "Form and Variation in Balinese Village Structure." *American Anthropologist* 61(6): 991–1012.
1963 *Agricultural Involution: The Process of Ecological Change in Indonesia*. Berkeley and Los Angeles: University of California Press.
1973 *The Interpretation of Cultures*. New York: Harper Colophon.
Ghindelli, Azzo
1971 "The Alimentation of the Maya." *Ethnos* 36:23–31.
Gladwin, Thomas
1970 *East Is a Big Bird: Navigation and Logic on Puluwat Atoll*. Cambridge, Mass.: Harvard University Press.
Gmelch, George
1994 "Ritual and Magic in American Baseball." In *Conformity and Conflict: Readings in Cultural Anthropology*, edited by James P. Spradley and David W. McCurdy, 351–361. 8th ed. New York: HarperCollins.
González, Alvaro, and María Teresa Prado
1992 "Marco introductorio." In *Etnias, desarrollo, recursos, y tecnologías en Oaxaca*, edited by Alvaro González and Marco Antonio Vásquez, 13–20. Oaxaca, Mexico: CIESAS.
González, Roberto J., Laura Nader, and C. Jay Ou
1995 "Between Two Poles: Bronislaw Malinowski, Ludwik Fleck, and the Anthropology of Science." *Current Anthropology* 36(5):866–869.
Goodenough, Ward
1953 *Native Astronomy in the Central Carolines*. Philadelphia: University of Pennsylvania Museum.
1996 "Navigation in the Western Carolines: A Traditional Science." In *Naked Science: Anthropological Inquiry into Boundaries, Power, and Knowledge*, edited by Laura Nader, 29–42. New York: Routledge.
Goodman, David, and Michael Redclift
1991 *Refashioning Nature: Food, Ecology, and Culture*. New York and London: Routledge.
Greenberg, James B.
1989 *Blood Ties: Life and Violence in Rural Mexico*. Tucson: University of Arizona Press.

Greenberg, Russell
1994 "Phenomena, Comment and Notes." *Smithsonian* 25(8):24-26.
Guadarrama Olivera, Fernando
1996 "La autosuficiencia subversiva." *La Hora* (Oaxaca, Mexico) 236:5-6.
1997 "Forest Resources and People in the Sierra Juárez." In *Indigenous Peoples and Sustainability: Cases and Actions*, edited by Darrell A. Posey and Graham Dutfield, pp. 315-321. Utrecht, The Netherlands: IUCN Inter-Commission Task Force on Indigenous Peoples.
Gudeman, Stephen
1978 *The Demise of a Rural Economy: From Subsistence to Capitalism in a Latin American Village*. Boston and London: Routledge and Kegan Paul.
Guillow, Eulogio G.
1994 "Idolatrías en Caxonos." In *Los Zapotecos de la Sierra Norte de Oaxaca*, edited by
[1889] Manuel Ríos Morales, 167-184. Oaxaca, Mexico: CIESAS.
Guiteras Holmes, Calixto
1961 *Perils of the Soul: The World View of a Tzotzil Indian*. New York: Free Press.
Gusterson, Hugh
1995 *Nuclear Rites: A Weapons Laboratory at the End of the Cold War*. Berkeley: University of California Press.
1996 "Nuclear Weapons Testing: Scientific Experiment as Political Ritual." In *Naked Science: Anthropological Inquiry into Boundaries, Power, and Knowledge*, edited by Laura Nader, 131-147. New York: Routledge.
Habermas, Jürgen
1971 *Toward a Rational Society*. Translated by Jeremy J. Shapiro. London: Heinemann.
Hamnett, Brian
1971 *Politics and Trade in Southern Mexico, 1750-1821*. Cambridge: Cambridge University Press.
Harding, Sandra
1998 *Is Science Multicultural? Postcolonialisms, Feminisms, and Epistemologies*. Bloomington: Indiana University Press.
Harlan, Jack R.
1975 *Crops and Man*. Madison, Wis.: American Society of Agronomy.
Harris, Marvin
1964 *Patterns of Race in the Americas*. New York: W. W. Norton.
Hartmann, Karl M., and W. Nezadal
1990 "Photocontrol of Weeds without Herbicides." *Naturwissenschaften* 77:158-163.
Harvey, Neil
1992 "La Unión de Uniones de Chiapas y los retos políticos del desarrollo de base." In *Autonomía y nuevos sujetos sociales en el desarrollo rural*, edited by Julio Moguel, Carlota Botey, and Luis Hernández, 219-234. Mexico City: Siglo XXI.
Hatch, John K.
1974 *The Corn Farmers of Motupe: A Study of Traditional Farming Practices in Northern Coastal Peru*. Madison: University of Wisconsin, Land Tenure Center.
Hernández Castillo, Rosalva Aída
1997 "Confronting Racism: The Indian Movement in Chiapas and Its Quest for Autonomy." Presented at the Annual Meetings of the American Anthropological Association, Washington, D.C., November 16-22.

Hernández Castillo, Rosalva Aída, and Ronald Nigh
1998   "Global Processes and Local Identity: Indians of the Sierra Madre of Chiapas and the International Organic Coffee Market." *American Anthropologist* 100(3):136-147.
Hernández Díaz, Jorge
1987   *El café amargo: Los procesos de diferenciación y cambio social entre los Chatinos.* Oaxaca, Mexico: UABJO.
Hernández Navarro, Luis
1992   "Cafetaleros: Del adelgazamiento estatal a la guerra del mercado." In *Autonomía y nuevos sujetos sociales en el desarrollo rural*, edited by Julio Moguel, Carlota Botey, and Luis Hernández, 78-97. Mexico City: Siglo XXI.
1994   "El café y la guerra." *La Jornada* (Mexico City), January 30, 48.
Hernández Navarro, Luis, and Fernando Célis Callejas
1994   "Solidarity and the New Campesino Movements: The Case of Coffee Production." In *Transforming State-Society Relations in Mexico*, edited by Wayne A. Cornelius, Ann L. Craig, and Jonathan Fox, 217-231. San Diego: UCSD Center for U.S.-Mexican Studies.
Hernández Xolocotzl, Efraím
1977   *Agroecosistemas de México: Contribuciones a la enseñanza, investigación, y divulgación agrícola.* Chapingo, Mexico: Colegio de Posgraduados.
Hewitt de Alcántara, Cynthia (editor)
1994   *Economic Restructuring and Rural Subsistence in Mexico: Corn and the Crisis of the 1980s.* San Diego: UCSD Center for U.S.-Mexican Studies.
Hightower, Jim
1972   *Hard Tomatoes, Hard Times: The Failure of the U.S. Land Grant College Complex.* Washington, D.C.: Agriculture Accountability Project.
1975   *Eat Your Heart Out: How Food Profiteers Victimize the Consumer.* New York: Crown Publishers.
Hirabayashi, Lane
1993   *Cultural Capital: Mountain Zapotec Migrant Associations in Mexico City.* Tucson: University of Arizona Press.
Ho, Ping-ti
1959   *Studies on the Population of China, 1368-1953.* Cambridge, Mass.: Harvard University Press.
Horton, Robin
1967   "African Traditional Thought and Western Science." *Africa* 37:50-71, 155-187.
Horton, Robin, and Ruth Finnegan
1973   *Modes of Thought.* London: Faber and Faber.
Hurt, R. Douglas
1981   *The Dust Bowl: An Agricultural and Social History.* Chicago: Nelson-Hall.
Institute of Development Studies, Sussex University (IDS)
1979   "Rural Development: Whose Knowledge Counts?" *IDS Bulletin* 10(2):1-8.
Instituto Nacional de Estadística, Geografía e Informática (INEGI)
1991   *Censo general de población y vivienda, 1990 — Oaxaca: Datos por localidad (integración territorial).* Vol. 11. Aguascalientes, Mexico: INEGI.
Jackson, Wes, Wendell Berry, and Bruce Colman (editors)
1984   *Meeting the Expectations of the Land: Essays in Sustainable Agriculture and Stewardship.* San Francisco: North Point Books.

Janofsky, Michael
1997 "Twenty-five Million Pounds of Beef Is Recalled." *New York Times*, August 22, A1.
Jasanoff, Sheila, Gerald E. Markle, James C. Petersen, and Trevor Pinch (editors)
1995 *Handbook of Science and Technology Studies*. Thousand Oaks, Calif.: Sage Publications.
Jaspan, M. A.
1965 "Agricultural Involution: The Process of Ecological Change in Indonesia" (Review). *Man* 65:134.
Jiménez, Michael
1995 "From Plantation to Cup: Coffee and Capitalism in the United States, 1830-1930." In *Coffee, Society, and Power in Latin America*, edited by William Roseberry, Lowell Gudmundson, and Mario Samper Kutschbach, 38-64. Baltimore: Johns Hopkins University Press.
Johannes, Robert E.
1981 *Words of the Lagoon: Fishing and Marine Lore in the Palau District of Micronesia*. Berkeley and Los Angeles: University of California Press.
Johannes, Robert E. (editor)
1989 *Traditional Ecological Knowledge: A Collection of Essays*. Gland, Switzerland: International Union for the Conservation of Nature.
Kaplan, Lawrence
1995 "*Phaseolus* Beans: Accelerator Dates in the Americas." Paper presented at the annual meetings of the Society for American Archaeology, Minneapolis, Minnesota, 1994.
Katz, Esther
1992 "Yosotato: La definición de los espacios, de lo natural a lo cultural." In *Etnias, desarrollo, recursos, y tecnologías en Oaxaca*, edited by Alvaro González and Marco Antonio Vásquez, 91-116. Oaxaca, Mexico: CIESAS.
Katz, S. H., M. L. Hediger, and L. A. Valleroy
1974 "Traditional Maize Processing Techniques in the New World." *Science* 184:765-773.
Kearney, Michael
1972 *The Winds of Ixtepeji: World View and Society in a Zapotec Town*. Prospect Heights, Ill.: Waveland Press.
1996 *Reconceptualizing the Peasantry: Anthropology in a Global Perspective*. Boulder, Colo.: Westview Press.
Kidd, R., and N. Colletta (editors)
1982 *Tradition for Development: Indigenous Structures and Folk Media in Non-formal Education*. Bonn, Germany: Foundation for International Development.
Kit Bond News
1999 "Bond, Respected Scientists Discuss Biotechnology at WTO Conference." See http://bond.senate.gov/Hot_Topics/Biotech/biotech.html.
Kloppenburg, Jack
1988 *First the Seed: The Political Economy of Plant Biotechnology, 1492-2000*. Cambridge: Cambridge University Press.
Kroeber, Alfred L.
1917 "The Superorganic." *American Anthropologist* 19:163-213.
1948 *Anthropology*. New York: Harcourt and Brace.
Kuhn, Thomas
1962 *The Structure of Scientific Revolutions*. Chicago: University of Chicago Press.

Kuroda, Etsuko
1984   *Under Mt. Zempoaltépetl: Highland Mixe Society and Ritual.* Senri Ethnological Studies (Osaka, Japan), vol. 12.
Lappé, Frances Moore
1971   *Diet for a Small Planet.* New York: Ballantine Books.
Lappé, Frances Moore, and Joseph Collins
1977   *Food First.* New York: Ballantine Books.
Levins, Richard, and Richard C. Lewontin
1985   *The Dialectical Biologist.* Cambridge, Mass.: Harvard University Press.
Lévi-Strauss, Claude
1966   *The Savage Mind.* Chicago: University of Chicago Press.
Lévy-Bruhl, Lucien
1923   *Primitive Mentality.* Translated by Lilian A. Clare. Boston: Beacon Press.
Lipp, Frank J.
1991   *The Mixe of Oaxaca: Religion, Ritual, and Healing.* Austin: University of Texas Press.
Lockeretz, W., G. Shearer, and D. H. Kohl
1981   "Organic Farming in the Corn Belt." *Science* 211:540–547.
Long, A., B. F. Benz, D. J. Donohue, A. J. T. Jull, and L. J. Toolin
1989   "First Direct AMS Dates on Early Maize from Tehuacán, Mexico." *Radiocarbon* 31(3): 1035.
Losey, John E., Linda S. Rayor, and Maureen E. Carter
1999   "Transgenic Pollen Harms Monarch Larvae." *Nature,* May 20, 214.
Lovelock, James
1979   *Gaia: A New Look at Life on Earth.* Oxford and New York: Oxford University Press.
1991   *Healing Gaia: Practical Medicine for the Planet.* New York: Harmony Books.
1995   *The Ages of Gaia: A Biography of Our Living Earth.* New York: Norton.
MacVean, Charles
1996   "The Case of the Missing Migrants." *Science* 274:1299–1301.
Madgwick, R. B.
1936   "Coral Gardens and Their Magic (Review)." *Oceania* 7(1):140–142.
Magee, Michelle
1996   "Japanese Fattening Up, U.S.-Style." *San Francisco Chronicle,* July 17, A1.
Malinowski, Bronislaw
1935   *Coral Gardens and Their Magic.* New York: Dover.
1948   *Magic, Science, Religion, and Other Essays.* Garden City, N.J.: Doubleday Anchor
[1925] Books.
Malinowski, Bronislaw, and Julio de la Fuente
1982   *Malinowski in Mexico: The Economics of a Mexican Market System.* London: Routledge
[1940] and Kegan Paul.
Mangelsdorf, Paul C.
1974   *Corn: Its Origin, Evolution, and Development.* Cambridge, Mass.: Harvard University Press.
Mannheim, Karl
1960   "Types of Rationality and Organized Insecurity." In *Images of Man: The Classic Tradition in Sociological Thinking,* edited by C. Wright Mills, 508–528. New York: George Braziller.

Manzano, Crisanto
1993   *Don Chendo* (video). Oaxaca, Mexico: INI Centro de Video y Posproducción.
Marcus, Joyce
1983   "Zapotec Religion." In *The Cloud People: Divergent Evolution of Zapotec and Mixtec Civilizations*, edited by Kent Flannery and Joyce Marcus, 345-351. New York: Academic Press.
Marcus, Joyce, and Kent Flannery
1996   *Zapotec Civilization: How Urban Society Evolved in Mexico's Oaxaca Valley*. London: Thames and Hudson.
Marcuse, Herbert
1964   *One-Dimensional Man*. Boston: Beacon Press.
Martínez-Alier, Juan
1987   *Ecological Economics: Energy, Environment, and Society*. Oxford: Basil Blackwell.
Martínez Luna, Jaime
1995a "¿Es la comunidad nuestra identidad?" *Ojarasca* (Mexico City) 42-43:34-38.
1995b "Resistencia comunitaria y cultura popular." In *Culturas populares y política cultural*, edited by Guillermo Bonfil Batalla, 53-64. Mexico City: Consejo Nacional para la Cultura y las Artes.
Marx, Karl, and Friedrich Engels
1960   "On Alienation." In *Images of Man: The Classic Tradition in Sociological Thinking*,
[1844] edited by C. Wright Mills, 486-507. New York: George Braziller.
Mathews, Jay
1994   "Increasingly, Coffee Isn't Our Cup of Tea." *Washington Post*, November 4, C1.
Mauss, Marcel
1954   *The Gift*. Translated by I. Cunnison. New York: Free Press.
[1924]
McKinsey, Kitty
1997   "Hungarians' Lifestyle Is Killing Them." *San Francisco Chronicle*, October 6, A1.
McWilliams, Carey
1971   *Factories in the Field*. Santa Barbara, Calif.: Peregrine Publishers.
Mead, Margaret
1928   *Coming of Age in Samoa*. Washington, D.C.: American Museum of Natural History.
Meadows, Donella H. (editor)
1972   *The Limits to Growth: A Report for the Club of Rome's Project on the Predicament of Mankind*. New York: Universe Books.
Mendieta y Núñez, Lucio (editor)
1949   *Los Zapotecos*. Mexico City: UNAM.
Merchant, Carolyn
1980   *The Death of Nature: Women, Ecology, and the Scientific Revolution*. New York: Harper and Row.
Merriam, Mansfield, and H. A. Hazen
1892   "The Influence of the Moon on Rainfall—A Symposium." *Science* o.s. 20:310-311.
Messer, Ellen
1972   "Patterns of 'Wild' Plant Consumption in Oaxaca, Mexico." *Ecology of Food and Nutrition* 1:325-332.
Mintz, Sidney W.
1979   "Time, Sugar, and Sweetness." *Marxist Perspectives* 2:56-73.

1985 *Sweetness and Power: The Place of Sugar in Modern History.* New York: Penguin.
1991 "Pleasure, Profit, and Satiation." In *Seeds of Change*, edited by Herman J. Viola and Carolyn Margolis, 112–129. Washington, D.C.: Smithsonian Institution.
Mitchell, Donald O., Merlinda D. Ingco, and Ronald C. Duncan
1997 *The World Food Outlook.* Cambridge: Cambridge University Press.
Moguel, Julio
1992 "La lucha por la apropiación de la vida social en la economía cafetalera: La experiencia de la CNOC 1990–1991." In *Autonomía y nuevos sujetos sociales en el desarrollo rural*, edited by Julio Moguel, Carlota Botey, and Luis Hernández, 98–118. Mexico City: Siglo XXI.
Moguel, Julio, and Josefina Aranda
1992 "Los nuevos caminos en la construcción de la autonomía: La experiencia de la Coordinadora Estatal de Productores de Café de Oaxaca." In *Autonomía y nuevos sujetos sociales en el desarrollo rural*, edited by Julio Moguel, Carlota Botey, and Luis Hernández, 167–193. Mexico City: Siglo XXI.
Monsanto Company
1998 "The Challenge of the Future." See http://www.monsanto.com/ag/articles/sourcebook98/Hug.htm.
1999 "Monsanto Unveils 'Beautiful Science' Exhibit at Disney's Epcot Center." See http://www.monsanto.com/monsanto/mediacenter/99/99oct1_disney.html.
Moore, Molly
1996 "Guerrillas Put Mexico's Hope for Stability under Siege." *Washington Post*, August 31, A24.
Morales, Fidel
1997 "Propuestas para todos." *Ojarasca* (Mexico City) 47–48:50–57.
Mumford, Lewis
1963 *Technics and Civilization.* New York: Harcourt, Brace, and Jovanovich.
[1934]
1964 "Authoritarian and Democratic Technics." *Technology and Culture* 5(1):1–8.
Murray, Matt
1997 "A Bit of Prosperity and Some Fast Food Fatten Zimbabweans." *Wall Street Journal*, August 25, A1.
Nabhan, Gary Paul
1992 "Native Crops of the Americas: Passing Novelties or Lasting Contributions to Diversity?" In *Chilies to Chocolate: Food the Americas Gave the World*, edited by Nelson Foster and Linda S. Cordell, 143–162. Tucson: University of Arizona Press.
Nader, Laura
1964 *Talea and Juquila: A Comparison of Zapotec Social Organization.* Berkeley: University of California Press.
1966 *To Make the Balance* (film). Berkeley: University of California Media Extension Center.
1969 "The Zapotec of Oaxaca." In *Handbook of Middle American Indians*, edited by Robert Wauchope, vol. 7, part 1, 329–359. Austin: University of Texas Press.
1981 *Little Injustices: Laura Nader Looks at the Law* (film). Alexandria, Va.: PBS Video.
1990 *Harmony Ideology: Justice and Control in a Zapotec Mountain Village.* Stanford: Stanford University Press.
1996 "Introduction: Anthropological Inquiry into Boundaries, Power, and Knowledge."

In *Naked Science: Anthropological Inquiry into Boundaries, Power, and Knowledge*, edited by Laura Nader, 1-26. New York: Routledge.
1998   "The Harder Path: Shifting Gears." Paper presented at the Energy and Resources Group, University of California at Berkeley, April 2.
Nader, Ralph (editor)
1993   *The Case against Free Trade: GATT, NAFTA, and the Globalization of Corporate Power.* San Francisco: Earth Island Press.
Nagengast, Carol, and Michael Kearney
1990   "Mixtec Ethnicity: Social Identity, Political Consciousness, and Political Action." *Latin American Research Review* 25(2):31-61.
Nash, June
1970   *In the Eyes of the Ancestors: Belief and Behavior in a Maya Community.* Prospect Heights, Ill.: Waveland Press.
1992   "Global Integration and Subsistence Insecurity." *American Anthropologist* 96(1):7-30.
1994   "The Challenge of Trade Liberalization to Cultural Survival on the Southern Frontier of Mexico." *Indiana Journal of Global Legal Studies* 1(2):367-395.
National Research Council (NRC)
1989   *Alternative Agriculture.* Washington, D.C.: National Academy Press.
Nazarea, Virginia (editor)
1999   *Ethnoecology: Situated Knowledge/Located Lives.* Tucson: University of Arizona Press.
Netting, Robert McC.
1993   *Smallholders, Householders: Farm Families and the Ecology of Intensive, Sustainable Agriculture.* Stanford: Stanford University Press.
New York Times
1999   "U.N. Warning, as Population Nears 6 Billion." *New York Times*, September 22, A5.
Nietschmann, Bernard
1989   "Traditional Sea Territories, Resources, and Rights in the Torres Straits." In *A Sea of Small Boats: Cultural Survival Report 26*, edited by John Cordell, 60-93. Cambridge, Mass.: Cultural Survival.
Nigh, Ronald, and Nemesio J. Rodríguez
1995   *Territorios violados: Indios, medio ambiente y desarrollo en América Latina.* Mexico City: INI.
Nisbet, E. G.
1991   *The Living Earth: A Short History of Life and Its Home.* New York: HarperCollins.
Noble, David F.
1977   *America by Design: Science, Technology, and the Rise of Corporate Capitalism.* Oxford: Oxford University Press.
Nolasco, Margarita
1985   *Café y sociedad en México.* Mexico City: Centro de Ecodesarrollo.
Norman, D. W.
1974   "The Rationalisation of a Crop Mixture Strategy Adopted by Farmers under Indigenous Conditions: The Example of Northern Nigeria." *Journal of Development Studies* 11:3-21.
Novartis
2000   *Maize Is Maize: Why We Use Gene Technology.* See http://www.seeds.novartis.com/.

Nyerges, A. Endre (editor)
1996   *The Ecology of Practice: Studies of Food Crop Production in Sub-Saharan West Africa.* Amsterdam: Gordon and Breach Publishers.

Ohnuki-Tierney, Emiko
1993   *Rice as Self: Japanese Identities through Time.* Princeton: Princeton University Press.

Opler, Morris E.
1959   "Component, Assemblage, and Theme in Cultural Integration and Differentiation." *American Anthropologist* 61(6):955-964.

Ortiz de Montellano, Bernard R.
1990   *Aztec Medicine, Health, and Nutrition.* New Brunswick and London: Rutgers University Press.

Ou, C. Jay
1996   "Native Americans and the Monitored Retrievable Storage Plan for Nuclear Wastes: Late Capitalism, Negotiation, and Controlling Processes." *Kroeber Anthropological Society Papers* 80:32-89.

Pacey, Arnold
1990   *Technology in World Civilization.* Cambridge, Mass.: MIT Press.

Papendick, R. I., P. A. Sánchez, and G. B. Triplett
1976   *Multiple Cropping.* ASA Special Publication No. 27. Madison, Wis.: ASA.

Paredes, Lorena Paz, and Remedio Cobo
1992   "El Proyecto Cafetalero de la Coalición de Ejidos de la Costa Grande de Guerrero." In *Autonomía y nuevos sujetos sociales en el desarrollo rural,* edited by Julio Moguel, Carlota Botey, and Luis Hernández, 144-156. Mexico City: Siglo XXI.

Paredes, Lorena Paz, Remedio Cobo, and Armando Bartra
1997   "La hora del café." *Ojarasca* (Mexico City) 46:27-49.

Parnell, Phillip
1989   *Escalating Disputes: Social Participation and Change in the Oaxacan Highlands.* Tucson: University of Arizona Press.

Parsons, Elsie Clews
1936   *Mitla: Town of the Souls.* Chicago: University of Chicago Press.

Partridge, William L., Antoinette Brown, and Jeffrey Nugent
1982   "The Papaloapan Dam and Resettlement Project: Human Ecology and Health Impacts." In *Involuntary Migration and Resettlement: The Problems and Responses of Dislocated People,* edited by Art Hansen and Anthony Oliver-Smith, 245-263. Boulder, Colo.: Westview Press.

Pascual López, Prócoro, and Neftalí Ortiz Medrano
1996   "UNOSJO: La escarpada producción." *Ojarasca* (Mexico City) 35-36:42-43.

Passell, Peter
1997   "Cutting Waste Can Be a Waste." *New York Times,* September 25, C1.

Paz, Octavio
1973   *In Praise of Hands: Contemporary Crafts of the World.* Greenwich, Conn.: New York Graphic Society.

Pelto, Gretel H.
1987   "Social Class and Diet in Contemporary Mexico." In *Food and Evolution: Toward a Theory of Human Food Habits,* edited by Marvin Harris and Eric B. Ross, 517-540. Philadelphia: Temple University Press.

Pérez García, Rosendo
1956　*La Sierra Juárez.* 2 vols. Mexico City: Gráfica Cervantina.
Poleman, Thomas T.
1964　*The Papaloapan Project: Agricultural Development in the Mexican Tropics.* Stanford: Stanford University Press.
Ponting, Clive
1991　*A Green History of the World: The Environment and the Collapse of Great Civilizations.* New York: Penguin Books.
Posey, Darrel
1984　"Ethnoecology as Applied Anthropology in Amazonian Development." *Human Organization* 43(2):95–107.
Pratt, Jeff
1994　*The Rationality of Rural Life: Economic and Cultural Change in Tuscany.* Chur, Switzerland: Harwood Academic Publishers.
Press, Eyal, and Jennifer Washburn
2000　"The Kept University." *Atlantic Monthly* 285(3):39.
Preston, Julie
1996　"Stung by Attacks, Mexico Intensifies Search for Rebels." *New York Times*, August 31, A1.
Proctor, Robert
1995　*Cancer Wars: How Politics Shapes What We Know and Don't Know about Cancer.* New York: Basic Books.
Rahnema, Majid
1986　"Under the Banner of Development." *Development* 36:1–23.
Ramales, Rosa
1997　"Si gana PRD, dialogaremos con el Congreso de la Unión: EPR." *Noticias* (Oaxaca, Mexico), April 20, 16A.
Randall, Laura (editor)
1996　*Reforming Mexico's Agrarian Reform.* Armonk, N.Y.: M. E. Sharpe.
Rappaport, Roy
1968　*Pigs for the Ancestors.* New Haven: Yale University Press.
Redfield, Robert
1930　*Tepoztlán: A Mexican Village.* Chicago: University of Chicago Press.
Rees, Martha W.
1996　"Ethnicity and Community in Oaxaca: Nursery, Hospital, and Retirement Home." *Reviews in Anthropology* 25:107–123.
Reganold, J. P., R. I. Papendick, and J. F. Parr
1990　"Sustainable Agriculture." *Scientific American* 262(6):112–120.
Reichel-Dolmatoff, Gerardo
1976　"Cosmology as Ecological Analysis: A View from the Rain Forest." *Man* 11(3):307–318.
Renner, Michael
1997　"Transforming Security." In *State of the World 1997*, edited by Lester Brown, 115–131. New York: Norton.
Restivo, Sal
1988　"Modern Science as a Social Problem." *Social Problems* 35:206–225.

Ricard, Robert
1966 *The Spiritual Conquest of Mexico*. Translated by Lesley Byrd Simpson. Berkeley and
[1933] Los Angeles: University of California Press.
Richards, Paul
1979 "Community Environmental Knowledge and African Rural Development." *IDS Bulletin* 10(2):28-36.
1985 *Indigenous Agricultural Revolution: Ecology and Food Production in West Africa*. Boulder, Colo.: Westview Press.
Ríos Morales, Manuel (editor)
1994 *Los Zapotecos de la Sierra Norte de Oaxaca*. Oaxaca, Mexico: CIESAS.
Robertson, James A.
1927 "Some Notes on the Transfer by Spain of Plants and Animals to Its Colonies Overseas." *Studies in Hispanic American History* 19(2):8-21.
Rodríguez, Nemesio J.
1989 "¿Desarrollo para quién?" *México Indígena* 5(27):21-22.
Romero Frizzi, María de los Angeles
1990 *Economía y vida de los españoles en la Mixteca Alta, 1519-1720*. Mexico City: INAH.
1996 *El Sol y la Cruz: Los pueblos indios de Oaxaca colonial*. Mexico City: CIESAS/INI.
Roseberry, William
1996 "The Rise of Yuppie Coffees and the Reimagination of Class in the United States." *American Anthropologist* 98(4):762-775.
Roush, R. T., and A. M. Shelton
1997 "Assessing the Odds: The Emergence of Resistance to Bt Transgenic Plants." *Nature Biotechnology* 15(9):816-817.
Rubel, Arthur J.
1964 "The Epidemiology of a Folk Illness: Susto in Hispanic America." *Ethnology* 3:268-283.
Ruiz Arrazola, Víctor
1997a "Empieza a militarizarse la Sierra Juárez de Oaxaca." *La Jornada* (Mexico City), April 23, 46.
1997b "Instalan soldados y policías una base en tierras comunales." *La Jornada* (Mexico City), June 28, 6.
Ruiz Cervantes, Francisco José
1988 "De la bola a los primeros repartos." In *Historia de la cuestión agraria mexicana, Estado de Oaxaca*, edited by Leticia Reina, 331-424. Mexico City: Juan Pablos Editor.
Ruiz Lombardo, Andrés
1991 *Cafeticultura y economía en una comunidad totonaca*. Mexico City: INI.
Rus, Jan, and Robert Wasserstrom
1980 "Civil-Religious Hierarchies in Central Chiapas: A Critical Perspective." *American Ethnologist* 7:466-478.
Sahlins, Marshall
1972 *Stone Age Economics*. Chicago: Aldine-Atherton.
Samuelson, Robert J.
1989 "The Coffee Cartel: Brewing Up Trouble." *Washington Post*, July 26, A25.
Sanderson, Steven
1986 *The Transformation of Mexican Agriculture: International Structure and the Politics of Rural Change*. Princeton: Princeton University Press.

Sandstrom, Alan
1991   *Corn Is Our Blood: Culture and Ethnic Identity in a Contemporary Aztec Indian Village.* Norman: University of Oklahoma Press.
Schmelzer, Paul
1998   "Label Loopholes: When Organic Isn't." *Progressive* (May), 28-29.
Schmieder, Oscar
1930   *The Settlements of the Tzapotec and Mije Indians.* Berkeley: University of California Press.
Schultz, Stacey, and Josh Fischman
1999   "Why We're Fat." *U.S. News and World Report* (November 8), 82.
Schumacher, E. F.
1973   *Small Is Beautiful: Economics As If People Mattered.* New York: Harper and Row.
Sclove, Richard
1995   *Democracy and Technology.* New York: Guilford Press.
Scott, Colin
1996   "Science for the West, Myth for the Rest? The Case of James Bay Cree Knowledge Construction." In *Naked Science: Anthropological Inquiry into Boundaries, Power, and Knowledge,* edited by Laura Nader, 69-86. New York: Routledge.
Scott, James
1976   *The Moral Economy of the Peasant.* New Haven: Yale University Press.
1998   *Seeing Like a State: How Certain Schemes to Improve the Human Condition Have Failed.* New Haven: Yale University Press.
Semmens, Elizabeth S.
1923   "The Effect of Moonlight on the Germination of Seed." *Nature* 111:49-50.
Shaffer, John A., Randall S. Cerveny, and Robert C. Balling, Jr.
1997   "Polar Temperature Sensitivity to Lunar Forcing." *Geophysical Research Letters* 24(1): 29-32.
Shiva, Vandana
1991   "The Green Revolution in the Punjab." *Ecologist* 21(2):57-60.
Shweder, Richard A.
1997   "Ancient Cures for Open Minds." *New York Times,* October 26, sec. 4, 6.
Smith, Bruce D.
1998   "The Initial Domestication of *Cucurbita pepo* in the Americas 10,000 Years Ago." *Science* 276:932-934.
Stavenhagen, Rodolfo
1965   "Classes, Colonialism, and Acculturation." *Studies in Comparative International Development* 1(4):53-77.
1971   "Decolonizing Applied Social Science." *Human Organization* 30(4):333-357.
Steingraber, Sandra
1997   *Living Downstream: An Ecologist Looks at Cancer and the Environment.* Reading, Mass.: Addison-Wesley.
Stephen, Lynn
1991   *Zapotec Women.* Austin: University of Texas Press.
Strange, Marty
1988   *Family Farming: A New Economic Vision.* Lincoln: University of Nebraska Press.

Swadesh, Morris
1949 "El idioma de los Zapotecos." In *Los Zapotecos*, edited by Lucio Mendieta y Núñez, 417-448. Mexico City: Imprenta Universitaria.
Szpir, Michael
1996 "Lunar Phases and Climactic Puzzles." *American Scientist* 84(2):59-64.
Tambiah, Stanley
1990 *Magic, Science, Religion, and the Scope of Rationality*. Cambridge: Cambridge University Press.
Thiesenhusen, William C.
1995 "Mexican Land Reform, 1934-1991: Success or Failure?" In *Reforming Mexico's Agrarian Reform*, 35-48. Armonk, N.Y.: M. E. Sharpe.
Thompson, E. P.
1963 *The Making of the English Working Class*. New York: Vintage.
Todaro, Michael P.
1989 *Economic Development in the Third World*. 4th ed. New York: Longman.
Toledo, Víctor M.
1980 "La ecología del modo campesino de producción." *Antropología y Marxismo* 3:33-55.
1995 *Ecología y campesinado: La vía no-occidental al desarrollo sustentable*. Mexico City: Siglo XXI.
Toledo, Víctor M., J. Carabias, C. Mapes, and C. Toledo
1985 *Ecología y autosuficiencia alimentaria*. Mexico City: Siglo XXI.
Toledo, V. M., J. Carabias, C. Toledo, and C. González-Pacheco
1989 *La producción rural en México: Alternativas ecológicas*. Mexico City: Fundación Universo Veintiuno.
Toulmin, Stephen
1990 *Cosmopolis: The Hidden Agenda of Modernity*. New York: Free Press.
Traffano, Daniela
1998 "Mujeres e idolatría en dos regiones de Oaxaca, siglos XVII y XVIII." *Cuadernos del Sur* (Oaxaca, Mexico) 13:7-24.
Traweek, Sharon
1987 *Beamtimes and Lifetimes: The World of High-Energy Physicists*. Cambridge, Mass.: Harvard University Press.
Tricks, Henry, and Andrea Mandel-Campbell
1999 "Mexico's Farming Habits under Pressure from Transgenics: Genetically Modified Seeds May Threaten the Country's Biodiversity." *Financial Times*, October 12.
Trouillot, Michel-Rolph
1991 "Anthropology and the Savage Slot: The Poetics and Politics of Otherness." In *Recapturing Anthropology: Working in the Present*, edited by Richard G. Fox, 17-44. Santa Fe: School of American Research Press.
Trujillo Arriaga, Javier
1987 "The Agroecology of Maize Production in Tlaxcala, Mexico: Cropping Systems Effects on Arthropod Communities." Ph.D. dissertation, University of California at Berkeley.
Tylor, Edward B.
1871 *Primitive Culture*. London: J. Murray.

Tyrtania, Leonardo
1992   *Yagavila: Un ensayo en ecología cultural.* Mexico City: UAM-Iztapalapa.
Ukers, William H.
1935   *All About Coffee.* 2nd ed. New York: Tea and Coffee Trade Journal Company.
United States Department of Agriculture (USDA)
1973   *Monoculture in Agriculture: Extent, Causes, and Problems—Report of the Task Force on Spatial Heterogeneity in Agricultural Landscapes and Enterprises.* Washington, D.C.: USDA.
1980   *Report and Recommendations on Organic Farming.* Washington, D.C.: USDA.
United States Department of Commerce
1961   *Coffee Consumption in the United States, 1920–1960.* Washington, D.C.: Government Printing Office.
United States Senate Committee on Nutrition and Human Needs
1978   *Dietary Goals for the United States.* Washington, D.C.: Government Printing Office.
Vásquez y de los Santos, Elena, and Yanga Villagómez Velázquez
1993   "La UCIRI, el café orgánico, y la experiencia de un proyecto campesino autosugestivo en la producción." *Cuadernos del Sur* (Oaxaca, Mexico) 2(5):121–137.
Visvanathan, Shiv
1998   "A Celebration of Difference: Science and Democracy in India." *Science* 280:41–42.
Wallerstein, Immanuel
1974   *The Modern World-System.* New York: Academic Press.
Wall Street Journal
1999   "EU Panel Clarifies Food-Labeling Rules." *Wall Street Journal,* October 22, sec. o, A16.
Warman, Arturo
1988   *La historia de un bastardo: Maíz y capitalismo.* Mexico City: Fondo de Cultura Económica.
1994   "La reforma al artículo 27 constitucional." *Perfil de la Jornada,* special supplement to *La Jornada* (Mexico City), March 8, 1.
Warman, Arturo (editor)
1970   *De eso que llaman antropología mexicana.* Mexico City: Editorial Nuestro Tiempo.
Warren, Dennis M.
1984   "Linking Scientific and Indigenous Agricultural Systems." In *The Transformation of International Agricultural Research and Development,* edited by J. L. Compton, 132–147. Boulder, Colo.: Westview Press.
Weatherford, Jack
1988   *Indian Givers: What the Native Americans Gave to the World.* New York: Crown.
Weatherwax, Paul
1954   *Indian Corn in Old America.* New York: Macmillan Company.
Weber, Max
1946   *From Max Weber: Essays in Sociology.* Translated and edited by H. H. Gerth and C. Wright Mills. New York: Oxford University Press.
1958   *The Protestant Ethic and the Spirit of Capitalism.* New York: Charles Scribner's Sons.
Weiss, Rick
1998   "Strategy Worries Crop Up in Biotechnology's War on Pests." *Washington Post,* September 21, A3.

Weitlaner, Roberto J.
1965 "Supervivencias de la religión y magia prehispánicas en Guerrero y Oaxaca." *Proceedings of the Thirty-fifth International Congress of Americanists*, 557-563. Mexico City: UNAM.
Weitlaner, Roberto J., and Gabriel de Cicco
1958 "La jerarquía de los dioses zapotecos del sur." *Proceedings of the Thirty-fourth International Congress of Americanists*, pp. 695-710. Vienna, Austria: F. Berger.
Wiegner, Kathleen
1995 "One Man's Problem Is Another's Delicacy." *Los Angeles Times*, July 12, D5.
Wilken, Gene
1987 *Good Farmers: Traditional Agricultural Resource Management in Mexico and Central America.* Berkeley and Los Angeles: University of California Press.
Williams, Aubrey A.
1973 "Dietary Patterns in Three Mexican Villages." In *Man and His Foods: Studies in the Ethnobotany of Nutrition*, edited by C. Earle Smith, 51-73. Birmingham: University of Alabama Press.
Williams, Nancy M., and Graham Baines (editors)
1993 *Traditional Ecological Knowledge: Wisdom for Sustainable Development.* Canberra, Australia: Centre for Resource and Environmental Studies, Australian National University.
Williams, Raymond
1985 *Keywords: A Vocabulary of Culture and Society.* New York: Oxford University Press.
Winner, Langdon
1987 *The Whale and the Reactor: The Search for Limits in an Age of High Technology.* Chicago: University of Chicago Press.
Wolf, Eric R.
1955 "Types of Latin American Peasantry: A Preliminary Discussion." *American Anthropologist* 57:452-471.
1959 *Sons of the Shaking Earth.* Chicago: University of Chicago Press.
1966 *Peasants.* Englewood Cliffs, N.J.: Prentice-Hall.
1982 *Europe and the People without History.* Berkeley and Los Angeles: University of California Press.
Wright, Angus
1984 "Innocents Abroad: American Agricultural Research in Mexico." In *Meeting the Expectations of the Land: Essays in Sustainable Agriculture and Stewardship*, edited by Wes Jackson, Wendell Berry, and Bruce Colman, 135-151. San Francisco: North Point Press.
1990 *The Death of Ramón González: The Modern Agricultural Dilemma.* Austin: University of Texas Press.
Yoon, Carol Kaesuk
1998 "Chocoholics Take Note: Beloved Bean in Peril." *New York Times*, May 4, A1.
1999a "Reassessing Biological Risks of Genetically Altered Crops." *New York Times*, November 3, A1.
1999b "Studies Raise Concerns about Genetically Engineered Crops." *New York Times*, November 3, A22.

2000   "E.P.A. Announces New Rules on Genetically Altered Corn." *New York Times*, January 17, A14.

Young, Kate

1976   "The Social Setting of Migration: Factors Affecting Migration from a Sierra Zapotec Village in Oaxaca, Mexico." Ph.D. dissertation, University of London.

1982   "The Creation of a Relative Surplus Population: A Case Study from Mexico." In *Women and Development: The Sexual Division of Labor in Rural Societies*, edited by Lourdes Benería, 149–177. New York: Praeger.

Zilberman, María Cristina

1966   "Idolatrías de Oaxaca en el siglo XVIII." In *Proceedings of the Thirty-sixth International Congress of Americanists*, 111–123. Seville, Spain: ECESA.

INDEX

acupuncture, 23, 240, 246-247, 279n.2
Adas, Michael, 18
AgrEvo, 249
agribusiness. *See* transnational corporations
agricultural implements: customization of, 76, 235-236; design of, 80-96; disappearance of, 95; ergonomic design of, 76-77, 235-236; manufacture of, 80-96; Zapotec adoption of Spanish, 72-73, 101
Agricultural Involution, 12
agriculture: during colonial era, 50-52, 70-73; commercial, 128-129, 225-227; Indonesian, 12-13; industrialized—*see* factory farming; fallow, 130, 283n.1, 288n.13; of the Hanunóo, 12; organic, 242, 244-246; persistence of, in Talea, 103; rituals, 109-113; shifting (swidden), 12; slash-and-burn, 71, 130, 283n.1; subsistence, 128-129, 225-227; sustainable, 218-220, 236, 239, 241-244; yields, 16, 164-166; of the Zapotec, 1-3, 13-21, 25, 70-73, 258-259, 261-262
agroecologists, 23, 139, 170, 259

Agroecology: The Science of Sustainable Agriculture, 243
agronomists, 100, 147, 167, 220, 255
alcaldes mayores, 43, 44-45, 48, 49-50
Alianza del Campo, 147
altars, 97, 190-192
Altieri, Miguel, 242
Anaheim, California, 65
Anderson Clayton, 126, 248
Antequera. *See* Oaxaca City
Archer Daniels Midland, 234, 248
area, 78-80
art, 92
Article 27, 124, 126, 248, 258, 281n.9, 287n.19
Augustinians, 52, 179
axes, 82-83, 94, 97
Azande, 6
Aztecs, 1, 119, 172, 190, 285n.5

barrios, 131
Barry, Tom, 100
Beadle, George, 118
Betao, 54
Betaza, 179, 201
bhni gui'a, 17, 42

"Big Science," 9, 10, 22
biodiversity, 218–220, 222, 256
biotechnology, 127, 238, 249–254, 293n.1.
  See also genetically modified crops
Bond, Kit, 253
Borlaug, Norman, 251
Bracero Program, 57, 65, 164
Brewster, Leticia, 245
bridewealth, 116
Brown, Lester, 239
Bt corn, 249–250
bulls. See oxen

Cacalotepec, 179, 199
cacao, 219, 291n.14
cafetales. See coffee plantations
campesinos: appropriation of coffee by, 204; children of, 92, 99; and crafts, 236; definition of, 279n.1; differences between rich and poor, 208–209, 212–213; as efficient consumers, 172–173; experiments conducted by, 3, 207–209, 220–221, 233, 235, 241, 254–255, 259, 284n.3; intelligence of, 95; knowledge of, 130, 279n.2; patron saint of, 106–109; pooling of agricultural labor by, 143, 176; pride taken in work by, 95; scientific style of, 254–255; as scientists, 3, 220–221; use of scrap materials by, 88–90, 235; views of coffee, 216–217, 223–227; views of development, 257–258; views of food, 162, 176–177; views of maize as living being, 102–106, 120, 129, 236; views of progress, 257–258; views of "science," 257–258
CAMPO (Centro de Apoyo al Movimiento Popular de Oaxaca), 292n.19
Cancian, Frank, 26
Cañada, 216
Capulalpan, 51
Cárdenas, Lázaro, 125, 197, 290n.1
cargo system, 49–50, 71, 281n.11
Carson, Rachel, 242
cash cropping. See agriculture, commercial
Catholicism: during colonial era, 52–54, 55; "folk," 54; and Todos Santos celebration, 190–193; and Zapotec religion, 52–54, 110–113
Catholic missionaries, 50, 65, 68, 111
Catholic saints, 16, 17, 106–109, 110, 111–112, 129, 160
CEPCO (State Network of Coffee Producers of Oaxaca), 229–230
Chambers, Robert, 25, 242
Chance, John, 55
The Changing American Diet, 245
Chevalier, François, 51
Chiapas, 7, 69, 197
Chichén Itzá, 2
children: cultivation by, 161; formal schooling of, 99, 161, 284n.8; tasks done by, 92, 136, 223
chiles, 159, 289n.19
China, 122
chinampas, 119
Chinantecs, 43, 62–63, 112, 282n.19
Choapan, 63, 198
civil-religious hierarchy. See cargo system
Club of Rome, 242
CNOC (National Network of Coffee Producers), 228
Coca-Cola, 21, 291n.13
cochineal: as commodity in colonial era, 45–46, 47; production of, 24, 71, 235; trade, 44–45
Cocijo, 109
coffee: appropriation of, by Taleans, 222; as basis for identity, 227, 233, 237; and children, 223–224; cooperatives, 228–233, 237; cultivation of, 204–217, 235; demand for, 198; global consumption of, 201–202, 203; history of, 195–204, 237; for household consumption, 217–218; meanings of, 222–225; in Mexico, 197–198; in Northern Sierra, 198–200; organic, 228, 230, 232–233; as a plantation crop, 196–198; politics of, 227–233; prices, 201, 203–204, 206, 224–225, 227–228, 230–231, 237, 280n.10, 284n.9, 291n.16; quality of, 216; in Rincón, 199–200; seeds, 205;

Index

in Soconusco, 197, 205; in Talea, 57–58, 95, 200–201; trading, 201–204; as a "traditional" crop, 195, 222–227, 237; varieties of, 207–208; yields, 214. *See also* coffee plantations
coffee beans: depulping, 214–215; fermenting, 215; grinding, 217–218; harvesting, 209–213; sun-drying, 216; toasting, 217–218; transporting, 216; washing, 215–216
coffee plantations, 128–129, 197, 214; conversion of maize land to, 204, 206; cultivation of, 205–209, 212–213; and diseases, 209; experimentation with, 207; shade trees for, 205–206, 218–220, 233; sun, 219–220; weeding, 209
Cohen, Jeffrey, 26
Community Food Councils, 229
community studies, 26
composting, 242
CONASUPO (National Company for Popular Subsistence), 5, 9, 125, 127, 128, 157, 176, 204, 218
Conklin, Harold, 12
Conquest, 42–46
convergent evolution, 240–241
Coral Gardens and Their Magic, 11, 12
corn smut, 150, 288n.16
corregidores, 45, 46
corregimientos, 45
Cortés, Hernán, 43, 54
cotton: as commodity, 45–47; production of, 24, 71, 235; trading, 44
coyotes (intermediaries), 216
crafts, 92–96, 236
craftwork, 90–92, 236
Cree, 10, 95, 279n.2
Crosby, Alfred, 23, 120
Crouch, Martha, 253–254
Cruz Verde (Green Cross), 112
customization, 76, 95, 235

death, 190–194
de Echarrí, Juan Francisco, 48
de la Fuente, Julio, 26, 75
Delicias, 213

de Maldonado, Angel, 47
de Olmedo, Bartolomé, 52, 55
development. *See* economic development
Díaz, Bernal, 43
Diego, Juan, 150
Diet for a Small Planet, 244
Dominicans, 51–52
Dow, 249
DuPont, 249
Durkheim, Emile, 5

E. coli, 245
earth: as female force, 110; as living being, 17, 106–113, 244, 247; as mechanical object, 18; personification of, 106–113, 129, 244, 247; post-Enlightenment views of, 18; sacrifices made to, 109–111
economic development: agents of, 23, 102, 123, 238, 243, 249; critiques of, 25, 244; imposition of, 24; initiated by Mexican government, 203–204, 247–249; of Papaloapan River basin, 24, 32, 40, 58, 59, 282n.19; planners, 16, 18; projects, 113
economic growth, 16
ejidos, 124–125, 126, 127, 258, 281n.9, 287n.19
El Guindi, Fadwa, 13
encomenderos, 45, 46
encomiendas, 45
Engels, Friedrich, 92
epidemic diseases, 45, 71, 72, 177, 258
EPR (Popular Revolutionary Army), 68, 231, 282n.14, 283n.27, 293n.22
ergonomic design: by U.S. military, 77; by Zapotec campesinos, 76–77, 95, 235
Eurocentric diffusionism, 24
Evans-Pritchard, E. E., 6
EZLN (Zapatista Army of National Liberation). *See* Zapatistas

factory farming, 24, 102, 130, 174, 238, 241–242, 248, 249, 258, 294n.6
fertilizers, 102, 103, 123, 125, 128, 140, 145–147, 165, 169–170, 204, 207, 218, 221, 230, 233, 236, 244, 258, 288n.14

Feyerabend, Paul, 10, 249, 259
Finnegan, Ruth, 4
Fleck, Ludwik, 6
flowers, 160–161
food: chains, 245–246; contamination, 245–246; crisis, 238–239; as gift, 114–117, 176, 190–194; organic, 245, 280n.12; quality, 4, 14, 19–20, 158, 167, 173–174, 176–177, 189–190, 216, 244–246; security, 24–25, 238–239; waste, 239
Food First, 244
Food First Institute, 241
Ford Foundation, 287n.17
forests, 41–42, 218–220
Foster, George, 20–21, 26
Franciscans, 52
Frankfurt School, 7
Frazer, James, 5

García, Agustín, 58, 60, 282n.18
Geertz, Clifford, 12
General Agreement on Tariffs and Trade (GATT), 24
gene revolution. *See* genetically modified crops
genetically modified crops, 24, 246, 249–254, 258
genetic contamination, 250–251
genetic diversity: loss of, 24, 103, 255
Glickman, Dan, 239
Goodenough, Ward, 70
Good Farmers, 243
Gore, Albert, 239
gozona, 16, 98, 113–117, 131, 143, 149, 164, 176, 182, 190, 194, 212, 258
Great Depression, 201
Green Cross (Cruz Verde), 112
Green Revolution, 9, 24, 102–103, 123, 125, 126, 145, 244, 249, 251, 254, 255, 256, 287n.17
Guanajuato, 46
guelaguetza. *See gozona*
Guelatao de Juárez, 40, 63, 230
Guilá Naquitz, 118

Habermas, Jürgen, 7
Hanunóo, 12
Harding, Sandra, 21
herbicides, 102, 123, 125, 246, 258
Hernández, Fidencio, 199–200, 205, 227, 283n.26, 290n.2
Hidalgo, Miguel, 37
Hightower, Jim, 242
Hirabayashi, Lane, 66
hoes, 80–82, 94, 97, 179
Hong Kong, 7
Horton, Robin, 4, 6
hot/cold concept, 4, 14, 20–21, 123, 131, 147, 167, 189, 220, 237, 240, 247, 280n.8, 288n.14
huitlacoche. *See* corn smut
humoral theory. *See* hot/cold concept
hybrid seeds, 102, 123, 125, 165

Ibarra, Isaac, 283n.26
IEEPO (Oaxaca State Institute of Public Education), 60
IFE (Federal Electoral Institute), 67
implements. *See* agricultural implements
"Indian Republics," 43, 44, 49–50
Indigenous Agricultural Revolution, 243
Industrial Revolution, 256
intercropping, 139, 165, 167–168, 236, 242, 243
International Coffee Agreement, 203
International Coffee Organization, 228
INI (National Indigenous Institute), 59
INMECAFE (National Coffee Institute), 60, 203–204, 207, 215, 220–221, 227–228, 229, 230, 232, 293n.23
integrated pest management, 242
intermediaries, 216, 227
Isthmus of Tehuantepec, 52, 63, 66, 282n.14
Ixtepeji, 18, 47, 63
Ixtlán de Juárez, 34, 39, 51, 68, 111, 199

Jacobson, Michael, 245
James Bay Cree, 10–11
Japan, 128, 158

Johannes, Robert, 10-11, 99
Josaa, 199

Kearney, Michael, 18, 279n.1
kitchen garden, 160-161
Koslow, Stephen, 247
Kuhn, Thomas, 6

labor. *See* work
La Chachalaca, 34
Lachatao, 63
Lachirioag, 50, 179
Lachixila, 34, 50, 63
Lalopa, 49, 179
land reform, 124-127, 197
Lappé, Frances Moore, 244
Las Palmas, 34
learning by working, 26-28
length, 74-77
Lévi-Strauss, Claude, 5, 13
Lévy-Bruhl, Lucien, 5
López Portillo, José, 125
Los Angeles, 65, 66, 150, 225, 226
lunar patterns, 135, 141, 143, 170-173, 289n.21

machines, 90-91
Magic, Science, Religion and the Scope of Rationality, 4
maize: and agricultural revolutions, 117-123; applying fertilizer to, 145-147; cleaning, 151-153; collecting fallen ears of, 148; from CONASUPO, 127, 157-158, 162, 173; converting to food, 154-162; diet based on, 159, 172-173; domestication of, 117-120; genetic contamination of, 250-251; genetic modification of, 249-254, 258; as gift, 113-117; global diffusion of, 120-123; grinding, 154; harvesting, 148-149; husking of, 149-151; imports and exports, 124-125, 229, 250-251; as living matter, 102-117, 164; and local markets, 162-163; meanings of, 103-117, 129, 258; milling, 154; and nutrition, 172-173; pests affecting, 144, 152; planting, 140-143; as plant-person, 106, 120, 164, 258; prices, 127-128; and reciprocity, 105-106, 129, 236; removing spikes from, 147-148; replanting, 143; seeds, 140-141, 143, 151; separating, 149-153; as social relations, 113-117; soul of, 102-117, 129, 258; sun-drying, 153-154, threshing, 151-153, 288n.17; transporting, 153; varieties of, 120, 256, 288n.11; views of, among Rincón Zapotec, 103-117, 129, 164, 236, 258; weeding, 144-145; yields, 164-166, 284n.1
maize mounds, 168-169
Malinowski, Bronislaw, 5, 11
Mangelsdorf, Paul, 118
Mannheim, Karl, 7
mantenimiento (maintenance), 4, 14, 15-16, 48-49, 103, 129, 163, 174, 225-227, 236, 237, 243, 258, 280n.2
Marcus, Joyce, 52
Marcuse, Herbert, 7
markets: in ancient Mesoamerica, 61; and ethnic groups, 64-65; in the Rincón, 61-62
Marx, Karl, 19, 92-93
matelacihua (*mal aire*), 17, 280n.11
Mauss, Marcel, 5, 16-17
Mayans, 1, 7, 25, 64, 110, 197, 247, 285n.5
Mazatecs, 282n.19
McDonald's, 293n.5
measles, 45, 71
measures. *See* units of measure
Mesoamericans, ancient: cultivation by, 1-2, 117-118; diet of, 172, 244; imperial expansion of, 166
mestizo, 56, 64-65, 200
metric system, 284n.4
Mexican Food System, 125-129, 229
Mexican government: agronomists, 2; bureaucrats, 19; land reform policies, 124-126; presence in Rincón, 68; programs, 145, 203-204; structural adjustment policies, 228

Mexican revolution of 1910, 66, 100, 124-125, 281n.9, 282n.21
Mexico: agriculture in, 24, 247-249; factory farms in, 24, 248; food policy in, 124-129; genetically modified crops in, 250-251; genetic contamination of maize in, 250-251
Mexico City, 44, 46, 47, 54, 65, 66, 68, 89, 117, 150, 212, 218, 226
MICHIZA (Mixe-Chinantec-Zapotec Organization), 230
Micronesia, 7. *See also* Palau district
migration, 65-66, 213, 226, 280n.10, 283n.25, 284n.9
mills: for grinding maize, 155, 156; for grinding sugarcane, 181-185, 194, 215
mining, 24, 48-49, 56-57, 200, 235, 281n.3
Mintz, Sidney, 19, 175
Mitla, 2, 289n.3
Mixes, 32, 43, 48, 49, 50, 55, 62, 63-64, 110, 111, 112, 198, 201, 216, 232, 285n.2
Mixtecs, 1, 42, 51, 65, 240
Moctezuma, 43
Modes of Thought, 4
mole, 190
Monsanto, 249, 252
Monte Albán, 2, 37
mountain spirits, 109-112
Mount Zempoaltéptl, 32, 48
multicropping. *See* intercropping
Mumford, Lewis, 90

Nader, Laura, 4
NAFTA (North American Free Trade Agreement), 24, 126, 248, 250, 292n.18
Nahuatl, 110
Naked Science, 4, 9
National Institutes of Health, 246
Nejapa, 47
Nescafé, 217
Nestlé, 127, 217
New Spain: alcaldes mayores of, 72; colonists' attitudes about work, 19; commodities, 45-49; crops, 50-52; history, 43-54; rebellions, 50; sugarcane, 178-179; tribute policies, 71; uprisings, 50
New York City, 65
Nexitzo Zapotec. *See* Zapotec, Rincón
Norgaard, Richard, 8
Northern Sierra (Oaxaca): colonial commodities in, 45-49; effects of coffee prices in, 201-204; ethnic groupings in, 62-65; ethnographic studies of, 25-26; geography of, 32-34; history of, 42-60; history of coffee in, 198-200; introduction of new crops in, 50-52; magical rites performed in, 109-113; money in, 202; political economy of, 44-50; politics in, 66-69, 227-233; population of, 72; rebellion in, 50; religious conquest of, 52-54; roads in, 202-203, 282n.22; uprisings in, 50
Novartis, 234, 249

Oaxaca City, 40, 44, 58, 59, 64, 65, 66, 68, 111, 149, 158, 202, 212, 216
Office of Alternative Medicine (National Institutes of Health), 246
Ohnuki-Tierney, Emiko, 128
Olmecs, 1
organic composts, 147
Ortiz de Montellano, Bernard, 172
Otatitlán de Morelos, 49, 179, 231
oxen, 135-138, 140, 146, 148, 150, 194, 288n.7

Palau district (Micronesia), 10, 95, 99-100, 101
panela: complementarity between coffee and, 189; conversion of sugarcane into, 188; as gift, 114-115; household consumption of, 176; meanings of, 190-194; as offering to deceased, 190-193; preference for, in Talea, 176, 180, 189-190, 236; production of, 186-190, 236; wrapping, 188-189
Papaloapan Commission, 24, 40, 58, 59, 204, 207, 220, 282n.19
Parsons, Elsie Clews, 26, 75

participant observation, 26-28
Partido de la Revolución Democrática
  (PRD), 67
Partido Revolucionario Institucional
  (PRI), 41, 66-68, 95, 232
Paz, Octavio, 92
Pepsico, 126, 293n.5
Pérez García, Rosendo, 200
personification: of earth, 17, 51-52, 106-
  113, 244, 247; of maize, 51-52, 106,
  164, 236, 258; of nonhumans and
  supernatural actors, 4, 14, 17-18, 51-52
pesticides, 102, 103, 123, 125, 144, 151, 165,
  221, 230, 242, 244, 246, 258
Pharmacia & Upjohn, 252
plows, 70-73, 83-88, 90, 94, 101, 134-137,
  139-140, 236
Porfiriato, 124, 198
Porvenir, 34, 213
PRD, 67
PRI. See Partido Revolucionario Institu-
  cional
PROCAMPO, 126, 147
Puebla, 46, 226
Pueblos Unidos (of the Rincón of the
  Sierra Juárez), 230, 282n.22, 292n.20

quelites, 139, 159

Ralston Purina, 126, 248, 293n.5
Rangel, Rodrigo, 55
rationality: anthropological debates
  about, 3-7; economic, 176; formal, 7;
  substantial, 7
rationalization, 7
reciprocal exchange, 4, 14, 16-17, 98, 105,
  190-194. *See also gozona*
reciprocity. *See* reciprocal exchange
Redfield, Robert, 26
Rees, Martha, 63
repartimientos de efectos, 44-45, 49-50
repartimientos de labor, 48, 49-50, 71
Ricard, Robert, 52, 190
rice, 128, 158
Richards, Paul, 25, 242

Rincón: boundaries of, 32-34; decline
  of craft industries in, 57-59; ethnic
  groupings in, 62-65; geographical features of, 32-34; history of, 24; history
  of coffee in, 199-200; location of, 32;
  markets in, 61-62; meanings of maize
  in, 103-117, 128-129; money in, 201;
  politics in, 66-69; roads of, 57-59;
  sociolinguistic groupings of, 62-65;
  sugarcane work in, 178-189
Rockefeller Foundation, 287n.17
Romero, María de los Angeles, 51

Salinas de Gortari, Carlos, 228, 248,
  292n.18
SAM. *See* Mexican Food System
San Andrés Yaa, 112
San Francisco Cajonos, 50
San Isidro Labrador, 106-109, 111, 112,
  285n.6
San Juan Juquila Vijanos, 39, 40, 41, 189,
  201
San Juan Tabaa, 41, 50, 51, 63, 104, 105,
  153
San Mateo Cajonos, 179
San Miguel Amatlán, 51
San Pablo Chiltepec, 111
San Pedro Yolox, 68
Santa Ana, 65
Santa Gertrudis, 48, 50, 56-57, 62, 281n.3,
  283n.22
Santa Monica, 65
The Savage Mind, 5
Schmieder, Oscar, 75
Schumacher, E. F., 242
science: anthropology of, 4-6, 21-22;
  boundaries between magic, religion,
  and, 4-5, 9, 90; campesinos as practitioners of, 220-221, 259; contextual-
  ization of, 254, 260; cosmopolitan, 3,
  6, 9, 10, 11, 95, 166-167, 237, 240, 242,
  256, 258, 259, 260, 279n.2; democratic
  approaches to, 257-262; funding of,
  253-254, 260; historical meaning of,
  22; in Holland, 261; in India, 260-

261; local, 3, 7-13, 240, 241, 242, 248, 254-257, 258, 279n.2; "normal," 6-7, 241, 259; plurality of, 261-262; politics of, 253-254; as practice, 23; range of meanings of, 22; Zapotec agriculture as, 3, 21-23, 130, 258, 259
Scott, James, 8, 10, 11, 279n.2
scrap materials, 88-90, 101, 235
Shapiro, Robert, 252
Sierra Juárez. *See* Northern Sierra
smallpox, 45, 71
soils, 131-132, 135, 139, 140, 146, 168-169, 206, 214, 218, 220, 242
Solaga, 50, 63, 201
Solidaridad, 67
Southern Sierra (Oaxaca), 216
spices, 160-161
Stavenhagen, Rodolfo, 56
Stephen, Lynn, 26
stone idols, 109, 110, 112
The Structure of Scientific Revolutions, 6
sugarcane: as cash crop, 190; conversion into panela, 181, 186-189, 236; coordinating work for production of, 181; cultivation, 179-185; cutting, 184; danger associated with, 175; and death, 190-194; gathering firewood for processing, 183; history of, 177-179, 236; and industrial work, 185; labor arrangements, 181-182, 236; persistence of, 190, 236; planting, 179-180; and slavery, 176, 177, 236; as small-scale crop, 176; as staple food in Talea, 175; varieties of, 180; weeding, 180-181
susto, 109
syncretism, 4, 53-54, 111-112, 261

Talea de Castro: architectural features of, 35-37, 40-41; cash crops of, 124; children of, 92, 99; and construction of road, 57-58; cultivated fields of, 42; effects of coffee prices in, 128-129, 201-203; effects of food policies in, 127-129; founding of, 55, 235; geographical features of, 34-41; "harmony ideology" in, 66; history of, 54-60, 235; history of coffee in, 200-203, 235; infrastructure in, 39-40; markets in, 61-62, 283n.22; musical groups of, 27-28, 287n.2; persistence of maize farming in, 103; politics in, 67-69; population of, 34; preference for panela in, 189-190; ranch houses in, 96-98; rich and poor in, 208-209, 212-213; stereotypes of, 216; subsistence crops in, 2, 124; sugarcane work in, 178-189; sustainable agriculture in, 218-220; technology and change in, 56-61; and Todos Santos, 190-193
Tambiah, Stanley, 4
Tanetze de Zaragoza, 49, 61, 62, 128, 145, 283n.1
Taxco, 46
Technics and Civilization, 90
technology, 92, 95-96
Tehuacán Valley, 117
Tenochtitlán, 2, 43, 119
teosinte, 118-119, 285n.5, 286n.14
Teotihuacán, 61, 119
Teotlaxco, 63, 199, 200
Tepanzacoalco, 199
Tepatitlán, 65
tequio, 30, 281n.5
Tijuana, 65, 68, 89, 226
Tiltepec, 43, 50
Tlacolula, 201
Todos Santos, 189, 190-193
Toltecs, 1, 119
Tonantzín, 109
Totonacs, 226
Totontepec, 48
tools, 90-91. *See also* agricultural implements
tortillas: consumption of, 155-157; preferences for, 155-158; preparation of, 154-159; and relishes, 161-162; and sauces, 161-162; symbolism of, 161; "throwing" of, 154-155; "tossing" of, 154-155
trade: during colonial era, 44-50; of coffee, 201-204; between Valley and Sierra Zapotec, 62

traditional ecological knowledge (TEK), 7-9
transgenic crops. *See* genetically modified crops
transnational corporations, 126-127, 222, 232, 238, 241, 242, 245, 248-249, 260, 287n.18, 293nn.23,5
tribute, 44-45, 72
Trobriand Islanders, 5, 12
Tuxtepec, 40, 198
Tylor, Edward, 5
typhus, 45, 71
Tyson Farms, 126

UCI-100 (Union of Indigenous Communities "One Hundred Years of Solitude"), 228
UCIRI (Union of Indigenous Communities of the Isthmus Region), 228
UCIZONI (Union of Indigenous Communities of the Northern Isthmus Zone), 228
Unilever, 127
United States: agriculture in, 164-165, 286n.16; alternative medicine in, 246-247; coffee consumption of, 201-203; Congress of, 253, 260; consumers in, 246, 253; diet in, 244-246; food supply of, 249-250; food system of, 245-246; genetically modified crops in, 249-254; health problems in, 244-245; life cycle events in, 16-17; organic foods in, 280n.12; prevalence of obesity and overweight in, 245
United States Department of Agriculture, 20, 239, 242, 293n.4
United States Environmental Protection Agency, 250
United States Food and Drug Administration, 246
United States National Institute of Mental Health, 247
United States Senate Committee on Nutrition and Human Needs, 244-245
units of measure: custom-made implements and, 76; in the Northern Sierra, 73-80; political implications of, 79-80; standardization of, 79-80
UNOSJO (Union of Organizations of the Sierra Juárez of Oaxaca), 230, 231

Veracruz, 32, 33, 178, 180, 197
Villa Alta, 18, 34, 39, 41, 43, 44, 45, 46-47, 50, 51, 52, 56, 146, 179, 198, 199
Virgen de Guadalupe, 106, 109, 150
volume, 77-78

Warman, Arturo, 118
Weber, Max, 6-7
Wilken, Gene, 25, 242
Wolf, Eric, 50
women: as experts in coffee processing, 217-218; as experts in the conversion of maize into food, 159; monetary contribution of, 163; as participants in weekly market, 162-163; responsibility of, for animals, 158; responsibility of, for maize, 153-163; role of, in coffee harvest, 210, 223; role of, in rituals, 110
work: attitudes toward, among colonists in New Spain, 19; on coffee plantations, 197-198, 210; inevitability and normality of, 4, 14, 18-19, 77, 95; pooling of, 143-176; reciprocal—*see* gozona; shortage of, 213; for wages, 208; women's, 47, 153, 158, 162-163, 210
world system, 12, 233
World Trade Organization, 253, 292n.18
Worldwatch Institute, 238-239
Wright, Angus, 8, 102

XEGLO, 59

Yaeé, 43, 49, 61, 62, 128, 145, 179, 215, 282n.22, 283n.25
Yagalaxi, 34, 63
Yagavila, 43, 50, 51, 62, 63, 111, 146, 199, 200, 204, 205
Yagila, 62, 199
Yalálag, 50, 110, 179, 198, 201
Yaneri, 199
Yatoni, 99, 179, 213, 225

Yohueche, 152
Yojobi, 50, 105, 106, 153, 213
Yotao, 199
Yovego, 146

Zacatecas, 46
Zapatistas, 69
Zapotec: agricultural system during colonial era, 70–73; ancient trade routes, 42, 61; Cajonos, 62, 105, 110, 112, 137, 152, 179, 200; as cultivators in ancient times, 1, 25; deities, 109; dialects, 283n.23; education of children, 52; as efficient food consumers, 24–25; ethnic grouping of, 62–65; ethnic identity of, 65–66; experimentation with new technologies by, 96, 241; history of, 1, 43; Isthmus, 64; migrant associations, 65–66; myth and ritual, 13–14; religion, 52, 54, 110–112, 285n.5; resistance, 43, 282n.14; Rincón, 55, 62–64, 104, 106, 110, 113, 167, 172–173, 234, 235, 240, 244, 257, 259, 261, 262; as scientists, 259; Sierra, 62–63, 200, 232; views of maize held by, 103–117, 129
Zoochila, 179
Zoogochi, 63, 199
Zoogocho, 137, 179, 201, 203

Lightning Source UK Ltd.
Milton Keynes UK
UKHW010125270123
416049UK00001B/24